T0185538

Raspberry Pi GPU Audio Video Programming

Jan Newmarch

Apress®

Raspberry Pi GPU Audio Video Programming

Jan Newmarch
Oakleigh, Victoria
Australia

ISBN-13 (pbk): 978-1-4842-2471-7 ISBN-13 (electronic): 978-1-4842-2472-4
DOI 10.1007/978-1-4842-2472-4

Library of Congress Control Number: 2016961518

Managing Director: Welmoed Spahr
Lead Editor: Steve Anglin
Technical Reviewer: Chaim Krause
Editorial Board: Steve Anglin, Pramila Balan, Laura Berendson, Aaron Black, Louise Corrigan, Jonathan Gennick, Robert Hutchinson, Celestin Suresh John, Nikhil Karkal, James Markham, Susan McDermott, Matthew Moodie, Natalie Pao, Gwenan Spearing
Coordinating Editor: Mark Powers
Copy Editor: Kim Wimpsett
Compositor: SPi Global
Indexer: SPi Global
Artist: SPi Global

Distributed to the book trade worldwide by Springer Science+Business Media New York, 233 Spring Street, 6th Floor, New York, NY 10013. Phone 1-800-SPRINGER, fax (201) 348-4505, e-mail `orders-ny@springer-sbm.com`, or visit `www.springeronline.com`. Apress Media, LLC is a California LLC and the sole member (owner) is Springer Science + Business Media Finance Inc (SSBM Finance Inc). SSBM Finance Inc is a **Delaware** corporation.

For information on translations, please e-mail `rights@apress.com`, or visit `www.apress.com`.

Apress and friends of ED books may be purchased in bulk for academic, corporate, or promotional use. eBook versions and licenses are also available for most titles. For more information, reference our Special Bulk Sales–eBook Licensing web page at `www.apress.com/bulk-sales`.

Any source code or other supplementary materials referenced by the author in this text are available to readers at `www.apress.com`. For detailed information about how to locate your book's source code, go to `www.apress.com/source-code/`. Readers can also access source code at SpringerLink in the Supplementary Material section for each chapter.

Printed on acid-free paper

To my wife, Linda Cai, and my daughter, Kathryn, who let me play with my computing toys without complaint.

Contents at a Glance

Contents

About the Author

Jan Newmarch is the head of the ICT (Higher Education) at Box Hill Institute, an adjunct professor at Canberra University, and an adjunct lecturer in the School of Information Technology, Computing, and Mathematics at Charles Sturt University. He is interested in more aspects of computing than he has time to pursue, but the major thrust over the last few years has developed from user interfaces under Unix into Java, to the Web, and then into general distributed systems. Jan has developed a number of publicly available software systems in these areas. For the last few years, he has been looking at sound for Linux systems and programming the Raspberry Pi's GPU. He is now exploring aspects of the IoT. He lives in Melbourne, Australia, and enjoys the food and culture there but is not so impressed by the weather.

About the Technical Reviewer

Chaim Krause currently lives in Leavenworth, Kansas, where the U.S. Army employs him as a simulation specialist. In his spare time, he likes to play PC games and occasionally develops his own. He has recently taken up the sport of golf to spend more time with his significant other, Ivana. Although he holds a bachelor's of science degree in political science from the University of Chicago, Chaim is an autodidact when it comes to computers, programming, and electronics. He wrote his first computer game in BASIC on a Tandy Model I Level I and stored the program on a cassette tape. Amateur radio introduced him to electronics, while the Arduino and the Raspberry Pi provided a medium to combine computing, programming, and electronics into one hobby.

Acknowledgments

This book was written with the aid of the extensive documentation on all topics related to computing and the helpful nature of the many thousands of programmers contributing to the Web as a knowledge source. The book is based on the engineering and development work of those behind the Raspberry Pi and of course of the organizations creating the hardware, the software, and the APIs to run programs on GPUs.

Introduction

The Raspberry Pi was created to meet a need to help younger people become involved in the IT field. As a low-cost computer, it can be used, experimented with, broken, and replaced. Initially expected to sell perhaps a few thousand, it has now sold more than 10 million units, and it is used for all sorts of activities, right up to building supercomputers held together with LEGO blocks.

A surprising and early development was to use the Raspberry Pi as a media center, where the GPU was used by software such as `omxplayer` to render 1080p movies to HDMI screens. This has led to software such as Kodi and OpenElec. My partner is Chinese, so for entertainment we sing karaoke. We have several karaoke players, and they all have different features. I wanted to combine all these features into a single application running on low-cost hardware, and the Raspberry Pi was a natural choice. But it turned out to be really, really *hard*. Some documentation about programming the Raspberry Pi's GPU was available but very inaccessible. So, as I explored what the GPU was capable of, I documented it all, and it turned into this book.

Who This Book Is For

This book is aimed at programmers familiar with C programming who want to write programs using the Raspberry Pi's GPU. It does not assume a prior background in graphics programming, but it does assume that you understand enough about the Unix/Linux programming environment to install and build software from source using tools such as `make`. Of course, it assumes you have a Raspberry Pi and are comfortable setting it up and installing software packages.

What This Book Covers

There are a multitude of APIs and programming styles for programming graphics systems. The Raspberry Pi supports a few of these: OpenGL ES for 3D programming, OpenVG for 2D vector graphics programming, and OpenMAX for video and audio. This book attempts to deal with all of these. The book deals with the basic concepts of each style of programming and illustrates them with many complete, working programs. You are fortunate in that although the Raspberry Pi has been through a number of hardware revisions, the GPU has remained the same, so the programs run unchanged on all versions of the hardware.

OpenGL ES is a well-documented API. I deal with it here in only enough depth to give simple programs showing the basic concepts of working on the Raspberry Pi. Then you can move on to other resources. The OpenMAX and OpenVG systems have not been documented anywhere outside of formal specifications and multiple PowerPoint presentations. Describing those systems form the major part of the book. In particular, the boundaries between CPU and GPU programming are discussed in detail, such as where text handling or audio decoding should occur.

The final chapter looks at the interplay between these different APIs: drawing dynamic text using OpenVG on top of a video played by OpenMAX, for example.

By the end of this book, you will be familiar with the current APIs using the Raspberry Pi's GPU. Source code for all the programs is accessible via the book's Apress.com product page, which is at `www.apress.com/9781484224717`.

CHAPTER 1

Introduction to the Raspberry Pi

The Raspberry Pi (RPi) is a low-cost Linux computer developed with the intention of improving the background of students entering university computer science courses by giving them a good, cheap environment in which to play. And it does! In fact, I've got a bunch of colleagues at work, well into middle age, who have leapt upon it to play with. So far their kids haven't found out, though....

Introduction

The RPi is a small, single-board computer developed originally for the purpose of teaching computer science in schools. It uses a system on a chip (SoC) and is characterized by a small form factor, low power, not much RAM, limited inputs and outputs, and disk storage provided either by an SD card or by external storage using a USB port.

There is a growing ecosystem of such computers, and sites such as Element14 Single Board Computers Line Card (`https://www.element14.com/community/docs/DOC-55875?ICID=sbc-featured-products`) and Wikipedia (`http://en.wikipedia.org/wiki/Single-board_computer`) list many of them.

Now this doesn't seem very exciting, but nevertheless the RPi has captured the imagination (and the pockets!), with more than ten million sold. In fact, Amazon lists more than 1,000 books related to the RPi.

So, what is it? Well, it is a single-board computer, about the size of a credit card. There are a series of models beginning with the Raspberry Pi 1 released in 2012 to the Model 3 Model B released in 2016 and the Pi Zero.

The Model B has 512 MB RAM, two USB ports, and an Ethernet port. It has HDMI and analog audio and video outputs. The RPi FAQ states that the models currently available are the Pi 2 Model B, the Pi 3 Model B, the Pi Zero, and the Pi 1 Model B+ and A+. The following is from the FAQ at `www.raspberrypi.org/faqs`:

> *The Model A+ is the low-cost variant of the Raspberry Pi. It has 256 MB RAM, one USB port, 40 GPIO pins, and no Ethernet port. The Model B+ is the final revision of the original Raspberry Pi. It has 512 MB RAM (twice as much as the A+), four USB ports, 40 GPIO pins, and an Ethernet port. In February 2015, it was superseded by the Pi 2 Model B, the second generation of the Raspberry Pi. The Pi 2 shares many specs with the Pi 1 B+, but it uses a 900 MHz quad-core ARM Cortex-A7 CPU and has 1 GB RAM. The Pi 2 is completely compatible with first-generation boards and is the model we recommend for use in schools, due to its flexibility for the learner. The Pi 3 Model B was launched in February 2016; it uses a 1.2 GHz 64-bit quad-core ARM Cortex-A53 CPU, has 1 GB RAM, integrated 802.11n wireless LAN, and Bluetooth 4.1. Finally, the Pi Zero is half the size of a Model A+, with a 1 GHz single-core CPU and 512 MB RAM, and mini-HDMI and USB On-the-Go ports.*

Electronic supplementary material The online version of this chapter (doi:10.1007/978-1-4842-2472-4_1) contains supplementary material, which is available to authorized users.

J. Newmarch, *Raspberry Pi GPU Audio Video Programming*, DOI 10.1007/978-1-4842-2472-4_1

Figure 1-1 shows what the RPi looks like graphically, and Figure 1-2 shows what it looks like in real life.

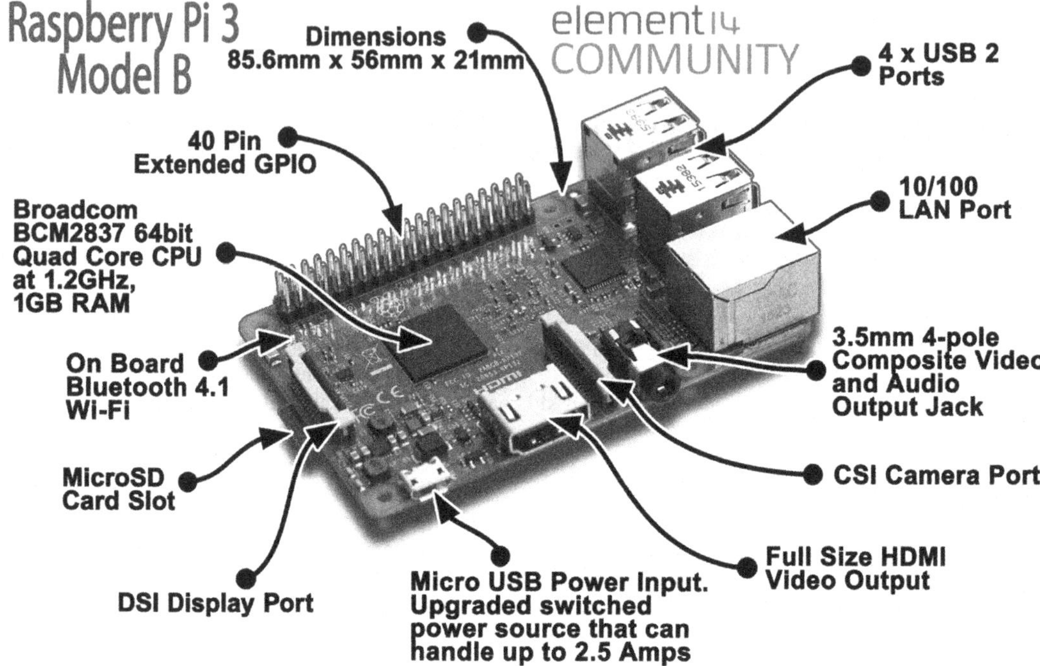

Figure 1-1. *Raspberry Pi Model 3B (https://www.element14.com/community)*

For its target audience, this is a wonderful computer. Its price point means that if it breaks, well, you just get another one. It runs a full-blown Linux system (the RPi 3B can run a very stripped down version "IoT" version of Windows) and can be used in a huge variety of situations (including as an array of RPis to build a supercomputer or hooked up to sensors as a participant in the Internet of Things). But if you want to use a single RPi for heavy-duty computation, well, it sucks. The following is from Wikipedia:

> *On the CPU level the performance of the RPi 2 is similar to a 300 MHz Pentium II of 1997–1999. The RPi 3 is substantially better but still not a world-shaker.*

The GPU, on the other hand, is quite reasonable. The following is from the RPi FAQ:

> *The GPU is capable of 1 Gpixel/s, 1.5 Gtexel/s, or 24 GFLOPs of general-purpose compute and features a bunch of texture filtering and DMA infrastructure.*
>
> *The GPU provides Open GL ES 2.0, hardware-accelerated OpenVG, and 1080p30 H.264 high-profile encode and decode.*

Now this isn't top-of-the-range GPU performance. A review of the NVidia GTX 980 gave 72 gigapixels per second, but then that single graphics card is ten times the cost of the whole RPi!

From the point of view of this book, an important factor is that while the CPU has changed in the different models, the GPU has remained the same. That means programs written to the GPU will run unchanged across all the different models.

A number of Linux images are available from the Raspberry Pi site, and others are being developed elsewhere. I use the Debian-based Raspbian image, which essentially comes in two forms: with or without a GUI. It doesn't matter which one you use since in this book you will write directly to the GPU and bypass any running window system.

ELinux.org maintains a list of RPi distributions. There are many standard Linux distros included, such as Fedora, Debian, Arch, SUSE, Gentoo, and others. The RPi has gained traction as a media center based on the XBMC media center, and this is represented by distros such as OSMC (formerly Raspbmc) and OpenElec.

It doesn't really matter which distro you use or even whether you use the X Window System or a command line. You will be bypassing those for this book.

There don't seem to be any books that go into detail about the Broadcom VideoCore GPU used on the RPi, besides several tutorials and a set of examples from Broadcom. I bumped into a need to do this kind of programming while trying to turn the RPi into a karaoke machine: with the CPU busting its gut rendering MIDI files, there was nothing left for showing images such as karaoke lyrics except for the GPU, and nothing really to tell me how to do it.

This book scratches my itch. Since I've had to learn a lot about RPi GPU programming, I might as well share it. Let's be honest: much of this is about sharing the pain. What started as a side issue turned into a full-blown project of its own, and this stuff is *hard*.

I will cover OpenGL ES, OpenMAX, and OpenVG. Along the way I will cover some Dispmanx. A version of the full OpenGL has just been released in beta, but this book will not cover that.

If you can't program in C, then go away. If you can't do concurrent programming using threads, then you will probably get glassy-eyed at various points. You need to be able to read formal specifications and pay attention to all the finicky details or your application won't work. And you will need to spend a lot of time experimenting and working through unexplained "features," buggy behavior, and not-yet-implemented core elements. Get used to staring at your screen in a totally bemused fashion, with a debugger running and the specification open on one side and a half-dozen web sites open on the other. I've found that a good bottle of red wine helps, not with the programming but for handling the frustration!

Programming Style

There are many different C programming styles that can be adopted. Regrettably this book uses most of them. I have taken code and specifications from the following places:

- The OpenMAX specification
- The OpenGL ES specification
- The OpenVG specification
- The OpenGL ES Programming Guide
- The Broadcom IL Client library

My preferred brace style for C programming is the "one true brace" with the opening brace on the same line as its conditional statement, as shown here:

```
if (x) {
    ...
}
```

The IL Client library uses the Allman style (brace on a new line), as shown here:

```
if (x)
{
    ...
}
```

The Programming Guide uses the Whitesmiths style (brace on a new line, indented), as shown here:

```
if (x)
  {
    ...
  }
```

For variable names, I prefer delimiting words with an underscore (_), as in `my_vbl`, though after many years of Java programming, I am also happy with CamelCase (`myVbl`). The IL Client library uses _ separation; the Programming Guide uses CamelCase.

The OpenMAX specification uses a simplified form of Hungarian notation for struct field types such as `pNativeRender` but prefixes all function calls with `OMX_`, and the struct names themselves are all uppercase such as `OMX_AUDIO_PORTDEFINITIONTYPE`.

The OpenGL ES specification uses CamelCase for its function names, all beginning with `gl` such as `glValidateProgram`.

It isn't possible to get one consistent style throughout the book, so I haven't even tried for this edition. Please be tolerant.

Conclusion

This chapter has given you a brief overview of the RPi and what this book will attempt to cover.

Resources

- *Raspberry Pi home page*: `https://www.raspberrypi.org/`
- *RPi OpenMAX forum*: `https://www.raspberrypi.org/forums/viewforum.php?f=70`
- *RPi Graphics forum*: `https://www.raspberrypi.org/forums/viewforum.php?f=67`
- *RPi OpenGL ES forum*: `https://www.raspberrypi.org/forums/viewforum.php?f=68`
- *Raspberry Pi VideoCore APIs*: `http://elinux.org/Raspberry_Pi_VideoCore_APIs`
- *Linux Gizmos reports on many SoCs including the RPi*: `http://hackerboards.com/`
- *Tutorial: VLC with hardware acceleration on Raspberry Pi*: `http://intensecode.blogspot.com.au/2013/10/tutorial-vlc-with-hardware-acceleration.html`
- *How to draw 2D images using OpenGL, in SDL*: `http://gamedev.stackexchange.com/questions/46640/how-to-draw-2d-images-using-opengl-in-sdl`
- *Hardware Accelerated Qt Multimedia Backend for Raspberry Pi and OpenGL Shaders on Video*: `http://thebugfreeblog.blogspot.com.au/2013/04/hardware-accelerated-qtmultimedia.html`
- *Decoding and Rendering Compressed Images with OpenMAX on Raspberry Pi*: `http://thebugfreeblog.blogspot.com.au/2012/12/decoding-and-rendering-compressed.html`

CHAPTER 2

Khronos Group

According to its web site, "The Khronos Group is a not-for-profit industry consortium creating open standards for the authoring and acceleration of parallel computing, graphics, dynamic media, computer vision, and sensor processing on a wide variety of platforms and devices."

Role of Khronos

The Khronos Group is concerned with devising and setting standards for graphics and video processing, among other things. The standards it is involved in include the following:

- OpenGL
- OpenGL ES
- OpenMAX
- EGL
- OpenVG
- OpenCL

What relevance does this have to the RPi? Well, the RPi supports OpenGL ES, OpenVG, and OpenMAX, and, by the structure of the standards, EGL. So, this is the primary source site for information about the video and graphics programming APIs used by the RPi. Having said that, it may not be the best site to learn what these APIs are. You can read this book instead.

OpenGL ES

According to its web site, "OpenGL ES is a royalty-free, cross-platform API for full-function 2D and 3D graphics on embedded systems, including consoles, phones, appliances, and vehicles. It consists of well-defined subsets of desktop OpenGL, creating a flexible and powerful low-level interface between software and graphics acceleration. OpenGL ES includes profiles for floating-point and fixed-point systems and the EGL specification for portably binding to native windowing systems. OpenGL ES 1.X: fixed function hardware offering acceleration, image quality, and performance. OpenGL ES 2.X: enables full programmable 3D graphics."

What distinguishes OpenGL ES from OpenGL is its emphasis on low-capability embedded systems, and the RPi falls into this category, although it is not necessarily embedded. OpenGL ES is a simplified and tidied-up form of OpenGL but still has the same capabilities as OpenGL. The emphasis with this API is on 2D and 3D graphics, suitable for games programming and other high-demand graphics applications.

J. Newmarch, *Raspberry Pi GPU Audio Video Programming*, DOI 10.1007/978-1-4842-2472-4_2

It will work in cooperation with other windowing systems, such as the X Window System or Wayland, or where no other windowing system is running.

OpenVG

According to its web site, "OpenVG is a royalty-free, cross-platform API that provides a low-level hardware acceleration interface for vector graphics libraries such as Flash and SVG. OpenVG is targeted primarily at handheld devices that require portable acceleration of high-quality vector graphics for compelling user interfaces and text on small-screen devices, while enabling hardware acceleration to provide fluidly interactive performance at very low power levels."

OpenVG is an alternative approach to graphics to OpenGL. While OpenGL renders onto textures, OpenVG is more concerned with drawing lines to form shapes and then rendering within those shapes.

EGL

According to its web site, "EGL is an interface between Khronos-rendering APIs such as OpenGL ES or OpenVG and the underlying native platform window system. It handles graphics context management, surface/buffer binding, and rendering synchronization and enables high-performance, accelerated, mixed-mode 2D and 3D rendering using other Khronos APIs."

EGL is the "glue" layer between the higher-level APIs and the hardware. It isn't used extensively by the application programmer, just enough to give the higher level the hooks into the lower level. Both OpenGL ES and OpenVG sit above EGL.

OpenMAX

According to its web site, "OpenMAX™ is a royalty-free, cross-platform API that provides comprehensive streaming media codec and application portability by enabling accelerated multimedia components to be developed, integrated, and programmed across multiple operating systems and silicon platforms. The OpenMAX API will be shipped with processors to enable library and codec implementers to rapidly and effectively make use of the full acceleration potential of new silicon, regardless of the underlying hardware architecture."

This is a bit more market-speak. But this API is designed for streaming media such as movies and music. Again, it will work with other windowing systems.

Conclusion

Khronos is the standards organization responsible for many of the graphics standards. This chapter has briefly introduced those relevant to the RPi.

Resources

- *Khronos Group*: `www.khronos.org/`

CHAPTER 3

Compiling Programs for the Raspberry Pi

Every system has its own quirks when it comes to compilation. The RPi is no exception. This chapter looks at the process of compiling C code for the RPi.

Generic Information

Here is some general information:

- The command `arch` gives the Linux idea of the CPU. On the RPi 2, it gives `armv7l`. According to the Arch Linux ARM (`http://archlinuxarm.org/platforms/armv7/broadcom/raspberry-pi-2`), this means "ARMv7l Cortex-A7 quad-core. [...] The SoC is a Broadcom BCM2836."
- The file `/etc/issue` contains the information `Raspbian GNU/Linux 7`.
- The file `/etc/os-release` contains general information about your operating system.

```
PRETTY_NAME="Raspbian GNU/Linux 7 (wheezy)"
NAME="Raspbian GNU/Linux"
VERSION_ID="7"
VERSION="7 (wheezy)"
ID=raspbian
ID_LIKE=debian
ANSI_COLOR="1;31"
HOME_URL="http://www.raspbian.org/"
SUPPORT_URL="http://www.raspbian.org/RaspbianForums"
BUG_REPORT_URL="http://www.raspbian.org/RaspbianBugs"
```

At the time of this writing, the RPi 3 gives a similar report, although the CPU is actually a different one, an ARMv8 Cortex A53. Apparently this is because of the current lack of a 64-bit version of Raspbian, so it reports the same as the 32-bit version for backward compatibility.

J. Newmarch, *Raspberry Pi GPU Audio Video Programming*, DOI 10.1007/978-1-4842-2472-4_3

Versions of gcc

Near the end of 2016, the Raspbian distros give gcc 4.8 as the default C compiler. This is fine. If you have an older version, say 4.6, you will want to upgrade your distro or gcc.

C99 or C11 Compilers

Based on language standards supported by gcc, gcc by default does not use the latest C standards. It uses a sort-of 1990 standard, using the default flag -std=gnu90. You can set it to use the C99 standard with the flag -std=c99 and to use C11 with the flag -std=c11.

But there is a problem. EGL, which forms the basis for OpenGL ES and OpenVG, does not support C99. If you try to use the C99 flag, then just including <EGL/egl.h> will throw a host of errors, mainly because it doesn't like the declaration of the function clock_gettime, even though the appropriate header <time.h> is included.

```
In file included from /opt/vc/include/interface/vcos/vcos.h:116:0,
                 from /opt/vc/include/interface/vmcs_host/vc_dispmanx.h:33,
                 from /opt/vc/include/EGL/eglplatform.h:110,
                 from /opt/vc/include/EGL/egl.h:36,
                 from tiny.c:2:
/opt/vc/include/interface/vcos/pthreads/vcos_platform.h: In function
'vcos_semaphore_wait_timeout':
/opt/vc/include/interface/vcos/pthreads/vcos_platform.h:297:4: warning:
implicit declaration of function 'clock_gettime' [-Wimplicit-function-declaration]
    if (clock_gettime(CLOCK_REALTIME, &ts) == -1)
    ^
/opt/vc/include/interface/vcos/pthreads/vcos_platform.h:297:22: error:
'CLOCK_REALTIME' undeclared (first use in this function)
    if (clock_gettime(CLOCK_REALTIME, &ts) == -1)
...
```

You can work around this by using the flag -std=gnu99 instead, but this still generates lots of warnings.

```
In file included from /opt/vc/include/interface/vcos/vcos.h:185:0,
                 from /opt/vc/include/interface/vmcs_host/vc_dispmanx.h:33,
                 from /opt/vc/include/EGL/eglplatform.h:110,
                 from /opt/vc/include/EGL/egl.h:36,
                 from tiny.c:2:
/opt/vc/include/interface/vcos/vcos_timer.h:112:6: warning: inline function
'vcos_timer_delete' declared but never defined [enabled by default]
 void vcos_timer_delete(VCOS_TIMER_T *timer);
      ^
/opt/vc/include/interface/vcos/vcos_timer.h:109:6: warning: inline function
'vcos_timer_reset' declared but never defined [enabled by default]
 void vcos_timer_reset(VCOS_TIMER_T *timer, VCOS_UNSIGNED delay);
...
```

This will compile nevertheless. Do you want to put up with the warnings or just use the default gcc mode?

gcc Compile Flags

The command

```
gcc -mcpu=native -march=native -Q --help=target
```

prints lots of interesting stuff on what gcc figures out. On the RPi 2 with gcc 4.8, it gives the following:

```
The following options are target specific:
-mabi=                                  aapcs-linux
-mabort-on-noreturn                     [disabled]
-mandroid                               [disabled]
-mapcs                                  [disabled]
-mapcs-float                            [disabled]
-mapcs-frame                            [disabled]
-mapcs-reentrant                        [disabled]
-mapcs-stack-check                      [disabled]
-march=                                 armv7-a
-marm                                   [enabled]
-mbig-endian                            [disabled]
-mbionic                                [disabled]
-mcallee-super-interworking             [disabled]
-mcaller-super-interworking             [disabled]
-mcpu=                                  cortex-a7
-mfix-cortex-m3-ldrd                    [enabled]
-mfloat-abi=                            hard
-mfp16-format=                          none
-mfpu=                                  vfp
-mglibc                                 [enabled]
-mhard-float
-mlittle-endian                         [enabled]
-mlong-calls                            [disabled]
-mpic-register=
-mpoke-function-name                    [disabled]
-msched-prolog                          [enabled]
-msingle-pic-base                       [disabled]
-msoft-float
-mstructure-size-boundary=              0x20
-mthumb                                 [disabled]
-mthumb-interwork                       [enabled]
-mtls-dialect=                          gnu
-mtp=                                   auto
-mtpcs-frame                            [disabled]
-mtpcs-leaf-frame                       [disabled]
-mtune=                                 [default]
-muclibc                                [disabled]
-munaligned-access                      [enabled]
-mvectorize-with-neon-double            [disabled]
-mvectorize-with-neon-quad              [enabled]
-mword-relocations                      [disabled]
-mwords-little-endian                   [disabled]
```

```
Known ARM ABIs (for use with the -mabi= option):
aapcs aapcs-linux apcs-gnu atpcs iwmmxt

Known ARM architectures (for use with the -march= option):
armv2 armv2a armv3 armv3m armv4 armv4t armv5 armv5e armv5t armv5te armv6
armv6-m armv6j armv6k armv6s-m armv6t2 armv6z armv6zk armv7 armv7-a armv7-m
armv7-r armv7e-m armv8-a iwmmxt iwmmxt2 native

Known __fp16 formats (for use with the -mfp16-format= option):
alternative ieee none

Known ARM FPUs (for use with the -mfpu= option):
crypto-neon-fp-armv8 fp-armv8 fpv4-sp-d16 neon neon-fp-armv8 neon-fp16
neon-vfpv4 vfp vfp3 vfpv3 vfpv3-d16 vfpv3-d16-fp16 vfpv3-fp16 vfpv3xd
vfpv3xd-fp16 vfpv4 vfpv4-d16

Valid arguments to -mtp=:
auto cp15 soft

Known floating-point ABIs (for use with the -mfloat-abi= option):
hard soft softfp

Known ARM CPUs (for use with the -mcpu= and -mtune= options):
arm1020e arm1020t arm1022e arm1026ej-s arm10e arm10tdmi arm1136j-s
arm1136jf-s arm1156t2-s arm1156t2f-s arm1176jz-s arm1176jzf-s arm2 arm250
arm3 arm6 arm60 arm600 arm610 arm620 arm7 arm70 arm700 arm700i arm710
arm7100 arm710c arm710t arm720 arm720t arm740t arm7500 arm7500fe arm7d
arm7di arm7dm arm7dmi arm7m arm7tdmi arm7tdmi-s arm8 arm810 arm9 arm920
arm920t arm922t arm926ej-s arm940t arm946e-s arm966e-s arm968e-s arm9e
arm9tdmi cortex-a15 cortex-a5 cortex-a7 cortex-a8 cortex-a9 cortex-m0
cortex-m0plus cortex-m1 cortex-m3 cortex-m4 cortex-r4 cortex-r4f cortex-r5
ep9312 fa526 fa606te fa626 fa626te fa726te fmp626 generic-armv7-a iwmmxt
iwmmxt2 marvell-pj4 mpcore mpcorenovfp native strongarm strongarm110
strongarm1100 strongarm1110 xscale

TLS dialect to use:
gnu gnu2
```

If you want to find out what this all means, look at the GCC Option Summary at `https://gcc.gnu.org/onlinedocs/gcc/Option-Summary.html`.

I built up an impressive set of options, basically by chucking in everything I saw in other people's code. They are now as follows:

```
        -mfloat-abi=hard -mcpu=arm1176jzf-s -fomit-frame-pointer -mabi=aapcs-linux
-mtune=arm1176jzf-s -mfpu=vfp -Wno-psabi -mno-apcs-stack-check -mstructure-size-boundary=32
-mno-sched-prolog -march=armv6zk
```

Some of these are specific to the RPi version 1, and some are redundant. They are discussed here:

- `-mabi=aapcs-linux`: This generates code for the specified ABI. Possible values are `apcs-gnu`, `atpcs`, `aapcs`, `aapcs-linux`, and `iwmmxt`.

  This seems to be primarily for linking to C++ libraries. For C only, can it be omitted?

- `-mfloat-abi=hard`: This specifies which floating-point ABI to use. Possible values are `soft`, `softfp`, and `hard`. For decent performance on the RPi, you will use the hard floating-point distro and use this flag in compilations. This seems to be the default anyway.
- `arch`, `cpu`, and `mtune`: These give information about the CPU. On later versions of `gcc`, it can be left as `native`, with one wrinkle. If all are left as native with gcc 4.6, you get the following:

```
warning: switch -mcpu=cortex-a7 conflicts with -march=armv7-a switch
```

and then the following:

```
Assembler messages:
Error: unknown cpu `native'
Error: unrecognized option -mcpu=native
cc1: error: bad value (native) for -mcpu switch
cc1: error: bad value (native) for -march switch
```

The following is according to Richard Earnshaw, at `https://gcc.gnu.org/bugzilla/show_bug.cgi?id=58869`:

> *If you specify -mcpu, you don't need to also specify the architecture; the compiler can work that out from the CPU.*
>
> *At a technical level, Cortex-A7 implements the ARMv7ve variant of ARMv7, which is why you get conflict messages (v7ve having the integer divide instructions); the diagnostic is telling you that the architecture derived from the CPU option is at variance with that coming from the -march option.*
>
> *If you really want to generate strict ARMv7 code that is tuned for cortex-a7, then use -march=armv7-a -mtune=cortex-a7.*

But then the value of `cortex-a7` isn't recognized by the assembler.

```
Assembler messages:
Error: unknown architecture `cortex-a7'

Error: unrecognized option -march=cortex-a7
cc1: error: bad value (cortex-a7) for -mtune switch
cc1: error: bad value (cortex-a7) for -march switch
```

This seems to be using the gcc version 4.6 assembler and is OK under the 4.8 version. The values of `arm1176jzf-s` and `armv6zk` are for the RPi 1.

- `-fomit-frame-pointer`: This is an optimization flag and may mess up debugging by omitting useful information.
- `-mfpu=vfp`: This specifies the floating-point h/w. This value is the default, so it could be omitted.
- `-mno-apcs-stack-check`: This is disabled anyway.

- `-mstructure-size-boundary=32`: This looks important; I got bitten once by assuming 64-bit architecture.
- `-mno-sched-prolog`: The default is enabled; maybe this is useful for debugging?

In summary, this looks reasonable:

```
        -mabi=aapcs-linux -mtune=cortex-a7 -march=armv7-a -mfpu=vfp -mstructure-size-
boundary=32 -mno-sched-prolog
```

This looks OK for some optimization while still allowing debugging.

Includes

The following are the includes.

Dispmanx

The Dispmanx include files are not organized very nicely. Ideally, only the top-level directory `/opt/vc/include` should be needed. Unfortunately, this includes files that don't have a base relative to this file. The following set may be needed:

```
        DMX_INC =  -I/opt/vc/include -I/opt/vc/include/interface/vmcs_host/ -I/opt/vc/
include/interface/vcos/pthreads -I/opt/vc/include/interface/vmcs_host/linux
        INCLUDES = $(DMX_INC)
```

EGL

This adds no new directories.

```
        DMX_INC =  -I/opt/vc/include/ -I /opt/vc/include/interface/vmcs_host/ -I/opt/vc/
include/interface/vcos/pthreads -I/opt/vc/include/interface/vmcs_host/linux
        EGL_INC =
        INCLUDES = $(DMX_INC) $(EGL_INC)
```

OpenMAX

I use both the OpenMAX API and the wrapper from Broadcom, the IL Client library. The IL Client library is intended for wider use than just the RPi, so it contains some conditional execution flags. The main one is to choose between 32- and 64-bit architectures since `time` to OpenMAX is 64 bits and on a 32-bit machine, that needs to be managed in a struct, not a simple type. This flag is `OMX_SKIP64BIT`.

I also use the define `RASPBERRY_PI` in my code where appropriate. There doesn't seem to be a default RPI define flag.

The appropriate includes and cflags are as follows:

```
        DMX_INC =  -I/opt/vc/include/ -I /opt/vc/include/interface/vmcs_host/ -I/opt/vc/
        include/interface/vcos/pthreads -I/opt/vc/include/interface/vmcs_host/linux
        EGL_INC =
        OMX_INC =  -I /opt/vc/include/IL
        OMX_ILCLIENT_INC = -I/opt/vc/src/hello_pi/libs/ilclient
```

```
INCLUDES = $(DMX_INC) $(EGL_INC) $(OMX_INC) $(OMX_ILCLIENT_INC)
CFLAGS =  -DOMX_SKIP64BIT -DRASPBERRY_PI $(INCLUDES)
```

OpenGL ES

I am using the esUtils utility from the OpenGL ES book by Munshi et al., and their files are in the subdirectory Common of my code. The includes reflect this, as shown here:

```
        DMX_INC =  -I/opt/vc/include/ -I /opt/vc/include/interface/vmcs_host/ -I/opt/vc/
include/interface/vcos/pthreads -I/opt/vc/include/interface/vmcs_host/linux
        EGL_INC =
        GLES_INC = -ICommon/
        INCLUDES = $(DMX_INC) $(EGL_INC) $(GLES_INC)
```

OpenVG

There are no additional include directories.

```
        DMX_INC =  -I/opt/vc/include/ -I /opt/vc/include/interface/vmcs_host/ -I/opt/vc/
include/interface/vcos/pthreads -I/opt/vc/include/interface/vmcs_host/linux
        EGL_INC =
        OPENVG_INC =
        INCLUDES = $(DMX_INC) $(EGL_INC) $(OPENVG_INC)
```

Libraries

The following are the libraries.

Dispmanx

The libraries required are as follows:

```
DMX_LIBS =  -L/opt/vc/lib/ -lbcm_host -lvcos -lvchiq_arm -lpthread
LDFLAGS = $(DMX_LIBS)
```

EGL

This pulls in the Dispmanx and EGL libraries. Unfortunately, it also requires the OpenGL ES library.

```
DMX_LIBS =  -L/opt/vc/lib/ -lbcm_host -lvcos -lvchiq_arm -lpthread
EGL_LIBS = -L/opt/vc/lib/ -lEGL -lGLESv2
LDFLAGS = $(DMX_LIBS) $(EGL_LIBS)
```

OpenMAX

The libraries are both the standard OpenMAX libraries and the IL Client libraries.

```
DMX_LIBS =  -L/opt/vc/lib/ -lbcm_host -lvcos -lvchiq_arm -lpthread
EGL_LIBS = -L/opt/vc/lib/ -lEGL
OMX_LIBS = -lopenmaxil
OMX_ILCLIENT_LIBS = -L/opt/vc/src/hello_pi/libs/ilclient -lilclient

LDFLAGS = $(DMX_LIBS) $(EGL_LIBS) $(OMX_LIBS) $(OMX_ILCLIENT_LIBS)
```

OpenGL ES

In addition to the Munshi objects, the math library has to be added. The `EGL_LIBS` library already has the `GLESv2` library.

```
DMX_LIBS =  -L/opt/vc/lib/ -lbcm_host -lvcos -lvchiq_arm -lpthread
EGL_LIBS = -L/opt/vc/lib/ -lEGL -lGLESv2
LDFLAGS = $(DMX_LIBS) $(EGL_LIBS) \
Common/esUtil.o Common/esShader.o Common/esTransform.o\
-lm
```

OpenVG

There is just one additional OpenVG library.

```
DMX_LIBS =  -L/opt/vc/lib/ -lbcm_host -lvcos -lvchiq_arm -lpthread
EGL_LIBS = -L/opt/vc/lib/ -lEGL -lGLESv2
OPENVG_LIBS = -lOpenVG
```

Sample Makefiles

The following are the makefiles.

Dispmanx

This `Makefile` is for the directory `OpenMAX/Components` with the C file `info.c`. The preprocessor flags are for the RPi 2.

```
        DMX_INC =  -I/opt/vc/include -I/opt/vc/include/interface/vmcs_host/ -I/opt/vc/
include/interface/vcos/pthreads -I/opt/vc/include/interface/vmcs_host/linux

INCLUDES = $(DMX_INC)

CFLAGS = $(INCLUDES)
CPPFLAGS = -march=armv7-a -mtune=cortex-a7

DMX_LIBS =  -L/opt/vc/lib/ -lbcm_host -lvcos -lvchiq_arm -lpthread
LDFLAGS =  $(DMX_LIBS)

all: info
```

Running the command `make` will build all the executables.

EGL

This Makefile is for the directory OpenMAX/Components with the C files info.c, context, and window.c. The preprocessor flags are for the RPi 2.

```
        DMX_INC =  -I/opt/vc/include/ -I /opt/vc/include/interface/vmcs_host/ -I/opt/vc/
include/interface/vcos/pthreads -I/opt/vc/include/interface/vmcs_host/linux
        EGL_INC =
        INCLUDES = $(DMX_INC) $(EGL_INC)

        CFLAGS = $(INCLUDES)
        CPPFLAGS = -march=armv7-a -mtune=cortex-a7

        DMX_LIBS =  -L/opt/vc/lib/ -lbcm_host -lvcos -lvchiq_arm -lpthread
        EGL_LIBS = -L/opt/vc/lib -lEGL -lGLESv2

        LDFLAGS = $(DMX_LIBS) $(EGL_LIBS)

        all: context info window
```

Running the command make will build all the executables.

OpenMAX

This Makefile is for the directory OpenMAX/Components with the C files info.c, listcomponents.c, and portinfo.c. The preprocessor flags are for the RPi 2.

```
        DMX_INC =  -I/opt/vc/include/ -I /opt/vc/include/interface/vmcs_host/
-I/opt/vc/include/interface/vcos/pthreads
-I/opt/vc/include/interface/vmcs_host/linux
        EGL_INC =
        OMX_INC =  -I /opt/vc/include/IL
        OMX_ILCLIENT_INC = -I/opt/vc/src/hello_pi/libs/ilclient
        INCLUDES = $(DMX_INC) $(EGL_INC) $(OMX_INC) $(OMX_ILCLIENT_INC)

        CFLAGS=-g -DRASPBERRY_PI -DOMX_SKIP64BIT $(INCLUDES)
        CPPFLAGS = -march=armv7-a -mtune=cortex-a7

        DMX_LIBS =  -L/opt/vc/lib/ -lbcm_host -lvcos -lvchiq_arm -lpthread
        EGL_LIBS = -L/opt/vc/lib/ -lEGL -lGLESv2
        OMX_LIBS = -lopenmaxil
        OMX_ILCLIENT_LIBS = -L/opt/vc/src/hello_pi/libs/ilclient -lilclient

        LDFLAGS =  $(DMX_LIBS) $(EGL_LIBS) $(OMX_LIBS) $(OMX_ILCLIENT_LIBS)

        all: info listcomponents portinfo
```

Running the command make will build all the executables.

OpenVG

This Makefile is for the directory OpenVG with the C files simple_shape.c and ellipse.c. The preprocessor flags are for the RPi 2.

```
        DMX_INC =  -I/opt/vc/include/ -I /opt/vc/include/interface/vmcs_host/ -I/opt/vc/
include/interface/vcos/pthreads -I/opt/vc/include/interface/vmcs_host/linux
        EGL_INC =
        OPENVG_INC =
        INCLUDES = $(DMX_INC) $(EGL_INC) $(OPENVG_INC)

        CFLAGS= $(INCLUDES)
        CPPFLAGS = -march=armv7-a -mtune=cortex-a7

        DMX_LIBS =  -L/opt/vc/lib/ -lbcm_host -lvcos -lvchiq_arm -lpthread
        EGL_LIBS = -L/opt/vc/lib/ -lEGL -lGLESv2
        OPENVG_LIBS = -lOpenVG

        LDFLAGS =  $(DMX_LIBS) $(EGL_LIBS) $(OPENVG_LIBS)

        all: simple_shape ellipse
```

Using pkg-config

The tool pkg-config is useful for simplifying include and library options. For example, if you want to also use the Glib libraries, then rather than track down all of the include files and library files, you can use the following:

```
$(pkg-config --cflags glib-2.0)
$(pkg-config --libs glib-2.0)
```

These then expand into the following:

```
-I/usr/include/glib-2.0 -I/usr/lib/x86_64-linux-gnu/glib-2.0/include
-lglib-2.0
```

These work for nearly all the packages installed using a Debian package manager such as apt-get. (Red Hat systems use pkgconfig.)

The relevant package for the Raspberry Pi GPU include files and libraries is libraspberrypi-dev. Many package-related commands work fine for this package. For example, to show all the files in the package, dpkg-query -L libraspberrypi-dev lists the 241 files in the package. Unfortunately, pkg-config does *not* work, as the creators of this package neglected to include the file libraspberrypi-dev.pc, which would be required by pkg-config.

Heiher (https://github.com/heiher) includes a Git package called raspberrypi-firmware-pkgconfig that fills that gap. This includes these files:

```
bcm_host.pc
egl.pc
glesv2.pc
vg.pc
```

There is, however, a nuisance problem in using these files as they are given: the Mesa project also has the configuration files `egl.pc` and `glesv2.pc`, which are installed if you have installed the Mesa package `libegl1-mesa` or `libgles2-mesa`. Mesa has not followed the expected naming conventions (the EGL package descriptor should have been `libegl1-mesa.pc`, for example).

Download the package and copy the `.pc` files to `/usr/share/pkgconfig`. The different includes/libraries are then managed in `Makefiles` with the following:

- Dispmanx:

```
CFLAGS = $(shell pkg-config --cflags bcm_host)
LDLIBS = $(shell pkg-config --libs bcm_host)
```

- EGL:

```
CFLAGS = $(shell pkg-config --cflags egl-vc)
LDLIBS = $(shell pkg-config --libs egl-vc)
```

- OpenGL ES:

```
CFLAGS = $(shell pkg-config --cflags glesv2-vc)
LDLIBS = $(shell pkg-config --libs glesv2-vc)
```

- OpenVG:

```
CFLAGS = $(shell pkg-config --cflags vg-vc)
LDLIBS = $(shell pkg-config --libs vg-vc)
```

- OpenMAX:

```
CFLAGS = $(shell pkg-config --cflags openmax-vc)
LDLIBS = $(shell pkg-config --libs openmax-vc)
```

- OpenMAX with IL client library:

```
CFLAGS = $(shell pkg-config --cflags openmax--il-vc)
LDLIBS = $(shell pkg-config --libs openmax-il-vc)
```

Conclusion

This chapter is perhaps premature but summarizes the build process for the RPi.

CHAPTER 4

Dispmanx on the Raspberry Pi

Dispmanx is the lowest level of programming for the RPi's GPU. I don't go very deeply into it in this chapter, but you'll learn enough to get a native window for use by other toolkits.

Building Programs

The following is a `Makefile` to build the program in this chapter:

```
DMX_INC =  -I/opt/vc/include -I/opt/vc/include/interface/vmcs_host/
-I/opt/vc/include/interface/vcos/pthreads
-I/opt/vc/include/interface/vmcs_host/linux

INCLUDES = $(DMX_INC)

CFLAGS = $(INCLUDES)
CPPFLAGS = -march=armv7-a -mtune=cortex-a7

DMX_LIBS =  -L/opt/vc/lib/ -lbcm_host -lvcos -lvchiq_arm -lpthread
LDFLAGS =  $(DMX_LIBS)

all: info
```

You can use it by running `make` in the same directory as the `Makefile` and source file `info.c`.

Accessing the GPU

The lowest level of accessing the GPU seems to be by an API called Dispmanx. Now, people don't write their graphics applications using Dispmanx. Instead, they use Dispmanx to get a window that is then used by a framework such as EGL and from there by OpenGLES.

Just to give you a total lack of encouragement from using Dispmanx, there are hardly any examples and no serious documentation. However, that hasn't stopped AndrewFromMelbourne from developing a set of programs that can draw Mandelbrot figures, animate triangles, and so on. I'm not going to delve deeply into his programs; you can download them from raspidmx (`https://github.com/AndrewFromMelbourne/raspidmx`) if you want.

You just use Dispmanx to get some information about the display size and then to create a window that takes up the full screen. You won't do anything else in this chapter; this is just a building block on the way to using the Dispmanx window within a higher-level API. Again, if you want, explore AndrewFromMelbourne's examples.

J. Newmarch, *Raspberry Pi GPU Audio Video Programming*, DOI 10.1007/978-1-4842-2472-4_4

The following are the relevant Dispmanx calls to build a window:

```
    DISPMANX_ELEMENT_HANDLE_T dispman_element;
    DISPMANX_DISPLAY_HANDLE_T dispman_display;
    DISPMANX_UPDATE_HANDLE_T dispman_update;
    VC_RECT_T dst_rect;
    VC_RECT_T src_rect;

   success = graphics_get_display_size(0 /* LCD */,
            &screen_width, &screen_height);
    assert( success >= 0 );

    dst_rect.x = 0;
    dst_rect.y = 0;
    dst_rect.width = screen_width;
    dst_rect.height = screen_height;

    src_rect.x = 0;
    src_rect.y = 0;
    src_rect.width = screen_width << 16;
    src_rect.height = screen_height << 16;

    dispman_display = vc_dispmanx_display_open( 0 /* LCD */);
    dispman_update = vc_dispmanx_update_start( 0 );
    dispman_element =
        vc_dispmanx_element_add(dispman_update, dispman_display,
                                0/*layer*/, &dst_rect, 0/*src*/,
                                &src_rect, DISPMANX_PROTECTION_NONE,
                                0 /*alpha*/, 0/*clamp*/, 0/*transform*/);
```

At the end of this, you have a window stored in `dispman_element` that can be used as a native window object later.

The complete program is `info.c` and just prints the size of the screen, as shown here:

```
 /*
 * code stolen from openGL-RPi-tutorial-master/encode_OGL/
 */

#include <stdio.h>
#include <assert.h>

#include <bcm_host.h>

typedef struct
{
    uint32_t screen_width;
    uint32_t screen_height;
    DISPMANX_DISPLAY_HANDLE_T dispman_display;
} DISPMANX_STATE_T;

DISPMANX_STATE_T state, *p_state = &state;
```

```
void init_dispmanx(DISPMANX_STATE_T *state) {
    int32_t success = 0;

    bcm_host_init();

    DISPMANX_ELEMENT_HANDLE_T dispman_element;

    DISPMANX_UPDATE_HANDLE_T dispman_update;
    VC_RECT_T dst_rect;
    VC_RECT_T src_rect;

    success = graphics_get_display_size(0 /* LCD */,
                                        &state->screen_width,
                                        &state->screen_height);
    assert( success >= 0 );

    printf("Screen height %d, width %d\n",
           state->screen_height, state->screen_width);

    dst_rect.x = 0;
    dst_rect.y = 0;
    dst_rect.width = state->screen_width;
    dst_rect.height = state->screen_height;

    src_rect.x = 0;
    src_rect.y = 0;
    src_rect.width = state->screen_width << 16;
    src_rect.height = state->screen_height << 16;

    state->dispman_display = vc_dispmanx_display_open( 0 /* LCD */);
    dispman_update = vc_dispmanx_update_start( 0 );

    dispman_element =
        vc_dispmanx_element_add(dispman_update, state->dispman_display,
                                0/*layer*/, &dst_rect, 0/*src*/,
                                &src_rect, DISPMANX_PROTECTION_NONE,
                                0 /*alpha*/, 0/*clamp*/, 0/*transform*/);

    /* Now we have created a native Dispmanx window
     * Toolkits such as OpenGL ES will use this for
     * their own 'native window' type, with code like this:

    static EGL_DISPMANX_WINDOW_T nativewindow;

    nativewindow.element = dispman_element;
    nativewindow.width = state->screen_width;
    nativewindow.height = state->screen_height;
    vc_dispmanx_update_submit_sync( dispman_update );
    assert(vc_dispmanx_element_remove(dispman_update, dispman_element) == 0);
    */
}
```

```
int
main(int argc, char *argv[])
{
    init_dispmanx(p_state);

    assert( vc_dispmanx_display_close(p_state->dispman_display) == 0);

    return 0;
}
```

It can then be run easily by the following:

```
./info
```

The output on my RPi is as follows:

```
Screen height 1080, width 1920
```

Garbage Collection

C does not have automatic garbage collection like Java or other languages have. You need to do it yourself. This program does little else apart from opening a display. This can be closed by `vc_dispmanx_display_close(p_state->dispman_display)`. It doesn't actually seem to make a difference but is tidier.

Showing the free memory in RAM used by the CPU is common; tools such as `top` give a good idea of this, with `ps` drilling down deeper. Showing the memory used by the GPU is trickier, and there is a Broadcom tool `vcdbg` to show that. Even if nothing has been run, it shows some memory usage (with 512 MB of RAM devoted to the GPU).

```
$/opt/vc/lib /opt/vc/bin/vcdbg reloc

Relocatable heap version 4 found at 0x1f000000
total space allocated is 492M, with 492M relocatable, 0 legacy and 0 offline
0 legacy blocks of size 2359296

free list at 0x3d886520
489M free memory in 1 free block(s)
largest free block is 489M bytes

0x1f000000: free 489M
[    4] 0x3d886540: used  576 (refcount 1 lock count 0, size      512, align    4, data
0x3d886560, d0rual) 'ILCS VC buffer pool'
[    3] 0x3d886780: used 3.5M (refcount 1 lock count 8, size  3618816, align 4096, data
0x3d887000, d1rual) 'ARM FB'
[    2] 0x3dbfafa0: used  16K (refcount 1 lock count 0, size    16384, align   32, data
0x3dbfafc0, d0ruAl) 'audioplus_tmp_buf'
[    1] 0x3dbfefe0: used 4.0K (refcount 1 lock count 0, size        0, align 4096, data
0x3dbff000, d1rual) 'camera fast alloc arena'
small allocs not requested
```

This program is useful for showing what is going on when you *haven't* been doing garbage collection!

Screen Capture

AndrewFromMelbourne has written a program using Dispmanx called raspi2png to capture the RPi screen and save it as a PNG file. The source code demonstrates additional Dispmanx calls.

Conclusion

Dispmanx is the layer between EGL and the Broadcom GPU. You just looked at enough to supply that bridge.

Resources

- *raspidmx by AndrewFromMelbourne*: `https://github.com/AndrewFromMelbourne/raspi2png`
- *cherry: A GPU accelerated slide show daemon for Raspberry Pi*: `https://sanje2v.wordpress.com/tag/dispmanx/`

CHAPTER 5

EGL on the Raspberry Pi

EGL provides a base for graphics programming, common to many systems such as OpenCL, OpenGL, OpenGL ES, and OpenVG. It is system neutral and for the RPi is the layer above Dispmanx.

Building Programs

The following is a Makefile to build these programs:

```
DMX_INC =  -I/opt/vc/include/ -I /opt/vc/include/interface/vmcs_host/ -I/opt/vc/include/
interface/vcos/pthreads -I/opt/vc/include/interface/vmcs_host/linux
EGL_INC =
INCLUDES = $(DMX_INC) $(EGL_INC)

CFLAGS = $(INCLUDES)
CPPFLAGS = -march=armv7-a -mtune=cortex-a7

DMX_LIBS =  -L/opt/vc/lib/ -lbcm_host -lvcos -lvchiq_arm -lpthread
EGL_LIBS = -L/opt/vc/lib -lEGL -lGLESv2

LDFLAGS = $(DMX_LIBS) $(EGL_LIBS)

all: context info window
```

Overview

According to the EGL specification at www.khronos.org/registry/egl/specs/eglspec.1.5.pdf, "EGL [is] an interface between rendering APIs such as OpenCL, OpenGL, OpenGL ES, or OpenVG (referred to collectively as client APIs) and one or more underlying platforms (typically window systems such as X11)." It is not intended that application programmers write directly to EGL; instead, they should use one of the APIs such as OpenGL.

J. Newmarch, *Raspberry Pi GPU Audio Video Programming*, DOI 10.1007/978-1-4842-2472-4_5

Each underlying platform will have a means of rendering EGL surfaces. In this chapter, you will look at how EGL surfaces are linked to Dispmanx surfaces and how windows are built. You won't actually draw into any windows because that is best done using, for example, OpenGL. You will do that in Chapter 6.

Why EGL? The following is from the Wayland FAQ at `http://wayland.freedesktop.org/faq.html#heading_toc_j_12`:

> *EGL is the only GL binding API that lets us avoid dependencies on existing window systems, in particular X. GLX obviously pulls in X dependencies and only lets us set up GL on X drawables. The alternative is to write a Wayland-specific GL binding API, say, WaylandGL.*
>
> *A more subtle point is that libGL.so includes the GLX symbols, so linking to that library will pull in all the X dependencies. This means that we can't link to full GL without pulling in the client side of X, so we're using GLES2 for now. Longer term, we'll need a way to use full GL under Wayland.*

From the RPi viewpoint, EGL forms the link between native Dispmanx windows and the OpenGL ES API.

Initializing EGL

EGL has a display that it writes on. The display is built on a native display and is obtained by the call `eglGetDisplay`. The EGL platform is then initialized using `eglInitialize`.

Typically an EGL display will support a number of configurations. For example, a pixel may be 16 bits (5 red, 5 blue, and 6 green), 24 bits (8 red, 8 green, and 8 blue), or 32 bits (8 extra bits for alpha transparency). An application will specify certain parameters, such as the minimum size of a red pixel, and can then access the array of matching configurations using `eglChooseConfig`. The attributes of a configuration can be queried using `eglGetConfigAttrib`. One configuration should be chosen before proceeding.

The program `info.c` lists the possible configurations that are available to EGL.

```
/*
 * code stolen from openGL-RPi-tutorial-master/encode_OGL/
 */

#include <stdio.h>
#include <assert.h>
#include <math.h>
#include <unistd.h>

#include <EGL/egl.h>
#include <EGL/eglext.h>

typedef struct
{
    uint32_t screen_width;
    uint32_t screen_height;

    EGLDisplay display;
    EGLSurface surface;
    EGLContext context;
```

```
    EGLConfig config;
} EGL_STATE_T;

EGL_STATE_T state, *p_state = &state;

void printConfigInfo(int n, EGLDisplay display, EGLConfig *config) {
    int size;

    printf("Configuration %d is\n", n);

    eglGetConfigAttrib(display,
                       *config, EGL_RED_SIZE, &size);
    printf("  Red size is %d\n", size);
    eglGetConfigAttrib(display,
                       *config, EGL_BLUE_SIZE, &size);
    printf("  Blue size is %d\n", size);
    eglGetConfigAttrib(display,
                       *config, EGL_GREEN_SIZE, &size);
    printf("  Green size is %d\n", size);
    eglGetConfigAttrib(display,
                       *config, EGL_BUFFER_SIZE, &size);
    printf("  Buffer size is %d\n", size);

   eglGetConfigAttrib(display,
                       *config,  EGL_BIND_TO_TEXTURE_RGB , &size);
   if (size == EGL_TRUE)
       printf("  Can be bound to RGB texture\n");
   else
       printf("  Can't be bound to RGB texture\n");

   eglGetConfigAttrib(display,
                       *config,  EGL_BIND_TO_TEXTURE_RGBA , &size);
   if (size == EGL_TRUE)
       printf("  Can be bound to RGBA texture\n");
   else
       printf("  Can't be bound to RGBA texture\n");
}

void init_egl(EGL_STATE_T *state)
{
    EGLBoolean result;
    EGLint num_configs;

    EGLConfig *configs;

    // get an EGL display connection
    state->display = eglGetDisplay(EGL_DEFAULT_DISPLAY);

    // initialize the EGL display connection
    result = eglInitialize(state->display, NULL, NULL);
    if (result == EGL_FALSE) {
```

```
        fprintf(stderr, "Can't initialise EGL\n");
        exit(1);
    }

    eglGetConfigs(state->display, NULL, 0, &num_configs);
    printf("EGL has %d configs\n", num_configs);

    configs = calloc(num_configs, sizeof *configs);
    eglGetConfigs(state->display, configs, num_configs, &num_configs);

    int i;
    for (i = 0; i < num_configs; i++) {
        printConfigInfo(i, state->display, &configs[i]);
    }
}

int
main(int argc, char *argv[])
{
    init_egl(p_state);

    return 0;
}
```

The program is run by `info`.

The following is a portion of the output:

```
EGL has 28 configs
Configuration 0 is
  Red size is 8
  Blue size is 8
  Green size is 8
  Buffer size is 32
  Can't be bound to RGB texture
  Can be bound to RGBA texture
Configuration 1 is
  Red size is 8
  Blue size is 8
  Green size is 8
  Buffer size is 24
  Can be bound to RGB texture
  Can be bound to RGBA texture
```

Creating a Rendering Context

Each configuration will support one or more client APIs such as OpenGL. The API is usually requested through the configuration attribute `EGL_RENDERABLE_TYPE`, which should have a value such as `EGL_OPENGL_ES2_BIT`.

In addition to a configuration, each application needs one or more *contexts*. Each context defines a level of the API that will be used for rendering. Examples typically use a level of 2, and a context is created using `eglCreateContext`.

The following is typical code to perform these steps:

```
init_egl() {
    EGLint major, minor, count, n, size;
    EGLConfig *configs;
    int i;
    EGLint config_attribs[] = {
        EGL_SURFACE_TYPE, EGL_WINDOW_BIT,
        EGL_RED_SIZE, 8,
        EGL_GREEN_SIZE, 8,
        EGL_BLUE_SIZE, 8,
        EGL_RENDERABLE_TYPE, EGL_OPENGL_ES2_BIT,
        EGL_NONE
    };

    static const EGLint context_attribs[] = {
        EGL_CONTEXT_CLIENT_VERSION, 2,
        EGL_NONE
    };

   ...

    display->context =
        eglCreateContext(display->display,
                         display->config,
                         EGL_NO_CONTEXT, context_attribs);

}
```

The program `context.c` takes the first available configuration and then creates a context.

```
/*
 * code stolen from openGL-RPi-tutorial-master/encode_OGL/
 */

#include <stdio.h>
#include <assert.h>
#include <math.h>

#include <EGL/egl.h>
#include <EGL/eglext.h>

typedef struct
{
    uint32_t screen_width;
    uint32_t screen_height;
```

```
    EGLDisplay display;
    EGLSurface surface;
    EGLContext context;
    EGLConfig config;
} EGL_STATE_T;

EGL_STATE_T state, *p_state = &state;

void init_egl(EGL_STATE_T *state)
{
    EGLBoolean result;
    EGLint num_configs;

    static const EGLint attribute_list[] =
        {
            EGL_RED_SIZE, 8,
            EGL_GREEN_SIZE, 8,
            EGL_BLUE_SIZE, 8,
            EGL_ALPHA_SIZE, 8,
            EGL_SURFACE_TYPE, EGL_WINDOW_BIT,
            EGL_NONE
        };

    static const EGLint context_attributes[] =
        {
            EGL_CONTEXT_CLIENT_VERSION, 2,
            EGL_NONE
        };

    EGLConfig *configs;

    // get an EGL display connection
    state->display = eglGetDisplay(EGL_DEFAULT_DISPLAY);

    // initialize the EGL display connection
    result = eglInitialize(state->display, NULL, NULL);

    eglGetConfigs(state->display, NULL, 0, &num_configs);
    printf("EGL has %d configs\n", num_configs);

    configs = calloc(num_configs, sizeof *configs);
    eglGetConfigs(state->display, configs, num_configs, &num_configs);

    // get an appropriate EGL configuration - just use the first available
    result = eglChooseConfig(state->display, attribute_list,
                             &state->config, 1, &num_configs);
    assert(EGL_FALSE != result);

    // Choose the OpenGL ES API
    result = eglBindAPI(EGL_OPENGL_ES_API);
    assert(EGL_FALSE != result);
```

```
    // create an EGL rendering context
    state->context = eglCreateContext(state->display,
                                      state->config, EGL_NO_CONTEXT,
                                      context_attributes);
    assert(state->context!=EGL_NO_CONTEXT);
    printf("Got an EGL context\n");
}

int
main(int argc, char *argv[])
{
    init_egl(p_state);

    return 0;
}
```

The program is run by context and has the following output:

```
EGL has 28 configs
Got an EGL context
```

Creating an EGL Drawing Surface with Dispmanx

An EGL window needs to be created from a native window or surface. This is different for each system. For the RPi, you use a Dispmanx window, as created in the previous chapter. This is used to create an EGL drawing surface by eglCreateWindowSurface.

You take the EGL context creation from the previous program and the Dispmanx window creation from the previous chapter and build an EGL window surface with the following:

```
void egl_from_dispmanx(EGL_STATE_T *state,
                       EGL_DISPMANX_WINDOW_T *nativewindow) {
    EGLBoolean result;

    state->surface = eglCreateWindowSurface(state->display,
                                            state->config,
                                            nativewindow, NULL );
    assert(state->surface != EGL_NO_SURFACE);

    // connect the context to the surface
    result = eglMakeCurrent(state->display, state->surface, state->surface, state->context);
    assert(EGL_FALSE != result);
}
```

Drawing into this surface is generally done by an API such as OpenGL, and the choice of this has been set in the EGL context. This is set for the drawing surfaces by eglMakeCurrent.

Following this, drawing can take place. For now, you will draw only a big red rectangle occupying the screen. Once the drawing is complete, the EGL layer renders the surface by swapping the drawing buffer using eglSwapBuffers.

The program is window.c.

```
/*
 * code stolen from openGL-RPi-tutorial-master/encode_OGL/
 */

#include <stdio.h>
#include <assert.h>
#include <math.h>

#include <EGL/egl.h>
#include <EGL/eglext.h>
#include <GLES2/gl2.h>

#include <bcm_host.h>

typedef struct
{
    EGLDisplay display;
    EGLSurface surface;
    EGLContext context;
    EGLConfig config;

    EGL_DISPMANX_WINDOW_T nativewindow;
    DISPMANX_DISPLAY_HANDLE_T dispman_display;
} EGL_STATE_T;

EGL_STATE_T state = {
    .display = EGL_NO_DISPLAY,
    .surface = EGL_NO_SURFACE,
    .context = EGL_NO_CONTEXT
};
EGL_STATE_T *p_state = &state;

void init_egl(EGL_STATE_T *state)
{
    EGLint num_configs;
    EGLBoolean result;

    bcm_host_init();

    static const EGLint attribute_list[] =
        {
            EGL_RED_SIZE, 8,
            EGL_GREEN_SIZE, 8,
            EGL_BLUE_SIZE, 8,
            EGL_ALPHA_SIZE, 8,
            EGL_SURFACE_TYPE, EGL_WINDOW_BIT,
            EGL_NONE
        };

    static const EGLint context_attributes[] =
        {
```

```
            EGL_CONTEXT_CLIENT_VERSION, 2,
            EGL_NONE
        };

    // get an EGL display connection
    state->display = eglGetDisplay(EGL_DEFAULT_DISPLAY);

    // initialize the EGL display connection
    result = eglInitialize(state->display, NULL, NULL);

    // get an appropriate EGL frame buffer configuration
    result = eglChooseConfig(state->display, attribute_list, &state->config, 1,
    &num_configs);
    assert(EGL_FALSE != result);

    // Choose the OpenGL ES API
    result = eglBindAPI(EGL_OPENGL_ES_API);
    assert(EGL_FALSE != result);

    // create an EGL rendering context
    state->context = eglCreateContext(state->display,
                                      state->config, EGL_NO_CONTEXT,
                                      context_attributes);
    assert(state->context!=EGL_NO_CONTEXT);
}

void init_dispmanx(EGL_STATE_T *state) {
    EGL_DISPMANX_WINDOW_T *nativewindow = &p_state->nativewindow;
    int32_t success = 0;
    uint32_t screen_width;
    uint32_t screen_height;

    DISPMANX_ELEMENT_HANDLE_T dispman_element;
    DISPMANX_DISPLAY_HANDLE_T dispman_display;
    DISPMANX_UPDATE_HANDLE_T dispman_update;
    VC_RECT_T dst_rect;
    VC_RECT_T src_rect;

    bcm_host_init();

    // create an EGL window surface
    success = graphics_get_display_size(0 /* LCD */,
                                        &screen_width,
                                        &screen_height);
    assert( success >= 0 );

    dst_rect.x = 0;
    dst_rect.y = 0;
    dst_rect.width = screen_width;
    dst_rect.height = screen_height;
```

```
    src_rect.x = 0;
    src_rect.y = 0;
    src_rect.width = screen_width << 16;
    src_rect.height = screen_height << 16;

    dispman_display = vc_dispmanx_display_open( 0 /* LCD */);
    dispman_update = vc_dispmanx_update_start( 0 );
    state->dispman_display = dispman_display;

    dispman_element =
        vc_dispmanx_element_add(dispman_update, dispman_display,
                                0/*layer*/, &dst_rect, 0/*src*/,
                                &src_rect, DISPMANX_PROTECTION_NONE,
                                0 /*alpha*/, 0/*clamp*/, 0/*transform*/);

    // Build an EGL_DISPMANX_WINDOW_T from the Dispmanx window
    nativewindow->element = dispman_element;
    nativewindow->width = screen_width;
    nativewindow->height = screen_height;
    vc_dispmanx_update_submit_sync(dispman_update);
    assert(vc_dispmanx_element_remove(dispman_update, dispman_element) == 0);

    printf("Got a Dispmanx window\n");
}

void egl_from_dispmanx(EGL_STATE_T *state) {
    EGLBoolean result;

    state->surface = eglCreateWindowSurface(state->display,
                                            state->config,
                                            &p_state->nativewindow, NULL );
    assert(state->surface != EGL_NO_SURFACE);

    // connect the context to the surface
    result = eglMakeCurrent(state->display, state->surface, state->surface, state->context);
    assert(EGL_FALSE != result);
}

void cleanup(int s) {
    if (p_state->surface != EGL_NO_SURFACE &&
        eglDestroySurface(p_state->display, p_state->surface)) {
        printf("Surface destroyed ok\n");
    }
    if (p_state->context !=  EGL_NO_CONTEXT &&
        eglDestroyContext(p_state->display, p_state->context)) {
        printf("Main context destroyed ok\n");
    }
    if (p_state->display != EGL_NO_DISPLAY &&
        eglTerminate(p_state->display)) {
        printf("Display terminated ok\n");
    }
```

```
    if (eglReleaseThread()) {
        printf("EGL thread resources released ok\n");
    }
    if (vc_dispmanx_display_close(p_state->dispman_display) == 0) {
        printf("Dispmanx display rleased ok\n");
    }
    bcm_host_deinit();
    exit(s);
}

int
main(int argc, char *argv[])
{
    signal(SIGINT, cleanup);

    init_egl(p_state);

    init_dispmanx(p_state);

    egl_from_dispmanx(p_state);

    glClearColor(1.0, 0.0, 0.0, 1.0);
    glClear(GL_COLOR_BUFFER_BIT);
    glFlush();

    eglSwapBuffers(p_state->display, p_state->surface);

    sleep(4);

    cleanup(0);
    return 0;
}
```

The program is run by `window` and has the following output:

```
Got a Dispmanx window
Surface destroyed ok
Main context destroyed ok
Display terminated ok
EGL thread resources released ok
Dispmanx display rleased ok
```

Garbage Collection

C does not have automatic garbage collection like Java or other languages have. You have to do it yourself. The function `cleanup` manages this. It is also attached to a `SIG_INT` interrupt handler to clean up after Ctrl-C interrupts.

```
void cleanup(int s) {
    if (p_state->surface != EGL_NO_SURFACE &&
```

```
        eglDestroySurface(p_state->display, p_state->surface)) {
        printf("Surface destroyed ok\n");
    }
    if (p_state->context !=  EGL_NO_CONTEXT &&
        eglDestroyContext(p_state->display, p_state->context)) {
        printf("Main context destroyed ok\n");
    }
    if (p_state->display != EGL_NO_DISPLAY &&
        eglTerminate(p_state->display)) {
        printf("Display terminated ok\n");
    }
    if (eglReleaseThread()) {
        printf("EGL thread resources released ok\n");
    }
    if (vc_dispmanx_display_close(p_state->dispman_display) == 0) {
        printf("Dispmanx display released ok\n");
    }
    bcm_host_deinit();
    exit(s);
}
```

Garbage collection is not complete; there seems to be a number of Khronos calls that don't go away.

```
$/opt/vc/bin/vcdbg reloc

Relocatable heap version 4 found at 0x1f000000
total space allocated is 492M, with 490M relocatable, 0 legacy and 2.3M offline
0 legacy blocks of size 2359296

free list at 0x3bf78d20
461M free memory in 1 free block(s)
largest free block is 461M bytes

0x1f000000: offline 2.3M
0x1f240000: free 461M
[  31] 0x3bf78d40: used  96K (refcount 1 lock count 1, size    98304, align  256, data
0x3bf78e00, D1rual) 'khrn_hw_bin_mem'
[  30] 0x3bf90e60: used  96K (refcount 1 lock count 1, size    98304, align  256, data
0x3bf90f00, D1rual) 'khrn_hw_bin_mem'
0x3bf90e60: corrupt trailer (space 98592 != 98593)
[  29] 0x3bfa8f80: used  96K (refcount 0 lock count 0, size    98304, align  256, data
0x3bfa9000, D1ruAl) 'khrn_hw_bin_mem'
0x3bfc10a0: free 25M
[  11] 0x3d885700: used   96 (refcount 0 lock count 0, size       22, align    1, data
0x3d885720, d1rual) 'khrn_hw_null_render'
0x3d885760: free 3.5K
[   4] 0x3d886540: used  576 (refcount 1 lock count 0, size      512, align    4, data
0x3d886560, d0rual) 'ILCS VC buffer pool'
[   3] 0x3d886780: used 3.5M (refcount 1 lock count 8, size  3618816, align 4096, data
0x3d887000, d1rual) 'ARM FB'
```

```
[   2] 0x3dbfafa0: used  16K (refcount 1 lock count 0, size    16384, align   32, data
0x3dbfafc0, d0ruAl) 'audioplus_tmp_buf'
[   1] 0x3dbfefe0: used 4.0K (refcount 1 lock count 0, size        0, align 4096, data
0x3dbff000, d1rual) 'camera fast alloc arena'
heap corruption detected
small allocs not requested
```

This appears to be a fixed allocation that does not increase over multiple runs of programs.

Conclusion

This chapter has looked at EGL, which is the layer above Dispmanx and the layer below the major APIs of OpenGL ES, OpenVG, and so on.

Resources

- *eglIntro—introduction to managing client API rendering through the EGL API*: https://www.khronos.org/registry/egl/sdk/docs/man/html/eglIntro.xhtml
- *Khronos Native Platform Graphics Interface (EGL Version 1.5)*: www.khronos.org/registry/egl/specs/eglspec.1.5.pdf

CHAPTER 6

OpenGL ES on the Raspberry Pi

OpenGL ES is one of the means of drawing graphics on the RPi. It is an extremely sophisticated system that I introduce in this chapter but do not attempt to cover in detail. The major contribution of this chapter is to establish how you use Dispmanx to create an EGL surface so you can do OpenGL ES rendering. It describes a version of `esUtils.c` from the Munshi et al. book adapted to the RPi and gives some examples of its use. This gives the basics of what is required to run OpenGL ES programs on the RPi but is not intended to be more than an introduction to general OpenGL ES programming.

Building Programs

Download Benosteen's OpenGL ES files for the RPi.

```
git clone git://github.com/benosteen/opengles-book-samples.git
```

This will create a directory structure in the current directory, including the RPi files you will use in this chapter in the directory `opengles-book-samples/Raspi/Common`. For convenience, move this up to the current directory, as shown here:

```
mv opengles-book-samples/Raspi/Common.
```

Add the following `Makefile` to the directory `Common`:

```
#
# Raspbery Pi library for building 'OpenGL ES 2.0 Programming Guide' examples
#
LIBES=libRPes2pg.a

DMX_INC =  -I/opt/vc/include/ -I /opt/vc/include/interface/vmcs_host/ -I/opt/vc/include/
interface/vcos/pthreads -I/opt/vc/include/interface/vmcs_host/linux
EGL_INC =
GLES_INC =
INCLUDES = $(DMX_INC) $(EGL_INC) $(GLES_INC)
CFLAGS = $(INCLUDES) -DRPI_NO_X -DPI_DISPLAY=4
```


J. Newmarch, *Raspberry Pi GPU Audio Video Programming*, DOI 10.1007/978-1-4842-2472-4_6

```
LIBOBJ = esUtil.o esShader.o esShapes.o esTransform.o

$(LIBES): $(LIBOBJ)
	ar rc $(LIBES) $(LIBOBJ)
	ranlib $(LIBES)

$(LIBOBJ): Makefile
clean:
	rm -f $(LIBOBJ) $(LIBES)
```

The purpose of this `Makefile` is to create the library file `libRPes2pg.a`. Then the programs in this chapter can be built using the `Makefile`.

```
DMX_INC =  -I/opt/vc/include/ -I /opt/vc/include/interface/vmcs_host/ -I/opt/vc/include/
interface/vcos/pthreads -I/opt/vc/include/interface/vmcs_host/linux
EGL_INC =
GLES_INC = -ICommon/
INCLUDES = $(DMX_INC) $(EGL_INC) $(GLES_INC)
CFLAGS = $(INCLUDES)

DMX_LIBS =  -L/opt/vc/lib/ -lbcm_host -lvcos -lvchiq_arm -lpthread
EGL_LIBS = -L/opt/vc/lib/ -lEGL -lGLESv2
ESP_LIBS = -LCommon -lRPes2pg
LDLIBS = $(DMX_LIBS) $(EGL_LIBS) $(ESP_LIBS) -lm

SRC = Hello_Triangle Hello_TriangleColour Simple_Image Simple_Texture2D Rotate_Image Hello_
Square

all: $(SRC)

$(SRC): Common/libRPes2pg.a

Common/libRPes2pg.a:
    make -C Common
```

OpenGL ES

In this chapter, you actually get to draw things, using the OpenGL ES API supported by the RPi. OpenGL was first specified in 1992, arising out of earlier systems by Silicon Graphics. Obviously, being so old, it has accreted a certain amount of fluff. OpenGL ES is a simplified form, more suitable to low-profile computers such as the RPi. OpenGL ES is now up to version 3, but the RPi supports only version 2. (Actually, it also supports version 1, but we will ignore that.)

Note that the latest Raspbian image from February 2016 has experimental support for full OpenGL. This book will not discuss this; I would expect its main use would be to build in support for the forthcoming Vulkan graphics system.

OpenGL ES is a significantly complex system; generating 2D and 3D images is an exceedingly nontrivial task! The canonical reference for this API is the *OpenGL ES 2.0 Programming Guide* by Aaftab Munshi, Dan Ginsburg, and Dave Shreiner. It is nearly 500 pages and is packed with information about how to program OpenGL ES.

Obviously, I cannot hope to replicate all the material in that book. Nevertheless, I will address the issues of getting OpenGL ES running on the RPi and show you a number of simple examples.

The esUtil Functions

The Munshi et al. book tries to present a platform-independent view of OpenGL ES programming so that the example programs will run essentially unchanged across multiple platforms, including Linux, Windows, the iPhone, and Android. To do this, it has to abstract above the OS-dependent layers. It does so with a set of functions wrapped in files such as `esUtil.c`, one version for each OS.

The original book did not contain any version of this file for the RPi (it hadn't been invented then), but authors such as Benosteen have created versions. His version contains code that will run both under the X Window System and as stand-alone; I will cover the stand-alone version.

The `esUtil.c` file essentially defines four functions.

- `esInitContext` initializes their framework.
- `esCreateWindow` creates an `EGLContext` for drawing.
- `esRegisterDrawFunc` registers a suitable OpenGL ES drawing function.
- `esMainLoop` draws things in a loop.

The library uses a struct.

```
typedef struct _escontext
{
   void*       userData;
   GLint       width;
   GLint       height;
   EGLNativeWindowType  hWnd;
   EGLDisplay  eglDisplay;
   EGLContext  eglContext;
   EGLSurface  eglSurface;
   void (ESCALLBACK *drawFunc) ( struct _escontext * );
   void (ESCALLBACK *keyFunc) ( struct _escontext *, unsigned char, int, int );
   void (ESCALLBACK *updateFunc) ( struct _escontext *, float deltaTime );
} ESContext;
```

This gives OS-independent information needed for drawing OpenGL ES. The four fields of `EGLNativeWindowType`, `EGLDisplay`, `EGLContext`, and `EGLSurface` were all covered for the RPi in previous chapters and are defined for the RPi as follows:

```
///
// CreateEGLContext()
//
//    Creates an EGL rendering context and all associated elements
//
EGLBoolean CreateEGLContext  (EGLNativeWindowType hWnd, EGLDisplay* eglDisplay,
                              EGLContext* eglContext, EGLSurface* eglSurface,
                              EGLint attribList[])
{
   EGLint numConfigs;
   EGLint majorVersion;
   EGLint minorVersion;
   EGLDisplay display;
   EGLContext context;
   EGLSurface surface;
```

```
EGLConfig config;
EGLint contextAttribs[] = { EGL_CONTEXT_CLIENT_VERSION, 2, EGL_NONE, EGL_NONE };

// Get Display
display = eglGetDisplay(EGL_DEFAULT_DISPLAY);
if ( display == EGL_NO_DISPLAY )
{
   return EGL_FALSE;
}

// Initialize EGL
if ( !eglInitialize(display, &majorVersion, &minorVersion) )
{
   return EGL_FALSE;
}

// Get configs
if ( !eglGetConfigs(display, NULL, 0, νmConfigs) )
{
   return EGL_FALSE;
}

// Choose config
if ( !eglChooseConfig(display, attribList, &config, 1, &numConfigs) )
{
   return EGL_FALSE;
}

// Create a surface
surface = eglCreateWindowSurface(display, config, (EGLNativeWindowType)hWnd, NULL);
if ( surface == EGL_NO_SURFACE )
{
   return EGL_FALSE;
}

// Create a GL context
context = eglCreateContext(display, config, EGL_NO_CONTEXT, contextAttribs );
if ( context == EGL_NO_CONTEXT )
{
   return EGL_FALSE;
}

// Make the context current
if ( !eglMakeCurrent(display, surface, surface, context) )
{
   return EGL_FALSE;
}

*eglDisplay = display;
*eglSurface = surface;
*eglContext = context;
```

```
    return EGL_TRUE;
}

///
//  WinCreate()
//
//      This function initialized the native X11 display and window for EGL
//
EGLBoolean WinCreate(ESContext *esContext, const char *title)
{
    int32_t success = 0;
    uint32_t screen_width;
    uint32_t screen_height;

    EGL_DISPMANX_WINDOW_T *nativewindow = malloc(sizeof(EGL_DISPMANX_WINDOW_T));
    DISPMANX_ELEMENT_HANDLE_T dispman_element;
    DISPMANX_DISPLAY_HANDLE_T dispman_display;
    DISPMANX_UPDATE_HANDLE_T dispman_update;
    VC_RECT_T dst_rect;
    VC_RECT_T src_rect;

    bcm_host_init();

    // create an EGL window surface
    success = graphics_get_display_size(0 /* LCD */,
                                        &screen_width,
                                        &screen_height);
    assert( success >= 0 );

    dst_rect.x = 0;
    dst_rect.y = 0;
    dst_rect.width = screen_width;
    dst_rect.height = screen_height;

    src_rect.x = 0;
    src_rect.y = 0;
    src_rect.width = screen_width << 16;
    src_rect.height = screen_height << 16;

    dispman_display = vc_dispmanx_display_open( 0 /* LCD */);
    dispman_update = vc_dispmanx_update_start( 0 );

    dispman_element =
        vc_dispmanx_element_add(dispman_update, dispman_display,
                                0/*layer*/, &dst_rect, 0/*src*/,
                                &src_rect, DISPMANX_PROTECTION_NONE,
                                0 /*alpha*/, 0/*clamp*/, 0/*transform*/);

    // Build an EGL_DISPMANX_WINDOW_T from the Dispmanx window
    nativewindow->element = dispman_element;
    nativewindow->width = screen_width;
    nativewindow->height = screen_height;
```

```
    vc_dispmanx_update_submit_sync(dispman_update);

    printf("Got a Dispmanx window\n");

    esContext->hWnd = (EGLNativeWindowType) nativewindow;

    return EGL_TRUE;
}
```

These are used in a typical application like this, where UserData is some application-dependent structure:

```
ESContext esContext;
UserData  userData;

esInitContext ( &esContext );
esContext.userData = &userData;

esCreateWindow ( &esContext, "My application", 320, 240, ES_WINDOW_RGB );
```

Vertices

A key data structure for drawing is the vertex, which is a point in two- or three-dimensional space. Each vertex may have attributes associated with it such as a color. A line has two vertices, a triangle has three, a square has four, and a circle has as many points around the circumference as are needed to look like a circle.

Using three-dimensional (x, y, z) coordinates, you can specify the vertices of a triangle by doing the following:

```
GLfloat vVertices[] = {  0.0f,  0.5f, 0.0f,
                        -0.5f, -0.5f, 0.0f,
                         0.5f, -0.5f, 0.0f };
```

Note that the coordinate system (www.matrix44.net/cms/notes/opengl-3d-graphics/coordinate-systems-in-opengl) for OpenGL ES is

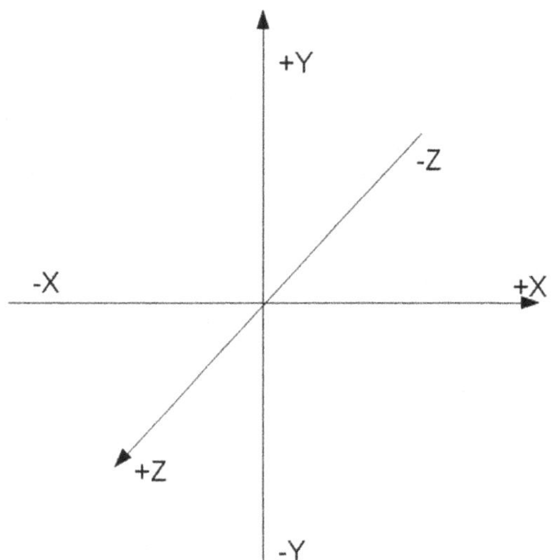

Figure 6-1. *OpenGL ES coordinate system*

so that this triangle is

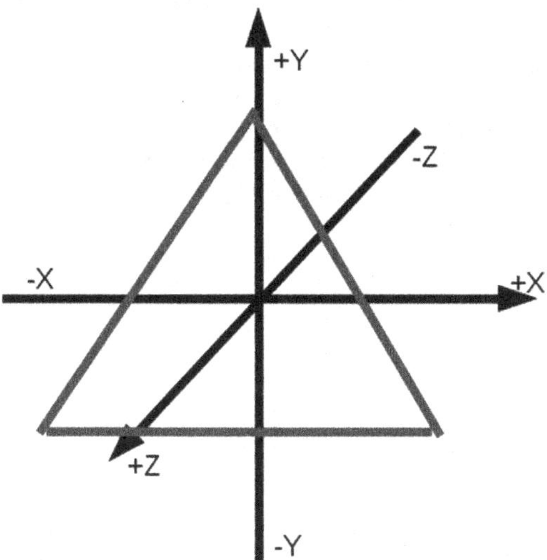

Figure 6-2. *Triangle in the coordinate system*

(but note, you haven't given the vertices any color yet, so it isn't going to appear red until you do.)

Shaders

Shaders are programs that run on the GPU rather than on the CPU. There is a communications pipeline from the CPU to the GPU: the CPU sends programs and data to the GPU, which then executes them. There are two types of programs that run on the GPU: vertex shaders and fragment shaders. Simplifying horribly dramatically the relationship between these is

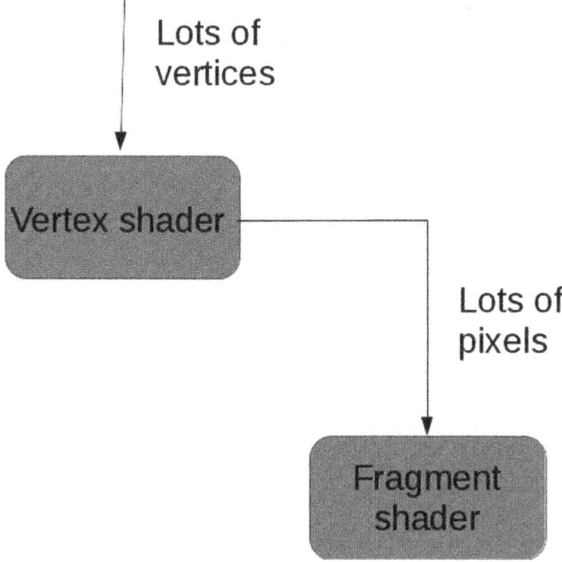

Figure 6-3. *Communications path*

The vertex shader is handed vertices one at a time and performs some operation such as moving the vertex to a new position. The OpenGL ES pipeline will then work out which pixels should be drawn and hand them one at a time to the fragment shader, which works out the color of the pixel.

How does OpenGL ES know which pixels are to be drawn? Because the application will have said, "Draw triangles incorporating these triplets of vertices and fill them in" (more details soon).

So, what do shaders look like? They are in a C-like language, custom designed for OpenGL ES. The language supports all the operations required for graphics and is quite extensive. I will need to cover some of it in the sequel. The specification is called the OpenGL ES Shading Language (`https://www.khronos.org/files/opengles_shading_language.pdf`).

Minimal Vertex Shader

The following is a really simple vertex shader:

```
attribute vec4 vPosition;
void main()
{
   gl_Position = vPosition;
}
```

This takes a parameter called `vPosition` and assigns it to the variable `gl_Position`. *Every vertex shader must assign a value to this variable.* This shader is taking an input value for the location of the vertex and assigning the final location of that vertex to the same value.

The data type of `gl_Position` is a four-dimensional vector (x, y, z, w). The first three components are the (x, y, z) values of the vertex, while the fourth is typically set to 1.

The shader is *not* C code; it is not compiled by the application. It is passed as a string to the GPU and compiled at runtime by the GPU.

Minimal Fragment Shader

A fragment shader is responsible for setting a color on a pixel. A color is also a four-dimensional vector, of the form (red, green, blue, alpha), with all values between 0 and 1. The alpha value is the amount of "transparency," where 1 means opaque and 0 means transparent.

A minimal fragment shader just sets the pixel to opaque red.

```
void main()
{
   gl_FragColor = vec4 ( 1.0, 0.0, 0.0, 1.0 );
}
```

The variable `gl_FragColor` *must* be assigned a value.

This also is not compiled by the application but is passed as a string to the GPU, which compiles it at runtime.

Loading the Shaders

The shaders are given as source code strings to OpenGL ES. You have to create a shader object, tell it the source code, compile the shader, and check for errors. The code for this is as follows:

```
GLuint LoadShader ( GLenum type, const char *shaderSrc )
{
   GLuint shader;
   GLint compiled;

   // Create the shader object
   shader = glCreateShader ( type );

   if ( shader == 0 )
        return 0;

   // Load the shader source
   glShaderSource ( shader, 1, &shaderSrc, NULL );

   // Compile the shader
   glCompileShader ( shader );

   // Check the compile status
   glGetShaderiv ( shader, GL_COMPILE_STATUS, &compiled );

   if ( !compiled )
   {
      GLint infoLen = 0;

      glGetShaderiv ( shader, GL_INFO_LOG_LENGTH, &infoLen );

      if ( infoLen > 1 )
      {
         char* infoLog = malloc (sizeof(char) * infoLen );
         char* infoLog = malloc (sizeof(char) * infoLen );

         glGetShaderInfoLog ( shader, infoLen, NULL, infoLog );
         esLogMessage ( "Error compiling shader:\n%s\n", infoLog );

         free ( infoLog );
      }

      glDeleteShader ( shader );
      return 0;
   }

   return shader;

}
```

Creating the Program Object

The OpenGL ES program to run in the GPU must be created and the shaders attached. This is a bit messy as you have to define the shaders, load the shaders, create the program that will run the shaders, attach the shaders to the program, add the information about the attributes (here `vPosition`), link the program, and take error correction action if it fails.

The typical code to do this is as follows:

```
int Init ( ESContext *esContext )
{
   esContext->userData = malloc(sizeof(UserData));

   UserData *userData = esContext->userData;
   GLbyte vShaderStr[] =
      "attribute vec4 vPosition;    \n"
      "void main()                  \n"
      "{                            \n"
      "   gl_Position = vPosition;  \n"
      "}                            \n";

   GLbyte fShaderStr[] =
       "void main()                                  \n"
      "{                                            \n"
      "  gl_FragColor = vec4 ( 1.0, 0.0, 0.0, 1.0 );\n"
      "}                                            \n";

   GLuint vertexShader;
   GLuint fragmentShader;
   GLuint programObject;
   GLint linked;

   // Load the vertex/fragment shaders
   vertexShader = LoadShader ( GL_VERTEX_SHADER, vShaderStr );
   fragmentShader = LoadShader ( GL_FRAGMENT_SHADER, fShaderStr );

   // Create the program object
   programObject = glCreateProgram ( );

   if ( programObject == 0 )
      return 0;

   glAttachShader ( programObject, vertexShader );
   glAttachShader ( programObject, fragmentShader );

   // Bind vPosition to attribute 0
   glBindAttribLocation ( programObject, 0, "vPosition" );

   // Link the program
   glLinkProgram ( programObject );
```

```
   // Check the link status
   glGetProgramiv ( programObject, GL_LINK_STATUS, &linked );

   if ( !linked )
   {
      GLint infoLen = 0;

      glGetProgramiv ( programObject, GL_INFO_LOG_LENGTH, &infoLen );

      if ( infoLen > 1 )
      {
         char* infoLog = malloc (sizeof(char) * infoLen );

         glGetProgramInfoLog ( programObject, infoLen, NULL, infoLog );
         esLogMessage ( "Error linking program:\n%s\n", infoLog );

         free ( infoLog );
      }

      glDeleteProgram ( programObject );
      return GL_FALSE;
   }

   // Store the program object
   userData->programObject = programObject;

   glClearColor ( 0.0f, 0.0f, 0.0f, 0.0f );
   return GL_TRUE;
}
```

Drawing Something

That's a lot of code so far, and we still haven't drawn anything! That's the purpose of the next function, which will set up the array of vertices, set a viewport in which to see the drawing, establish the vertices to be drawn, and finally draw them.

Establishing the vertices involves first setting the active program by `glUseProgram`; multiple programs can exist at any one time, but only one is active for drawing (you have defined only one so far).

The array of vertices given as C code must then be set so that OpenGL ES can hand each vertex to the attribute `vPosition`. The array is, for example, the set of triangle vertices that you already saw early on.

```
GLfloat vVertices[] = {  0.0f,  0.5f, 0.0f,
                        -0.5f, -0.5f, 0.0f,
                         0.5f, -0.5f, 0.0f };
```

The attribute that will be set to each vertex in turn is `vPosition` in the vertex shader. In the C code, you don't use the string `vPosition` to identify it, but instead use the integer index 0 that you set *once* when creating and linking the program with the following:

```
// Bind vPosition to attribute 0
glBindAttribLocation ( programObject, 0, "vPosition" );
```

The link is made with the following function:

```
VertexAttribPointer( uint index, int size, enum type,
                     boolean normalized, sizei stride, const
                     void *pointer );
```

where you specify 0 for the index attribute, 3 for the number of elements in each vertex (for x, y and z), GL_FLOAT for the data type of each array element, GL_FALSE for normalized data, and a stride of 0, meaning that you have only vertex data and nothing else in the vertex array.

```
glVertexAttribPointer ( 0, 3, GL_FLOAT, GL_FALSE, 0, vVertices );
```

One last little wrinkle and then you can draw: attributes by default are *disabled*, meaning that the C data will not be given to the attribute. The attribute (with index 0) must be enabled with the following:

```
glEnableVertexAttribArray ( 0 );
```

Now you draw using the function glDrawArrays, which takes three parameters: the drawing mode (you will use triangle mode for the single triangle), the initial vertex in the vertex array, and the number of vertices to be drawn.

```
glDrawArrays ( GL_TRIANGLES, 0, 3 );
```

The draw function is as follows:

```
void Draw ( ESContext *esContext )
{
   UserData *userData = esContext->userData;
   GLfloat vVertices[] = {  0.0f,  0.5f, 0.0f,
                           -0.5f, -0.5f, 0.0f,
                            0.5f, -0.5f, 0.0f };

   // Set the viewport
   glViewport ( 0, 0, esContext->width, esContext->height );

   // Clear the color buffer
   glClear ( GL_COLOR_BUFFER_BIT );

   // Use the program object
   glUseProgram ( userData->programObject );

   // Load the vertex data
   glVertexAttribPointer ( 0, 3, GL_FLOAT, GL_FALSE, 0, vVertices );
   glEnableVertexAttribArray ( 0 );

   glDrawArrays ( GL_TRIANGLES, 0, 3 );
}
```

Putting It Together

The main function brings it all together.

```
int main ( int argc, char *argv[] )
{
   ESContext esContext;
   UserData  userData;

   esInitContext ( &esContext );
   esContext.userData = &userData;

   esCreateWindow ( &esContext, "Hello Triangle", 320, 240, ES_WINDOW_RGB );

   if ( !Init ( &esContext ) )
      return 0;

   esRegisterDrawFunc ( &esContext, Draw );

   esMainLoop ( &esContext );
}
```

Drawing an Opaque Red Triangle

If you now take the set of functions from esUtil.c, the convenience functions defined earlier with the special values for arrays and shaders, then you can finally draw a red opaque triangle. The program is Hello_Triangle.c.

```
//
// Book:      OpenGL(R) ES 2.0 Programming Guide
// Authors:   Aaftab Munshi, Dan Ginsburg, Dave Shreiner
// ISBN-10:   0321502795
// ISBN-13:   9780321502797
// Publisher: Addison-Wesley Professional
// URLs:      http://safari.informit.com/9780321563835
//            http://www.opengles-book.com
//

// Hello_Triangle.c
//
//    This is a simple example that draws a single triangle with
//    a minimal vertex/fragment shader.  The purpose of this
//    example is to demonstrate the basic concepts of
//    OpenGL ES 2.0 rendering.
#include <stdlib.h>
#include "esUtil.h"
```

```
typedef struct
{
   // Handle to a program object
   GLuint programObject;

} UserData;

///
// Create a shader object, load the shader source, and
// compile the shader.
//
GLuint LoadShader ( GLenum type, const char *shaderSrc )
{
   GLuint shader;
   GLint compiled;

   // Create the shader object
   shader = glCreateShader ( type );

   if ( shader == 0 )
        return 0;

   // Load the shader source
   glShaderSource ( shader, 1, &shaderSrc, NULL );

   // Compile the shader
   glCompileShader ( shader );

   // Check the compile status
   glGetShaderiv ( shader, GL_COMPILE_STATUS, &compiled );

   if ( !compiled )
   {
      GLint infoLen = 0;

      glGetShaderiv ( shader, GL_INFO_LOG_LENGTH, &infoLen );

      if ( infoLen > 1 )
      {
         char* infoLog = malloc (sizeof(char) * infoLen );

         glGetShaderInfoLog ( shader, infoLen, NULL, infoLog );
         esLogMessage ( "Error compiling shader:\n%s\n", infoLog );

         free ( infoLog );
      }

      glDeleteShader ( shader );
      return 0;
   }

   return shader;

}
```

```
///
// Initialize the shader and program object
//
int Init ( ESContext *esContext )
{
   esContext->userData = malloc(sizeof(UserData));

   UserData *userData = esContext->userData;
   GLbyte vShaderStr[] =
      "attribute vec4 vPosition;    \n"
      "void main()                  \n"
      "{                            \n"
      "    gl_Position = vPosition; \n"
      "}                            \n";

   GLbyte fShaderStr[] =
      "precision mediump float;\n"\
      "void main()                                \n"
      "{                                          \n"
      "   gl_FragColor = vec4 ( 1.0, 0.0, 0.0, 1.0 );\n"
      "}                                          \n";

   GLuint vertexShader;
   GLuint fragmentShader;
   GLuint programObject;
   GLint linked;

   // Load the vertex/fragment shaders
   vertexShader = LoadShader ( GL_VERTEX_SHADER, vShaderStr );
   fragmentShader = LoadShader ( GL_FRAGMENT_SHADER, fShaderStr );

   // Create the program object
   programObject = glCreateProgram ( );

   if ( programObject == 0 )
      return 0;

   glAttachShader ( programObject, vertexShader );
   glAttachShader ( programObject, fragmentShader );

   // Bind vPosition to attribute 0
   glBindAttribLocation ( programObject, 0, "vPosition" );

   // Link the program
   glLinkProgram ( programObject );

   // Check the link status
   glGetProgramiv ( programObject, GL_LINK_STATUS, &linked );
```

```
   if ( !linked )
   {
      GLint infoLen = 0;

      glGetProgramiv ( programObject, GL_INFO_LOG_LENGTH, &infoLen );

      if ( infoLen > 1 )
      {
         char* infoLog = malloc (sizeof(char) * infoLen );

         glGetProgramInfoLog ( programObject, infoLen, NULL, infoLog );
         esLogMessage ( "Error linking program:\n%s\n", infoLog );

         free ( infoLog );
      }

      glDeleteProgram ( programObject );
      return GL_FALSE;
   }

   // Store the program object
   userData->programObject = programObject;

   glClearColor ( 0.0f, 0.0f, 0.0f, 0.0f );
   return GL_TRUE;
}

///
// Draw a triangle using the shader pair created in Init()
//
void Draw ( ESContext *esContext )
{
   UserData *userData = esContext->userData;
   GLfloat vVertices[] = {  0.0f,  0.5f, 0.0f,
                           -0.5f, -0.5f, 0.0f,
                            0.5f, -0.5f, 0.0f };

   // Set the viewport
   glViewport ( 0, 0, esContext->width, esContext->height );

   // Clear the color buffer
   glClear ( GL_COLOR_BUFFER_BIT );

   // Use the program object
   glUseProgram ( userData->programObject );

   // Load the vertex data
   glVertexAttribPointer ( 0, 3, GL_FLOAT, GL_FALSE, 0, vVertices );
   glEnableVertexAttribArray ( 0 );

   glDrawArrays ( GL_TRIANGLES, 0, 3 );
}
```

```
int main ( int argc, char *argv[] )
{
   ESContext esContext;
   UserData  userData;

   esInitContext ( &esContext );
   esContext.userData = &userData;

   esCreateWindow ( &esContext, "Hello Triangle", 320, 240, ES_WINDOW_RGB );

   if ( !Init ( &esContext ) )
      return 0;

   esRegisterDrawFunc ( &esContext, Draw );

   esMainLoop ( &esContext );
}
```

The program is run by calling `Hello_Triangle` and draws a small red triangle.

A Colored Triangle

In the previous example, you drew a triangle with a solid red color. This was done by setting the fragment shader variable `fragmentColor` to opaque red. But what if you wanted to get a color triangle?

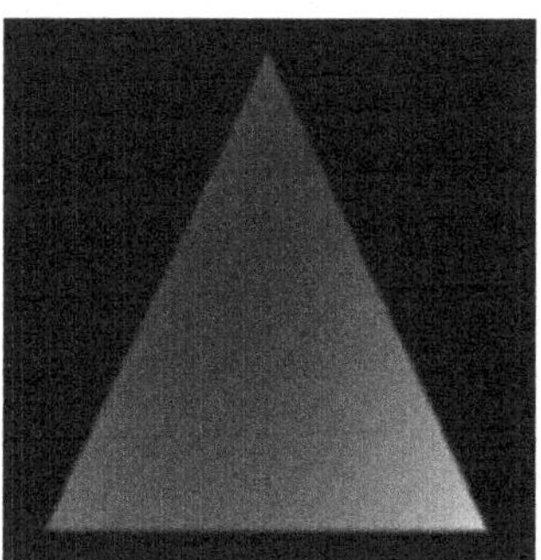

***Figure 6-4.** Coloured triangle*

Do you have to calculate all of the color values yourself?

Varyings

You could do it by yourself, but fortunately OpenGL ES can do the work for you. It has the concept of *varyings*, which are variables with values interpolated from the vertex shader data. For the colored triangle, the interpolated values are calculated using the vertex colors, averaged out based on the distance from each vertex.

To use them, you define a variable to be varying in both the vertex and the fragment shader, with the same name. You set a value in each vertex using a new attribute and then can access the interpolated value in each fragment shader. The vertex shader is as follows:

```
attribute vec4 vPosition;
attribute vec4 vColour;

varying vFragmentColour;

void main()
{
   gl_Position = vPosition;
   vFragmentColour = vColour;
}
```

The fragment shader is as follows:

```
varying vec4 vFragmentColour;

void main()
{
   gl_FragColor = vFragmentColour;
}
```

Passing Multiple Attributes to the Vertex Shader

In the simple triangle program, you passed a set of vertex values to the vertex shader by first setting the index of the attribute to 0 and then calling the vertex attribute pointer function. To add more attributes, the simplest way (but not the most efficient) is to set extra attributes to higher index values (one, two, and so on) and call the vertex attribute pointer function on these as well.

Here are the changes required to set colors for the color attribute:

```
GLfloat vColours[] = {1.0f, 0.0f, 0.0f,
                      0.0f, 1.0f, 0.0f,
                      0.0f, 0.0f, 1.0f,};

...

glVertexAttribPointer ( 1, 3, GL_FLOAT, GL_FALSE, 0, vColours );
glEnableVertexAttribArray ( 1 );
```

Here are more changes:

```
// Bind vColour to attribute 1
glBindAttribLocation ( programObject, 1, "vColour" );
```

The program is `Hello_TriangleColour.c`. It isn't worth including the full code here because only the previous lines differ from the `Hello_Triangle.c` program. It is run by `Hello_TriangleColour`.

Drawing Squares and Other Shapes

If you want to draw complex shapes such as squares, cubes, or even circles and spheres, you need to break them down into triangles. Squares and other regular shapes are easy to divide into triangles, but even then there can be many choices (divide on *this* diagonal or on *that* one?).

Far more complex, and more important, is how to label the vertices of these triangles, because that affects how you ask OpenGL ES to draw them. The OpenGL ES specification illustrates this with

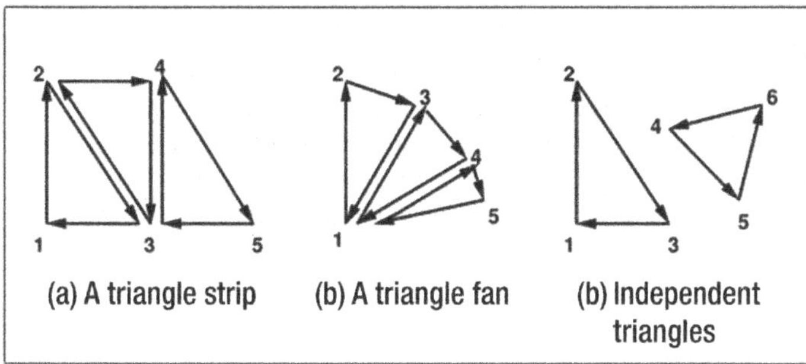

Figure 6-5. *Possible vertex labelling of figures*

Making separate triangles is easiest for the programmer: you just have to cut the polygons into any triangles with any labeling and call `glDrawArrays` where every three vertices define a separate triangle. But it may cost more in processing time.

A square, for example, can be divided into two separate triangles with a total of six vertices. Doing so would look like this:

```
GLfloat vVertices[] = { // first triangle
                        1.0f, -1.0f, 0.0f,
                        1.0f, 1.0f, 0.0f,
                        -1.0f, -1.0f, 0.0f,

                        // second triangle
                        -1.0f, 1.0f, 0.0f,
                        -1.0f, -1.0f, 0.0f,
                        1.0f, 1.0f, 0.0f
                       };
```

Here is the drawing code:

```
glVertexAttribPointer ( 0, 3, GL_FLOAT, GL_FALSE, 0, vVertices );
glEnableVertexAttribArray ( 0 );

glDrawArrays ( GL_TRIANGLES, 0, 6 );
```

But a square can also be divided into a triangle strip or even a triangle fan with only four vertices. But you have to be more careful. If you want a strip, the first vertex must not be shared with any other triangle and neither can the last. Here is one possibility for a square at (±1, ±1):

```
GLfloat vVertices[] = {
                        1.0f, -1.0f, 0.0f,
                        1.0f, 1.0f, 0.0f,
```

```
                    -1.0f, -1.0f, 0.0f,
                    -1.0f, 1.0f, 0.0f
                   };
```

The square can then be drawn with the following:

```
glVertexAttribPointer ( 0, 3, GL_FLOAT, GL_FALSE, 0, vVertices );
glEnableVertexAttribArray ( 0 );

glDrawArrays ( GL_TRIANGLE_STRIP, 0, 4 );
```

An alternative mechanism uses the vertex arrays as shown earlier but draws elements by explicit indexes into this array rather than the implicit indexes caused by the element ordering. This makes use of another array, this time of the indices. The vertex array must be used too, of course.

Even if an index array is used, there are still the choices of drawing individual triangles, strips, or fans. Using triangles, the code is as follows:

```
  GLfloat vVertices[] = {
                    1.0f, -1.0f, 0.0f,
                    1.0f, 1.0f, 0.0f,
                    -1.0f, -1.0f, 0.0f,
                    -1.0f, 1.0f, 0.0f,
                    -1.0f, -1.0f, 0.0f,
                    1.0f, 1.0f, 0.0f
                   };
  GLubyte vIndices[] = {
                   0, 1, 2,
                   3, 4, 5
                  };
```

(This is usually optimized by reusing some of the vertices as {0, 1, 2, 3, 2, 1} and omitting the last two entries of the vertices array.) The square can then be drawn with the following:

```
glVertexAttribPointer ( 0, 3, GL_FLOAT, GL_FALSE, 0, vVertices );
glEnableVertexAttribArray ( 0 );

glDrawElements ( GL_TRIANGLES, 6, GL_UNSIGNED_BYTE, vIndices );
```

Using strips, the code is as follows:

```
  GLfloat vVertices[] = {
                    1.0f, -1.0f, 0.0f,
                    1.0f, 1.0f, 0.0f,
                    -1.0f, -1.0f, 0.0f,
                    -1.0f, 1.0f, 0.0f
                   };
  GLubyte vIndices[] = {
                   0, 1, 2, 3
                  };
```

The square can then be drawn with the following:

```
glVertexAttribPointer ( 0, 3, GL_FLOAT, GL_FALSE, 0, vVertices );
glEnableVertexAttribArray ( 0 );

glDrawElements ( GL_TRIANGLE_STRIP, 4, GL_UNSIGNED_BYTE, vIndices );
```

These possibilities can all be seen in a revised Draw function, where the different choices described earlier can be made by setting TRIANGLES and ELEMENTS to 0 or 1:

```
void Draw ( ESContext *esContext )
{
#define TRIANGLES 1
#define ELEMENTS 1

   UserData *userData = esContext->userData;
   GLfloat vVertices[] = {
                          1.0f, -1.0f, 0.0f,
                          1.0f, 1.0f, 0.0f,
                          -1.0f, -1.0f, 0.0f,
                          -1.0f, 1.0f, 0.0f,
#if TRIANGLES
                          -1.0f, -1.0f, 0.0f,
                          1.0f, 1.0f, 0.0f
#endif
                         };

#if ELEMENTS
   GLubyte vIndices[] = {
                         0, 1, 2,
                         3,
#  if TRIANGLES
                         4, 5
#  endif
                        };
#endif

   // Set the viewport
   glViewport ( 0, 0, esContext->width, esContext->height );

   // Clear the color buffer
   glClear ( GL_COLOR_BUFFER_BIT );

   // Use the program object
   glUseProgram ( userData->programObject );

   // Load the vertex data
   glVertexAttribPointer ( 0, 3, GL_FLOAT, GL_FALSE, 0, vVertices );
   glEnableVertexAttribArray ( 0 );
```

```
#if TRIANGLES
#  if ELEMENTS
   glDrawElements ( GL_TRIANGLES, 6, GL_UNSIGNED_BYTE, vIndices );
#  else
   glDrawArrays ( GL_TRIANGLES, 0, 6 );
#  endif
#else
#  if ELEMENTS
   glDrawElements ( GL_TRIANGLE_STRIP, 4, GL_UNSIGNED_BYTE, vIndices );
#  else
   glDrawArrays ( GL_TRIANGLE_STRIP, 0, 4 );
#  endif
#endif
}
```

The full program is `Hello_Square.c` and is run by `Hello_Square`.

Which to Choose?

The choice of triangles, strip, or fan will depend on the geometry of the shape and any semantic context it might have; for example, drawing a wall of rectangles would use a strip, while drawing circles would use a fan.

The choice of drawing arrays or elements usually comes down in favor of elements for efficiency.

Textures

Textures are another means of filling triangles. A texture is image data that can be manipulated by the fragment shader. Textures can be two-dimensional rectangular images or the six sides of a cube. In this section, you will look only at two-dimensional images. Textures will typically come from image data stored in files or hard-coded into the program. The Munshi et al. book uses a simple hard-coded 2x2 image. Here you will take a file image.

TGA Files

There are many, many different file formats—some lossy, some lossless, some compressed, some not, some with metadata, some without. In this section, you will just use TGA (`www.fileformat.info/format/tga/egff.htm`), which is an uncompressed format with enough useful metadata that is simple to load.

TGA files can be created from, for example, JPEG files by using the `convert` utility from the Gimp drawing system. It is simple to convert a file from JPEG to TGA format. You just give the appropriate file extensions.

```
convert image.jpg image.tga
```

A TGA file has a header section that gives the width and height of the image from which its size can be calculated. The default format will be RGB, as in 24-bit pixels. Reading in such a file is just a matter of locating the dimensions, `malloc`'ing the right size buffer, skipping to the start of the image data, and reading it all in.

The `esUtil.c` file includes a function called `esLoadTGA` that will read in an image, returning the image and its dimensions.

The default is for the origin of the image to be the top-left corner, with the y-axis growing down. OpenGL ES, on the other hand, has the origin in the bottom-left corner with the y-axis growing up. So, the image will be upside down relative to the OpenGL coordinates. This can be fixed by reading the data in differently or by using an OpenGL ES reflection. You will map the texture upside down to give the correct orientation.

Mipmaps

An application may need to render an image in fine detail or, say, if it is far away, in only coarse detail. To avoid GPU load, multiple images can be given for one texture at varying levels of detail. I won't cover that here and will just set one image for all levels of detail.

Creating a Texture Object

Textures are stored in texture objects that have to be created using an image with a known format, dimensions, and pixel values. Assuming you have stored the width, height, and image in the `userData` field of an `esContext`, where the format is known to be RGB as unsigned bytes, the code is as follows:

```
GLuint CreateSimpleTexture2D(ESContext *esContext)
{
   // Texture object handle
   GLuint textureId;
   UserData *userData = esContext->userData;
   char *pixels = userData->image;

   // Use tightly packed data
   glPixelStorei ( GL_UNPACK_ALIGNMENT, 1 );

   // Generate a texture object
   glGenTextures ( 1, &textureId );

   // Bind the texture object
   glBindTexture ( GL_TEXTURE_2D, textureId );

   // Load the texture
   glTexImage2D ( GL_TEXTURE_2D, 0, GL_RGB,
                  userData->width, userData->height,
                  0, GL_RGB, GL_UNSIGNED_BYTE, pixels );

   // Set the filtering mode
   glTexParameteri ( GL_TEXTURE_2D, GL_TEXTURE_MIN_FILTER, GL_NEAREST );
   glTexParameteri ( GL_TEXTURE_2D, GL_TEXTURE_MAG_FILTER, GL_NEAREST );

   return textureId;

}
```

Texture Attributes and the Shaders

Two-dimensional textures will only need (x, y) values as coordinates, unlike vertex coordinates that are described by four coordinates (x, y, z, w). The vertices of the 2D texture will need to be fed into the vertex shaders in order that the texture can be properly located within the final rendering. This will need a parameter within the shader of type `vec2`. The texture coordinates are limited to the square with vertices (0, 0), (1, 0), (0, 1), and (1, 1).

The fragment shader will need a varying texture coordinate, which would be initialized by the vertex shader but then interpolated. This will need to be integrated with the image pixel values, and this is done by a fragment shader function `texture2D`.

The vertex shader program will typically look like this:

```
attribute vec4 a_position;
attribute vec2 a_texCoord;
varying vec2 v_texCoord;
void main()
{
   gl_Position = a_position;
   v_texCoord = a_texCoord;
}
```

The fragment shader will be as follows:

```
precision mediump float;
varying vec2 v_texCoord;
uniform sampler2D s_texture;
void main()
{
  gl_FragColor = texture2D( s_texture, v_texCoord );
}
```

These are all combined into a revised `Init` function as follows:

```
int Init ( ESContext *esContext )
{
    UserData *userData = esContext->userData;
    GLbyte vShaderStr[] =
      "attribute vec4 a_position;   \n"
      "attribute vec2 a_texCoord;   \n"
      "varying vec2 v_texCoord;     \n"
      "void main()                  \n"
      "{                            \n"
      "   gl_Position = a_position; \n"
      "   v_texCoord = a_texCoord;  \n"
      "}                            \n";

    GLbyte fShaderStr[] =
      "precision mediump float;                            \n"
      "varying vec2 v_texCoord;                            \n"
      "uniform sampler2D s_texture;                        \n"
      "void main()                                         \n"
      "{                                                   \n"
      "  gl_FragColor = texture2D( s_texture, v_texCoord );\n"
      "}                                                   \n";

   // Load the shaders and get a linked program object
   userData->programObject = esLoadProgram ( vShaderStr, fShaderStr );
```

```
    // Get the attribute locations
    userData->positionLoc = glGetAttribLocation ( userData->programObject, "a_position" );
    userData->texCoordLoc = glGetAttribLocation ( userData->programObject, "a_texCoord" );

    // Get the sampler location
    userData->samplerLoc = glGetUniformLocation ( userData->programObject, "s_texture" );

    // Load the texture
    userData->textureId = CreateSimpleTexture2D (esContext);

    glClearColor ( 0.0f, 0.0f, 0.0f, 0.0f );
    return GL_TRUE;
}
```

Drawing the Texture

To draw, you need to pass in the vertex coordinates. You will draw a square for the vertices. For the texture coordinates, you also need to specify their four coordinates. This could be done using two arrays as follows:

```
GLfloat vVertices[] = { -0.5f,  0.5f, 0.0f,  // Position 0
                        -0.5f, -0.5f, 0.0f,  // Position 1
                         0.5f, -0.5f, 0.0f,  // Position 2
                         0.5f,  0.5f, 0.0f,  // Position 3
                      };
 GLfloat tVertices[] = {
                         0.0f,  1.0f,        // TexCoord 0
                         0.0f,  0.0f,        // TexCoord 1
                         1.0f,  0.0f,        // TexCoord 2
                         1.0f,  1.0f         // TexCoord 3
                      };
```

More commonly, these arrays would be interleaved as follows:

```
GLfloat vVertices[] = { -0.5f,  0.5f, 0.0f,  // Position 0
                         0.0f,  1.0f,        // TexCoord 0
                        -0.5f, -0.5f, 0.0f,  // Position 1
                         0.0f,  0.0f,        // TexCoord 1
                         0.5f, -0.5f, 0.0f,  // Position 2
                         1.0f,  0.0f,        // TexCoord 2
                         0.5f,  0.5f, 0.0f,  // Position 3
                         1.0f,  1.0f         // TexCoord 3
                      };
```

(Note that you are rendering the texture upside down to compensate for it being upside down relative to OpenGL ES coordinates.)

The question then is how to get—separately—the two sets of vertices. This is where the stride parameter comes into play in the call to `glVertexAttribPointer`. Basically, you want to say the following:

```
Each vertex co-ordinate occurs in a set of three in every five values,
   starting at index zero
Each texture co-ordinate occurs in a set of two in every five values,
   starting at index three
```

The code for this is as follows:

```
// Load the vertex position
glVertexAttribPointer ( userData->positionLoc, 3, GL_FLOAT,
                        GL_FALSE, 5 * sizeof(GLfloat), &vVertices[0] );
// Load the texture coordinate
glVertexAttribPointer ( userData-gt;texCoordLoc, 2, GL_FLOAT,
                        GL_FALSE, 5 * sizeof(GLfloat), &vVertices[3] );
```

The drawing code is then as follows:

```
void Draw ( ESContext *esContext )
{
   UserData *userData = esContext->userData;
   GLfloat vVertices[] = { -0.5f,  0.5f, 0.0f,  // Position 0
                            0.0f,  1.0f,        // TexCoord 0
                           -0.5f, -0.5f, 0.0f,  // Position 1
                            0.0f,  0.0f,        // TexCoord 1
                            0.5f, -0.5f, 0.0f,  // Position 2
                            1.0f,  0.0f,        // TexCoord 2
                            0.5f,  0.5f, 0.0f,  // Position 3
                            1.0f,  1.0f         // TexCoord 3
                         };
   GLushort indices[] = { 0, 1, 2, 0, 2, 3 };

   // Set the viewport
   glViewport ( 0, 0, esContext->width, esContext->height );

   // Clear the color buffer
   glClear ( GL_COLOR_BUFFER_BIT );

   // Use the program object
   glUseProgram ( userData->programObject );

   // Load the vertex position
   glVertexAttribPointer ( userData->positionLoc, 3, GL_FLOAT,
                           GL_FALSE, 5 * sizeof(GLfloat), vVertices );
   // Load the texture coordinate
   glVertexAttribPointer ( userData->texCoordLoc, 2, GL_FLOAT,
                           GL_FALSE, 5 * sizeof(GLfloat), &vVertices[3] );

   glEnableVertexAttribArray ( userData->positionLoc );
   glEnableVertexAttribArray ( userData->texCoordLoc );

   // Bind the texture
   glActiveTexture ( GL_TEXTURE0 );
   glBindTexture ( GL_TEXTURE_2D, userData->textureId );
```

```
    // Set the sampler texture unit to 0
    glUniform1i ( userData->samplerLoc, 0 );

    glDrawElements ( GL_TRIANGLES, 6, GL_UNSIGNED_SHORT, indices );

}
```

Complete Code for Drawing an Image

The final code is `Simple_Image.c`. There appears to be a minor bug in the function esLoadTGA in the `Common/esUtil.c` file. It gets the colors of the image a little wrong.

```
// Authors:   Aaftab Munshi, Dan Ginsburg, Dave Shreiner
// Copyright: the authors and Jan Newmarch

// Simple_Image.c
//
//    This is a simple example that draws an image with a 2D
//    texture image.
//
#include <stdlib.h>
#include <stdio.h>
#include "esUtil.h"

typedef struct
{
   // Handle to a program object
   GLuint programObject;

   // Attribute locations
   GLint  positionLoc;
   GLint  texCoordLoc;

   // Sampler location
   GLint samplerLoc;

   // Texture handle
   GLuint textureId;

   GLubyte *image;
    int width, height;
} UserData;

///
// Create a simple 2x2 texture image with four different colors
//
GLuint CreateSimpleTexture2D(ESContext *esContext)
{
   // Texture object handle
   GLuint textureId;
   UserData *userData = esContext->userData;
```

```
   GLubyte *pixels = userData->image;
   userData->width = esContext->width;
   userData->height = esContext->height;

   // Use tightly packed data
   glPixelStorei ( GL_UNPACK_ALIGNMENT, 1 );

   // Generate a texture object
   glGenTextures ( 1, &textureId );

   // Bind the texture object
   glBindTexture ( GL_TEXTURE_2D, textureId );

   // Load the texture
   glTexImage2D ( GL_TEXTURE_2D, 0, GL_RGB,
                  userData->width, userData->height,
                  0, GL_RGB, GL_UNSIGNED_BYTE, pixels );

   // Set the filtering mode
   glTexParameteri ( GL_TEXTURE_2D, GL_TEXTURE_MIN_FILTER, GL_NEAREST );
   glTexParameteri ( GL_TEXTURE_2D, GL_TEXTURE_MAG_FILTER, GL_NEAREST );

   return textureId;

}

///
// Initialize the shader and program object
//
int Init ( ESContext *esContext )
{
    UserData *userData = esContext->userData;
    GLbyte vShaderStr[] =
      "attribute vec4 a_position;   \n"
      "attribute vec2 a_texCoord;   \n"
      "varying vec2 v_texCoord;     \n"
      "void main()                  \n"
      "{                            \n"
      "   gl_Position = a_position; \n"
      "   v_texCoord = a_texCoord;  \n"
      "}                            \n";

    GLbyte fShaderStr[] =
      "precision mediump float;                            \n"
      "varying vec2 v_texCoord;                            \n"
      "uniform sampler2D s_texture;                        \n"
      "void main()                                         \n"
      "{                                                   \n"
      "  gl_FragColor = texture2D( s_texture, v_texCoord );\n"
      "}                                                   \n";
```

```
    // Load the shaders and get a linked program object
    userData->programObject = esLoadProgram ( vShaderStr, fShaderStr );

    // Get the attribute locations
    userData->positionLoc = glGetAttribLocation ( userData->programObject, "a_position" );
    userData->texCoordLoc = glGetAttribLocation ( userData->programObject, "a_texCoord" );
    http://www.netmanners.com/e-mail-etiquette-tips/
    // Get the sampler location
    userData->samplerLoc = glGetUniformLocation ( userData->programObject, "s_texture" );

    // Load the texture
    userData->textureId = CreateSimpleTexture2D (esContext);

    glClearColor ( 0.0f, 0.0f, 0.0f, 0.0f );
    return GL_TRUE;
}

///
// Draw a triangle using the shader pair created in Init()
//
void Draw ( ESContext *esContext )
{
    UserData *userData = esContext->userData;
    GLfloat vVertices[] = { -0.5f,  0.5f, 0.0f,  // Position 0
                             0.0f,  1.0f,        // TexCoord 0
                            -0.5f, -0.5f, 0.0f,  // Position 1
                             0.0f,  0.0f,        // TexCoord 1
                             0.5f, -0.5f, 0.0f,  // Position 2
                             1.0f,  0.0f,        // TexCoord 2
                             0.5f,  0.5f, 0.0f,  // Position 3http://www.netmanners.com/
                                                    e-mail-etiquette-tips/
                             1.0f,  1.0f         // TexCoord 3
                          };
    GLushort indices[] = { 0, 1, 2, 0, 2, 3 };

    // Set the viewport
    glViewport ( 0, 0, esContext->width, esContext->height );

    // Clear the color buffer
    glClear ( GL_COLOR_BUFFER_BIT );

    // Use the program object
    glUseProgram ( userData->programObject );

    // Load the vertex position
    glVertexAttribPointer ( userData->positionLoc, 3, GL_FLOAT,
                            GL_FALSE, 5 * sizeof(GLfloat), vVertices );
    // Load the texture coordinate
    glVertexAttribPointer ( userData->texCoordLoc, 2, GL_FLOAT,
                            GL_FALSE, 5 * sizeof(GLfloat), &vVertices[3] );
```

```
   glEnableVertexAttribArray ( userData->positionLoc );
   glEnableVertexAttribArray ( userData->texCoordLoc );

   // Bind the texture
   glActiveTexture ( GL_TEXTURE0 );
   glBindTexture ( GL_TEXTURE_2D, userData->textureId );

   // Set the sampler texture unit to 0
   glUniform1i ( userData->samplerLoc, 0 );

   glDrawElements ( GL_TRIANGLES, 6, GL_UNSIGNED_SHORT, indices );

}

///
// Cleanup
//
void ShutDown ( ESContext *esContext )
{
   UserData *userData = esContext->userData;

   // Delete texture object
   glDeleteTextures ( 1, &userData->textureId );

   // Delete program object
   glDeleteProgram ( userData->programObject );

   free(esContext->userData);
}

int main ( int argc, char *argv[] )
{
   ESContext esContext;
   UserData  userData;

   int width, height;
   GLubyte *image;

   image = esLoadTGA("jan.tga", &width, &height);
   if (image == NULL) {
       fprintf(stderr, "No such image\n");
       exit(1);
   }
   printf("Width %d height %d\n", width, height);

   userData.image = image;
   userData.width = width;
   userData.height = height;

   esInitContext ( &esContext );
   esContext.userData = &userData;
```

```
   esCreateWindow ( &esContext, "Simple Texture 2D", width, height, ES_WINDOW_RGB );

   if ( !Init ( &esContext ) )
      return 0;

   esRegisterDrawFunc ( &esContext, Draw );

   esMainLoop ( &esContext );

   ShutDown ( &esContext );
}
```

Run this by `Simple_Image`. The image used is of me since that is possibly more interesting than a wall tile.

Animation: Rotating an Image

In this section, you will take a brief look at some of the issues and techniques in animating the graphics display. As you will have seen from multiple computer games and digital movies, the animations can be amazingly sophisticated and also involve much programmer time, program complexity, and CPU/GPU time.

You will do about the simplest: make the image of the previous section rotate around an axis.

Matrices

I cut my teeth on matrices years ago writing programs in Algol 60 to solve simultaneous linear equations. Now it is easier, with libraries to manage most calculations. You will use various ES functions from the Munshi et al. book to simplify things.

There are several primary operations you can carry out on graphic images to fit them into scenes: you can rotate them, move them nearer or further, apply perspective views so that the "parallel lines meet at the horizon," or do any other effects you may choose.

The operations are carried out on vectors such as the (x, y, z, w) coordinates of a vertex. The operations are represented by *matrices* that are square (or rectangular) arrays of numbers. The main operation is to *multiply* a *vector* by a *matrix*. Any book on linear algebra will give details on how to do this.

esUtil

These are the principal functions supplied by the `esUtil` package:

- Create an identity matrix with `esMatrixLoadIdentity`
- Multiply two matrices with `esMatrixMultiply`
- Translate a vector from one point to another with `esTranslate`
- Rotate a matrix through an angle around an axis with `esRotate`
- Scale a matrix by a factor with `esScale`
- Multiply a matrix by a perspective matrix with `esPerspective`

These may remove the need to explicitly specify actual matrices.

Invoking Animation

You already have a Draw function. You add to this an Update function to make the changes needed for the next Draw:

```
esRegisterDrawFunc ( &esContext, Draw );
esRegisterUpdateFunc ( &esContext, Update );
```

The Update function for a simple rotation can be as follows:

```
void Update ( ESContext *esContext, float deltaTime )
{
   UserData *userData = (UserData*) esContext->userData;

   // Compute a rotation angle based on time to rotate the cube
   userData->angle += ( deltaTime * 40.0f );
   if( userData->angle >= 360.0f )
      userData->angle -= 360.0f;

   // Generate an identity matrix before rotating the square
   esMatrixLoadIdentity( &userData->rotateMx );

   // Rotate the square about the (1, 0, 1) axis
   esRotate( &userData->rotateMx, userData->angle, 1.0, 0.0, 1.0 );
}
```

This uses a current rotation angle stored in userData and applies it to an identity matrix about a vector specified here as (x, y, z) = (1.0, 0.0, 1.0).

Uniform Parameters: The Rotation Matrix

If you are rotating an entire image, the rotation has to be applied to every pixel in the drawn image, and it is the same rotation matrix in each case. So, this matrix needs to be constant over all the vertex and fragment shaders. Such a matrix is a *uniform* parameter and is specified as such in both the vertex and fragment shaders.

Working out the value of a uniform parameter is something that needs to be done *once* in each drawing iteration; so, it is done in the application's Update C code, not in the shaders. It is then passed in as a parameter on each draw call by glUniformMatrix4fv.

Extra fields are added to UserData for the rotation angle and rotation matrix. Apart from that and the earlier changes, the standard program structure holds. The program to rotate an image loaded from a TGA file is Rotate_Image.c.

```
        //
// Book:      OpenGL(R) ES 2.0 Programming Guide
// Authors:   Aaftab Munshi, Dan Ginsburg, Dave Shreiner
// ISBN-10:   0321502795
// ISBN-13:   9780321502797
// Publisher: Addison-Wesley Professional
// URLs:      http://safari.informit.com/9780321563835
//            http://www.opengles-book.com
//
```

```
// Simple_Texture2D.c
//
//    This is a simple example that draws a quad with a 2D
//    texture image. The purpose of this example is to demonstrate
//    the basics of 2D texturing
//
#include <stdlib.h>
#include <stdio.h>
#include "esUtil.h"

typedef struct
{
   // Handle to a program object
   GLuint programObject;

   // Attribute locations
   GLint  positionLoc;
   GLint  texCoordLoc;

   // Uniform locations
   GLint  rotateLoc;

   // Sampler location
   GLint samplerLoc;

   // Texture handle
   GLuint textureId;

   GLubyte *image;
    int width, height;

   // Rotation angle
   GLfloat   angle;

   // rotate matrix
   ESMatrix  rotateMx;
} UserData;

///
// Create a simple 2x2 texture image with four different colors
//
GLuint CreateSimpleTexture2D(ESContext *esContext)
{
   // Texture object handle
   GLuint textureId;
   UserData *userData = esContext->userData;

#if 0
   // 2x2 Image, 3 bytes per pixel (R, G, B)
   GLubyte pixels[4 * 3] =
```

```
   {
      255,   0,    0, // Red
        0, 255,    0, // Green
        0,   0,  255, // Blue
      255, 255,    0  // Yellow
   };
   userData->width = 2;
   userData->height = 2;
#else
   GLubyte *pixels = userData->image;
   userData->width = esContext->width;
   userData->height = esContext->height;
#endif
   // Use tightly packed data
   glPixelStorei ( GL_UNPACK_ALIGNMENT, 1 );

   // Generate a texture object
   glGenTextures ( 1, &textureId );

   // Bind the texture object
   glBindTexture ( GL_TEXTURE_2D, textureId );

   // Load the texture
   glTexImage2D ( GL_TEXTURE_2D, 0, GL_RGB,
                  userData->width, userData->height,
                  0, GL_RGB, GL_UNSIGNED_BYTE, pixels );

   // Set the filtering mode
   glTexParameteri ( GL_TEXTURE_2D, GL_TEXTURE_MIN_FILTER, GL_NEAREST );
   glTexParameteri ( GL_TEXTURE_2D, GL_TEXTURE_MAG_FILTER, GL_NEAREST );

   return textureId;

}

///
// Initialize the shader and program object
//
int Init ( ESContext *esContext )
{
    UserData *userData = esContext->userData;
    GLbyte vShaderStr[] =
      "uniform mat4 u_rotateMx;                  \n"
      "attribute vec4 a_position;                \n"
      "attribute vec2 a_texCoord;                \n"
      "varying vec2 v_texCoord;                  \n"
      "void main()                               \n"
      "{                                         \n"
      "   gl_Position = u_rotateMx * a_position; \n"
      "   v_texCoord = a_texCoord;               \n"
      "}                                         \n";
```

```
    GLbyte fShaderStr[] =
      "precision mediump float;                          \n"
      "varying vec2 v_texCoord;                          \n"
      "uniform sampler2D s_texture;                      \n"
      "void main()                                       \n"
      "{                                                 \n"
      "  gl_FragColor = texture2D( s_texture, v_texCoord );\n"
      "}                                                 \n";

   // Load the shaders and get a linked program object
   userData->programObject = esLoadProgram ( vShaderStr, fShaderStr );

   // Get the attribute locations
   userData->positionLoc = glGetAttribLocation ( userData->programObject, "a_position" );
   userData->texCoordLoc = glGetAttribLocation ( userData->programObject, "a_texCoord" );
   userData->rotateLoc = glGetUniformLocation( userData->programObject, "u_rotateMx" );
   // Starting rotation angle for the square
   userData->angle = 0.0f;

   // Get the sampler location
   userData->samplerLoc = glGetUniformLocation ( userData->programObject, "s_texture" );

   // Load the texture
   userData->textureId = CreateSimpleTexture2D (esContext);

   glClearColor ( 0.0f, 0.0f, 0.0f, 0.0f );

   return GL_TRUE;
}

///
// Update rotate matrix based on time
//
void Update ( ESContext *esContext, float deltaTime )
{
   UserData *userData = (UserData*) esContext->userData;

   // Compute a rotation angle based on time to rotate the cube
   userData->angle += ( deltaTime * 40.0f );
   if( userData->angle >= 360.0f )
      userData->angle -= 360.0f;

   // Generate an identity matrix before rotating the square
   esMatrixLoadIdentity(  &userData->rotateMx );

   // Rotate the square
   esRotate(  &userData->rotateMx, userData->angle, 1.0, 0.0, 1.0 );
}
```

```
///
// Draw a triangle using the shader pair created in Init()
//
void Draw ( ESContext *esContext )
{
   UserData *userData = esContext->userData;
   GLfloat vVertices[] = { -0.5f,  0.5f, 0.0f,  // Position 0
                            0.0f,  1.0f,        // TexCoord 0
                           -0.5f, -0.5f, 0.0f,  // Position 1
                            0.0f,  0.0f,        // TexCoord 1
                            0.5f, -0.5f, 0.0f,  // Position 2
                            1.0f,  0.0f,        // TexCoord 2
                            0.5f,  0.5f, 0.0f,  // Position 3
                            1.0f,  1.0f         // TexCoord 3
                         };
   GLushort indices[] = { 0, 1, 2, 0, 2, 3 };

   // Set the viewport
   glViewport ( 0, 0, esContext->width, esContext->height );

   // Clear the color buffer
   glClear ( GL_COLOR_BUFFER_BIT );

   // Use the program object
   glUseProgram ( userData->programObject );

   // Load the vertex position
   glVertexAttribPointer ( userData->positionLoc, 3, GL_FLOAT,
                           GL_FALSE, 5 * sizeof(GLfloat), vVertices );
   // Load the texture coordinate
   glVertexAttribPointer ( userData->texCoordLoc, 2, GL_FLOAT,
                           GL_FALSE, 5 * sizeof(GLfloat), &vVertices[3] );

   glEnableVertexAttribArray ( userData->positionLoc );
   glEnableVertexAttribArray ( userData->texCoordLoc );

   // Load the rotate matrix
   glUniformMatrix4fv( userData->rotateLoc, // userData->mvpLoc,
                       1, GL_FALSE, (GLfloat*) &userData->rotateMx.m[0][0] );

   // Bind the texture
   glActiveTexture ( GL_TEXTURE0 );
   glBindTexture ( GL_TEXTURE_2D, userData->textureId );

   // Set the sampler texture unit to 0
   glUniform1i ( userData->samplerLoc, 0 );

   glDrawElements ( GL_TRIANGLES, 6, GL_UNSIGNED_SHORT, indices );

}
```

```
///
// Cleanup
//
void ShutDown ( ESContext *esContext )
{
   UserData *userData = esContext->userData;

   // Delete texture object
   glDeleteTextures ( 1, &userData->textureId );

   // Delete program object
   glDeleteProgram ( userData->programObject );

   free(esContext->userData);
}

int main ( int argc, char *argv[] )
{
   ESContext esContext;
   UserData  userData;

   int width, height;
   GLubyte *image;

   image = esLoadTGA("jan.tga", &width, &height);
   if (image == NULL) {
       fprintf(stderr, "No such image\n");
       exit(1);
   }
   printf("Width %d height %d\n", width, height);

   userData.image = image;
   userData.width = width;
   userData.height = height;

   esInitContext ( &esContext );
   esContext.userData = &userData;

   esCreateWindow ( &esContext, "Simple Texture 2D", width, height, ES_WINDOW_RGB );

   if ( !Init ( &esContext ) )
      return 0;

   esRegisterDrawFunc ( &esContext, Draw );
   esRegisterUpdateFunc ( &esContext, Update );

   esMainLoop ( &esContext );

   ShutDown ( &esContext );
}
```

Run this by Rotate_Image. It shows the picture of me in a rotating square.

Conclusion

OpenGL ES is an amazingly complex and sophisticated system that I have made a small attempt to explain. This chapter covered drawing on the EGL surface discussed in earlier chapters and showed a number of relatively simple cases. These have been standard OpenGL ES; the only RPi-specific part has been the EGL surface.

Resources

- OpenGL ES Shading Language: https://www.khronos.org/files/opengles_shading_language.pdf
- OpenGL ES Specification: https://www.khronos.org/registry/gles/specs/2.0/es_full_spec_2.0.25.pdf
- openGL-RPi-tutorial: https://github.com/peepo/openGL-RPi-tutorial
- OpenMAX+GLES2: Decode and render JPEG (with source code): jpeg_gles2: www.raspberrypi.org/forums/viewtopic.php?f=33&t=57721
- Using OpenGL ES 2.0 on the Raspberry Pi without X windows: http://benosteen.wordpress.com/2012/04/27/using-opengl-es-2-0-on-the-raspberry-pi-without-x-windows/
- KHR_image_base describes EGLImageKHR type: www.khronos.org/registry/egl/extensions/KHR/EGL_KHR_image_base.txt
- OpenGL Programming/Modern OpenGL Tutorial Text Rendering 01: http://en.wikibooks.org/wiki/OpenGL_Programming/Modern_OpenGL_Tutorial_Text_Rendering_01
- Using OpenGL ES to Accelerate Apps with Legacy 2D GUIs: https://software.intel.com/en-us/articles/using-opengl-es-to-accelerate-apps-with-legacy-2d-guis
- glEGLImageTargetTexture2DOES: http://e2e.ti.com/support/dsp/omap_applications_processors/f/447/t/81109.aspx
- Raspberry Pi VideoCore APIs: http://elinux.org/Raspberry_Pi_VideoCore_APIs
- OpenGL® ES 2.0 Programming Guide by Aaftab Munshi, Dan Ginsburg and Dave Shreiner: http://opengles-book.com/es2/index.html
- opengles-book-samples: https://code.google.com/p/opengles-book-samples/source/checkout
- TGA File Format Summary: www.fileformat.info/format/tga/egff.htm
- Raspberry Pi OpenGL ES forum: www.raspberrypi.org/forums/viewforum.php?f=68
- OpenGL ES 2.0 for iOS, Chapter 4 - Introducing the Programmable Pipeline: http://iphonedevelopment.blogspot.com.au/2010/11/opengl-es-20-for-ios-chapter-4.html

CHAPTER 7

OpenMAX on the Raspberry Pi Concepts

This chapter gives a brief overview of OpenMAX IL on the Raspberry Pi. It looks at the component model and the processing model. Further details are given in later chapters.

OpenMAX Overview

OpenMAX is designed to play audio and video. It consists of three layers.

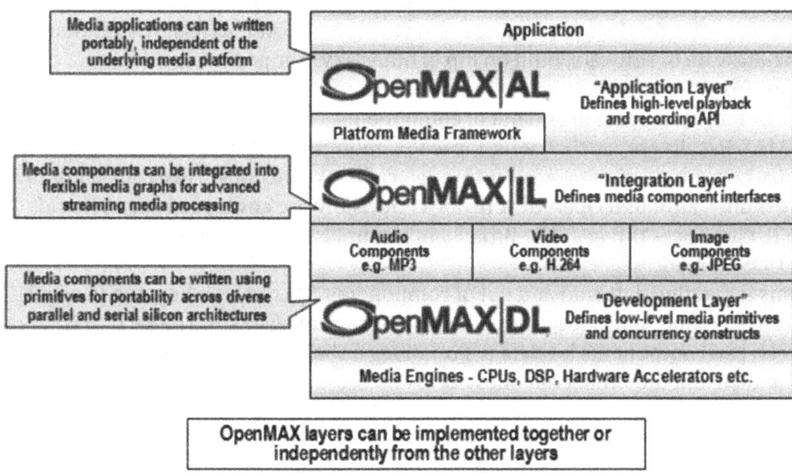

OpenMAX Media Portability Library

Figure 7-1. *OpenMAX APIs*

J. Newmarch, *Raspberry Pi GPU Audio Video Programming*, DOI 10.1007/978-1-4842-2472-4_7

The preferred programming layer is OpenMAX AL, the application layer. Unfortunately, the RPi does not support OpenMAX AL. What it does support is OpenMAX IL, the integration layer. Now this isn't nice to program to, as the following quotes from GitHub attest:

- According to jamesh, "OpenMAX is a complete and utter nightmare to use btw."
- According to dom, "I have written a fair bit of openmax client code and find it very hard. You have to get an awful lot right before you get anything useful out. Just lots of OMX_ErrorInvalidState and OMX_ErrorBadParameter messages if you are lucky. Nothing happening at all if you are not...."
- According to Twinkletoes, "I'm from a DirectShow background, and I thought that was badly documented...then I met OpenMax. Lots of ppts saying lovely things about it but no documentation or code examples I can find."

My own experience backs up these quotes, sadly. I've spent hours staring blankly at a screen, with the specification on one side and the debugger open on the other. It's a hard and rocky road ahead.

OpenMAX Concepts

The basic concept is of a *component*, which is an audio/video (or other) processing unit of some type, such as a volume control, a mixer, and an output device. Each component has zero or more input and output *ports*. Each port has a type, such as an audio port, a video port, or a timer port. The type is important: you manipulate different types with different structures. Each port can have one or more *buffers* that carry data either into or out of a component.

OpenMAX IL components use buffers to carry data. A component will usually process data from an input buffer and place it on an output buffer. This processing is not visible to the API, so it allows vendors to implement components in hardware or software, built on top of other A/V components, and so on. OpenMAX IL gives mechanisms for setting and getting parameters of components, for calling standard functions on the components, or for getting data in and out of components.

While some of the OpenMAX IL calls are synchronous, those that require possibly substantial amounts of processing are asynchronous, communicating the results through callback functions. This leads naturally to a multithreaded processing model, although OpenMAX IL does not visibly use any thread libraries and should be agnostic to how an IL client uses threads. The Broadcom examples for the Raspberry Pi use Broadcom's VideoCore OS (VCOS) threads.

There are two mechanisms for getting data into and out of components. The first is where the IL client makes calls on the component. All components are required to support this mechanism. The second is where a *tunnel* is set up between two components for data to flow along a shared buffer. A component is not required to support this mechanism, but the Broadcom ones do.

Once you have everything set up, the typical programming model for OpenMAX is as follows:

- Keep loading input buffers with data.
- Once an input buffer is loaded, ask the component to empty the buffer.
- Whenever a component empties the input buffer, it will call a function in the client, at which point the client can fill it again.
- Internal component processing will occur, and at some stage the component will fill one or more of its output buffers.

- When an output buffer is filled, the component will call another function in the client.
- The client will do something with the output buffer data and then return the buffer to the component to fill again.

There is, however, an immense amount of detail to each of these steps. It becomes more complex if a "pipeline" of components is involved, such as a decoder feeding into a format converter feeding into a rendering component. However, the principles are basically unchanged.

Conclusion

This chapter has given a brief overview of the architecture of OpenMAX IL. Later chapters will flesh this out in full detail.

Resources

- *Raspberry Pi Video on a Teapot Rendering OpenMAX onto an OpenGL surface*: `https://www.youtube.com/watch?v=-y3m_HFg4Do`
- *OpenMax rendering onto OpenGL texture*: `www.raspberrypi.org/forums/viewtopic.php?t=6577`
- *OpenMAX IL: The Standard for Media Library Portability*: `https://www.khronos.org/openmax/il/`
- *Decoding and Rendering to Texture H264 with OpenMAX on Raspberry Pi*: `http://thebugfreeblog.blogspot.com.au/2012/12/decoding-and-rendering-to-texture-h264.html`

CHAPTER 8

OpenMAX Components on the Raspberry Pi

OpenMAX uses components as the objects in its processing model. There is a large list of standard components. The Broadcom implementation on the Raspberry Pi has only a subset of these, plus some others. This chapter shows how to list the available components and get some of their properties.

Building Programs

The programs in this chapter can be built using the following Makefile:

```
DMX_INC = -I/opt/vc/include/ -I /opt/vc/include/interface/vmcs_host/ -I/opt/vc/include/
interface/vcos/pthreads -I/opt/vc/include/interface/vmcs_host/linux
EGL_INC =
OMX_INC =  -I /opt/vc/include/IL
INCLUDES = $(DMX_INC) $(EGL_INC) $(OMX_INC)

CFLAGS=-g -DRASPBERRY_PI -DOMX_SKIP64BIT $(INCLUDES)
CPPFLAGS =

DMX_LIBS =  -L/opt/vc/lib/ -lbcm_host -lvcos -lvchiq_arm -lpthread
EGL_LIBS = -L/opt/vc/lib/ -lEGL -lGLESv2
OMX_LIBS = -lopenmaxil

LDFLAGS =  $(DMX_LIBS) $(EGL_LIBS) $(OMX_LIBS)
SRC = info.c listcomponents.c portinfo.c

EXE = info listcomponents portinfo

all: $(EXE)
```

OpenMAX IL Components

OpenMAX IL in 1.1.2 lists a number of standard components, including a decoder, an encoder, a mixer, a reader, a renderer, a writer, a capturer, and a processor. An IL client gets such a component by calling `OMX_GetHandle()`, passing in the name of the component. This is a problem: the components do not have a standard name. The 1.1.2 specification says the following:

J. Newmarch, *Raspberry Pi GPU Audio Video Programming*, DOI 10.1007/978-1-4842-2472-4_8

> *Since components are requested by name, a naming convention is defined. OpenMAX IL component names are zero-terminated strings with the following format: "OMX.<vendor_name>.<vendor_specified_convention>." For example: OMX.CompanyABC.MP3Decoder.productXYZ.*
>
> *No standardization among component names is dictated across different vendors.*

The Broadcom OpenMAX IL components used on the Raspberry Pi are documented at the VMCS-X OpenMAX IL Components page (`http://home.nouwen.name/RaspberryPi/documentation/ilcomponents/index.html`). For example, that page shows a visualization component as look like this:

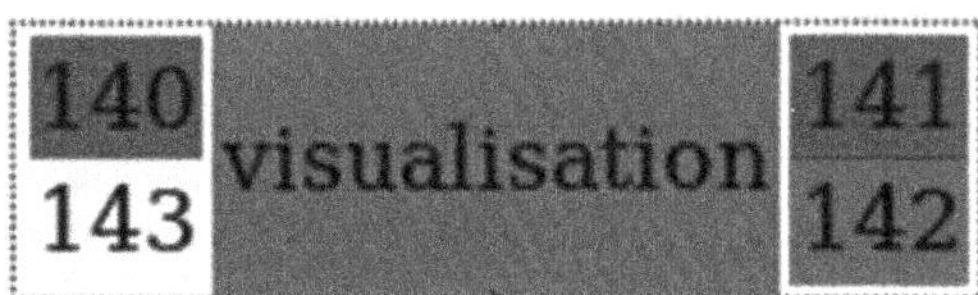

This has two audio ports (green), an input port 140 and an output port 141; a timer port (yellow), number 143; and a video output port (red), number 142. Clicking this component shows further details including the parameters that can be queried or set.

The detail view also shows the name of the component, which is used to look up the component. For the visualization component, its name is `OMX.broadcom.visualisation`.

Getting a List of Components

There is a simple way of programmatically getting the supported components. First initialize the OpenMAX system with `OMX_init()` and then make calls to `OMX_ComponentNameEnum()`. For successive index values, it returns a unique name each time, until it finally returns an error value of `OMX_ErrorNoMore`.

Each component may support a number of *roles*. These are given by `OMX_GetRolesOfComponent`. The Broadcom version has zero roles for each component, so it isn't useful for the RPi.

The program is `listcomponents.c`.

```
#include <stdio.h>
#include <stdlib.h>

#include <OMX_Core.h>

#include <bcm_host.h>

OMX_ERRORTYPE err;

void listroles(char *name) {
    int n;
    OMX_U32 numRoles;
    OMX_U8 *roles[32];

    /* get the number of roles by passing in a NULL roles param */
    err = OMX_GetRolesOfComponent(name, &numRoles, NULL);
    if (err != OMX_ErrorNone) {
        fprintf(stderr, "Getting roles failed\n", 0);
        exit(1);
    }
```

```
    printf("  Num roles is %d\n", numRoles);
    if (numRoles > 32) {
        printf("Too many roles to list\n");
        return;
    }

    /* now get the roles */
    for (n = 0; n < numRoles; n++) {
        roles[n] = malloc(OMX_MAX_STRINGNAME_SIZE);
    }
    err = OMX_GetRolesOfComponent(name, &numRoles, roles);
    if (err != OMX_ErrorNone) {
        fprintf(stderr, "Getting roles failed\n", 0);
        exit(1);
    }
    for (n = 0; n < numRoles; n++) {
        printf("    role: %s\n", roles[n]);
        free(roles[n]);
    }

    /* This is in version 1.2
    for (i = 0; OMX_ErrorNoMore != err; i++) {
        err = OMX_RoleOfComponentEnum(role, name, i);
        if (OMX_ErrorNone == err) {
            printf("   Role of omponent is %s\n", role);
        }
    }
    */
}

int main(int argc, char** argv) {

    int i;
    unsigned char name[OMX_MAX_STRINGNAME_SIZE];

    bcm_host_init();

    err = OMX_Init();
    if (err != OMX_ErrorNone) {
        fprintf(stderr, "OMX_Init() failed\n", 0);
        exit(1);
    }

    err = OMX_ErrorNone;
    for (i = 0; OMX_ErrorNoMore != err; i++) {
        err = OMX_ComponentNameEnum(name, OMX_MAX_STRINGNAME_SIZE, i);
        if (OMX_ErrorNone == err) {
            printf("Component is %s\n", name);
            listroles(name);
        }
    }
    printf("No more components\n");

    exit(0);
}
```

The following is the Makefile for the programs in this chapter:

```
DMX_INC =  -I/opt/vc/include/ -I /opt/vc/include/interface/vmcs_host/ -I/opt/vc/include/
interface/vcos/pthreads -I/opt/vc/include/interface/vmcs_host/linux
EGL_INC =
OMX_INC =  -I /opt/vc/include/IL
INCLUDES = $(DMX_INC) $(EGL_INC) $(OMX_INC)

CFLAGS=-g -DRASPBERRY_PI -DOMX_SKIP64BIT $(INCLUDES)
CPPFLAGS =

DMX_LIBS =  -L/opt/vc/lib/ -lbcm_host -lvcos -lvchiq_arm -lpthread
EGL_LIBS = -L/opt/vc/lib/ -lEGL -lGLESv2
OMX_LIBS = -lopenmaxil

LDLIBS =  $(DMX_LIBS) $(EGL_LIBS) $(OMX_LIBS)

SRC = info.c listcomponents.c portinfo.c
EXE = info listcomponents portinfo

all: $(EXE)
```

The listcomponents program is run with the following:

```
./listcomponents
```

The Raspberry Pi reports a large number of components but does not define a role for any of them.

```
Component is OMX.broadcom.audio_capture
  Num roles is 0
Component is OMX.broadcom.audio_decode
  Num roles is 0
Component is OMX.broadcom.audio_encode
  Num roles is 0
Component is OMX.broadcom.audio_render
  Num roles is 0
Component is OMX.broadcom.audio_mixer
  Num roles is 0
Component is OMX.broadcom.audio_splitter
  Num roles is 0
Component is OMX.broadcom.audio_processor
  Num roles is 0
Component is OMX.broadcom.camera
  Num roles is 0
Component is OMX.broadcom.clock
  Num roles is 0
Component is OMX.broadcom.coverage
  Num roles is 0
Component is OMX.broadcom.egl_render
  Num roles is 0
Component is OMX.broadcom.image_fx
```

```
  Num roles is 0
Component is OMX.broadcom.image_decode
  Num roles is 0
Component is OMX.broadcom.image_encode
  Num roles is 0
Component is OMX.broadcom.image_read
  Num roles is 0
Component is OMX.broadcom.image_write
  Num roles is 0
Component is OMX.broadcom.read_media
  Num roles is 0
Component is OMX.broadcom.resize
  Num roles is 0
Component is OMX.broadcom.source
  Num roles is 0
Component is OMX.broadcom.text_scheduler
  Num roles is 0
Component is OMX.broadcom.transition
  Num roles is 0
Component is OMX.broadcom.video_decode
  Num roles is 0
Component is OMX.broadcom.video_encode
  Num roles is 0
Component is OMX.broadcom.video_render
  Num roles is 0
Component is OMX.broadcom.video_scheduler
  Num roles is 0
Component is OMX.broadcom.video_splitter
  Num roles is 0
Component is OMX.broadcom.visualisation
  Num roles is 0
Component is OMX.broadcom.write_media
  Num roles is 0
Component is OMX.broadcom.write_still
  Num roles is 0
No more components
```

Getting a Handle to a Component

Before you can do anything with a component (apart from getting its name earlier), you need to get a *handle* to it. The call `OMX_GetHandle` takes four parameters.

- A pointer to the handle, which will be filled in by the call.
- The name of the component.
- Application callback data. This is used to pass information from this part of the program to later stages that are called asynchronously through events. Ignore it for now and set it to `NULL`.
- A pointer to a structure of callback functions. These are used extensively when doing event and buffer processing, but you aren't going to do that yet. For now you will set it to a struct of `NULL` functions.

```
OMX_CALLBACKTYPE callbacks  = { .EventHandler = NULL,
                                .EmptyBufferDone = NULL,
                                .FillBufferDone = NULL
};
```

Here's an example:

```
err = OMX_GetHandle(&handle, componentName,
                        // the next two fields are discussed later
                        NULL, &callbacks);
```

Port Information

Ports are the means by which a component communicates, where clients pass data into a port's input buffers and receive processed data back from a port's output buffers.

OpenMAX uses an `OMX_PARAM_PORTDEFINITIONTYPE` struct to keep port information. This is a complex structure, found in the `OMX_Component.h` file and in section 3.1.2.12 of the 1.1.2 specification.

All of the OpenMAX structures, just like this one, have as their first field the size of the structure and as the second field the version of OpenMAX that is being used. Before any struct is used, these must be initialized. It is also a good idea to ensure the structure is zeroed out.

```
struct OMX_PARAM_PORTDEFINITIONTYPE portdef;
memset(&portdef, 0, sizeof(OMX_PARAM_PORTDEFINITIONTYPE));
portdef.nSize = sizeof(OMX_PARAM_PORTDEFINITIONTYPE);
portdef.nVersion.nVersion = OMX_VERSION;
```

The next most important parameter is the index number of the port we are interested in.

One way of finding this is from the documentation mentioned earlier. For example, if you were querying the video port of the visualization component, you would specify this video port with the following:

```
portdef.nPortIndex = 142;
```

There are other ways, which will be discussed later when you look at particular port types.

Each port has buffers. Information about these buffers is stored in the next set of fields:

```
OMX_U32 nBufferCountActual;     /** The actual number of buffers allocated on this port */
OMX_U32 nBufferCountMin;        /** The minimum number of buffers this port requires */
OMX_U32 nBufferSize;            /** Size, in bytes, for buffers to be used for this channel
*/
```

Ports have a state flag as to whether they are enabled or disabled. Components also have state. There is interaction between the port state and component state, and it's not nice. This will be dealt with in more detail in a later chapter. For now, you often toggle this between `OMX_TRUE` and `OMX_FALSE`.

```
OMX_BOOL bEnabled;
```

The last important fields deal with the type of port. First, there is the direction: either OMX_DirInput or OMX_DirOutput.

```
OMX_DIRTYPE eDir;
```

Next is whether it is an audio (OMX_PortDomainAudio), video (OMX_PortDomainVideo), image (OMX_PortDomainImage), or other (OMX_PortDomainOther) port.

```
OMX_PORTDOMAINTYPE eDomain
```

Finally, there is a union corresponding to each of these domains:

```
union {
    OMX_AUDIO_PORTDEFINITIONTYPE audio;
    OMX_VIDEO_PORTDEFINITIONTYPE video;
    OMX_IMAGE_PORTDEFINITIONTYPE image;
    OMX_OTHER_PORTDEFINITIONTYPE other;
} format;
```

Getting and Setting Parameters

OpenMAX has a standard way of getting and setting parameter information for all types such as OMX_PARAM_PORTDEFINITIONTYPE shown earlier. A handle to a component is required; this will have come from a call to OMX_GetHandle() earlier. Then for each type of struct there is a corresponding index value. For OMX_PARAM_PORTDEFINITIONTYPE, the index is OMX_IndexParamPortDefinition, with a similar pattern for other data types. (Strangely, the index OMX_IndexParamPortDefinition doesn't seem to be individually documented in the 1.1.2 specification. Oh well, it works anyway.)

A call to set a parameter value is described in section 3.2.2.8 of the 1.1.2 specification and is as follows:

```
OMX_GetParameter(handle,
                 <index>,
                 <struct_address>);
```

Here is an example:

```
OMX_GetParameter(handle,
                 OMX_IndexParamPortDefinition,
                 &portdef);
```

Setting a parameter is described in section 3.2.2.9 of the 1.1.2 specification and is done with the following:

```
OMX_SetParameter(handle,
                 <index>,
                 <struct_address>);
```

Getting Port Information

You can now combine the last three sections to allow a program to get the information about a particular port on a component and print it out. The program is `portinfo.c`, which takes a component name and a port index as parameters.

```
#include <stdio.h>
#include <stdlib.h>

#include <OMX_Core.h>
#include <OMX_Component.h>

#include <bcm_host.h>

OMX_ERRORTYPE get_port_info(OMX_HANDLETYPE handle,
                            OMX_PARAM_PORTDEFINITIONTYPE *portdef) {
    return  OMX_GetParameter(handle,
                             OMX_IndexParamPortDefinition,
                             portdef);

}

void print_port_info(OMX_PARAM_PORTDEFINITIONTYPE *portdef) {
    char *domain;

    printf("Port %d\n", portdef->nPortIndex);
    if (portdef->eDir ==  OMX_DirInput) {
        printf("  is input port\n");
    } else {
        printf("  is output port\n");
    }

    switch (portdef->eDomain) {
    case OMX_PortDomainAudio: domain = "Audio"; break;
    case OMX_PortDomainVideo: domain = "Video"; break;
    case OMX_PortDomainImage: domain = "Image"; break;
    case OMX_PortDomainOther: domain = "Other"; break;
    }
    printf("  Domain is %s\n", domain);

    printf("  Buffer count %d\n", portdef->nBufferCountActual);
    printf("  Buffer minimum count %d\n", portdef->nBufferCountMin);
    printf("  Buffer size %d bytes\n", portdef->nBufferSize);
}

OMX_CALLBACKTYPE callbacks  = { .EventHandler = NULL,
                                .EmptyBufferDone = NULL,
                                .FillBufferDone = NULL

};
```

```
int main(int argc, char** argv) {

    int i;
    char componentName[128]; // min space required see /opt/vc/include/IL/OMX_Core.h
                             // thanks to Peter Maersk-Moller
    OMX_ERRORTYPE err;
    OMX_HANDLETYPE handle;
    OMX_PARAM_PORTDEFINITIONTYPE portdef;
    OMX_VERSIONTYPE specVersion, compVersion;
    OMX_UUIDTYPE uid;
    int portindex;

    if (argc < 3) {
        fprintf(stderr, "Usage: %s component-name port-index\n", argv[0]);
        exit(1);
    }
    strncpy(componentName, argv[1], 128);
    portindex = atoi(argv[2]);

    bcm_host_init();

    err = OMX_Init();
    if(err != OMX_ErrorNone) {
        fprintf(stderr, "OMX_Init() failed\n", 0);
        exit(1);
    }
    /** Ask the core for a handle to the component
     */
    err = OMX_GetHandle(&handle, componentName,
                        // the next two fields are discussed later
                        NULL, &callbacks);
    if (err != OMX_ErrorNone) {
        fprintf(stderr, "OMX_GetHandle failed\n", 0);
        exit(1);
    }

    // Get some version info
    err = OMX_GetComponentVersion(handle, componentName,
                                  &compVersion, &specVersion,
                                  &uid);
    if (err != OMX_ErrorNone) {
        fprintf(stderr, "OMX_GetComponentVersion failed\n", 0);
        exit(1);
    }
    printf("Component name: %s version %d.%d, Spec version %d.%d\n",
           componentName, compVersion.s.nVersionMajor,
           compVersion.s.nVersionMinor,
           specVersion.s.nVersionMajor,
           specVersion.s.nVersionMinor);
```

```
    memset(&portdef, 0, sizeof(OMX_PARAM_PORTDEFINITIONTYPE));
    portdef.nSize = sizeof(OMX_PARAM_PORTDEFINITIONTYPE);
    portdef.nVersion.nVersion = OMX_VERSION;
    portdef.nPortIndex = portindex;

    if (get_port_info(handle, &portdef) == OMX_ErrorNone) {
        print_port_info(&portdef);
    }

    exit(0);
}
```

When this is run, say with the following:

```
./portinfo  OMX.broadcom.image_decode 320
```

it results in the following:

```
Component name: OMX.broadcom.image_decode:12 version 0.0, Spec version 1.1
Port 320
  is input port
  Domain is Image
  Buffer count 3
  Buffer minimum count 2
  Buffer size 81920 bytes
```

(The component name has an integer field—here 12—that seems to increment by one each time it is run.)

Getting Information About Specific Types of Port

In the previous section, you specified the port of a component explicitly. You aren't likely to sit there with the specification handy just to input port numbers. You want to be able to determine valid port numbers programmatically.

Unfortunately, you can't just ask for a list of valid ports. What you can do is ask for a list of *audio* ports, a list of *video* ports, a list of *image* ports, and a list of valid *other* ports.

Getting such a list is done using a new data structure, the OMX_PORT_PARAM_TYPE. This is described in the file OMX_Core.h and is pretty simple.

```
typedef struct OMX_PORT_PARAM_TYPE {
    OMX_U32 nSize;                /** size of the structure in bytes */
    OMX_VERSIONTYPE nVersion;     /** OMX specification version information */
    OMX_U32 nPorts;               /** The number of ports for this component */
    OMX_U32 nStartPortNumber;     /** first port number for this type of port */
} OMX_PORT_PARAM_TYPE;
```

This structure is filled using a call to OMX_GetParameter as usual, but with an index appropriate to the port type you want. These are as follows:

- For audio ports, the index is OMX_IndexParamAudioInit.
- For video ports, the index is OMX_IndexParamVideoInit.

- For image ports, the index is OMX_IndexParamImageInit.
- For other ports, the index is OMX_IndexParamOtherInit.

Here's an example to get the video ports:

```
OMX_PORT_PARAM_TYPE param;

err = OMX_GetParameter(handle,
                       OMX_IndexParamVideoInit,
                       &param);
```

Note that within each type of port, the values are consecutive. For example, video_splitter has five video ports starting at 250, so the port numbers are 250 through 254.

Note that all that this does is to get the range of port numbers for a particular type of port. If the number of ports returned is zero, there are no ports of this type. Also, to get actual information about each port, you still have to make a call to the following:

```
OMX_PARAM_PORTDEFINITIONTYPE portdef;

OMX_GetParameter(handle,
                 OMX_IndexParamPortDefinition,
                 &portdef)
```

Detailed Audio Port Information

Each type of port (audio, video, image, or other) has a data structure for information specific to that type. For an audio port, the type is OMX_AUDIO_PORTDEFINITIONTYPE and is similar for the other types.

For an audio port, this information is returned as one of the union types in the OMX_PARAM_PORTDEFINITIONTYPE and can be filled in a call to the following, as in the previous program:

```
OMX_GetParameter(handle,
                 OMX_IndexParamPortDefinition,
                 &portdef)
```

The information you want is the audio port information within the general port information. This part of the structure is as follows:

```
portdef.format.audio
```

(Remember, the audio port information is in a union with tag format and field audio.)

The OMX_AUDIO_PORTDEFINITIONTYPE type is described in section 4.1.5 of the 1.1.2 specification and has two important fields.

```
OMX_STRING cMIMEType;
OMX_AUDIO_CODINGTYPE eEncoding;
```

On the RPi, MIMEType is NULL, which it shouldn't really be. I don't think it will get fixed.

The encoding is an enumerated data type. There are a large number of possible values, given in section 4.1.3 of the 1.1.2 specification. The types include PCM (pulse code modulated), AAC, MP3, Vorbis, and WMA. These are also defined in the header file /opt/vc/include/IL/OMX_Audio.h.

Here is where it gets a bit more interesting: that header file not only defines the "standard" OMX audio coding types but also defines an extra set: FLAC, DDP, DTS, and so on (none of which I have heard of). They are defined as part of the OMX_AUDIO_CODINGTYPE enumeration, in the section "Reserved region for introducing Vendor Extensions." In other words, it's perfectly legitimate but not standardized.

Using the OMX_PARAM_PORTDEFINITIONTYPE value you get from querying the port for general information, you can query for the audio types supported with the following:

```
void getAudioPortInformation(int nPort, OMX_PARAM_PORTDEFINITIONTYPE sPortDef) {
    printf("Port %d requires %d buffers\n", nPort, sPortDef.nBufferCountMin);
    printf("Port %d has min buffer size %d bytes\n", nPort, sPortDef.nBufferSize);

    if (sPortDef.eDir == OMX_DirInput) {
        printf("Port %d is an input port\n", nPort);
    } else {
        printf("Port %d is an output port\n",  nPort);
    }
    switch (sPortDef.eDomain) {
    case OMX_PortDomainAudio:
        printf("Port %d is an audio port\n", nPort);
        printf("Port mimetype %s\n",
               sPortDef.format.audio.cMIMEType);

        switch (sPortDef.format.audio.eEncoding) {
        case OMX_AUDIO_CodingPCM:
            printf("Port encoding is PCM\n");
            break;
        case OMX_AUDIO_CodingVORBIS:
            printf("Port encoding is Ogg Vorbis\n");
            break;
        case OMX_AUDIO_CodingMP3:
            printf("Port encoding is MP3\n");
            break;
        default:
            printf("Port encoding is not PCM or MP3 or Vorbis, is %d\n",
                   sPortDef.format.audio.eEncoding);
        }
        getSupportedAudioFormats(nPort);

        break;
        /* could put other port types here */
    default:
        printf("Port %d is not an audio port\n",  nPort);
    }
}
```

But now you can go much deeper. If the port can support MP3 data, what variations on MP3 does it support? For example, MP3 files can have different sampling rates (48 KHz, 44.1 KHz, and so on), number of bits per sample (16, 24, and so on), and number of channels (1, 2, 5, and so on). These can all be queried by—deep breath!—another set of data types.

The first type is OMX_AUDIO_PARAM_PORTFORMATTYPE (see section 4.1.6 of the 1.1.2 specification). It is quite simple.

```
typedef struct OMX_AUDIO_PARAM_PORTFORMATTYPE {
    OMX_U32 nSize;
    OMX_VERSIONTYPE nVersion;
    OMX_U32 nPortIndex;
    OMX_U32 nIndex;
    OMX_AUDIO_CODINGTYPE eEncoding;
} OMX_AUDIO_PARAM_PORTFORMATTYPE;
```

The first two fields are common to all OpenMAX structs: size and version. The third is the port index you are looking at. The fifth is the encoding: PCM, MP3, Vorbis, and so on. The new one is the fourth: the index.

You know from the OMX_PARAM_PORTDEFINITIONTYPE field portdef.format.audio.eEncoding that the port will support a single audio type such as PCM. But it may also support more, such as MP3 or AAC. That's what the index is for, for the different types of audio data. The index starts at zero and keeps increasing until you get an error trying to get parameters with that index value.

So, you start with a struct of type OMX_PARAM_PORTDEFINITIONTYPE and set the index to 0. You then keep asking for parameters with an OMX_IndexParamAudioPortFormat index to OMX_GetParameter. Each time round you increase the index by one until OMX_GetParameter returns an error. On each iteration, you get a new supported audio type.

For example, on the RPi, the audio_render component has one input port that suports PCM and DDT data. You will see the code in just a minute.

Well, if you now know that PCM is supported, you can go further and ask for what parameters it supports. Before doing that, here is the code to list the formats and call for this further level of detail:

```
void getSupportedAudioFormats(int indentLevel, int portNumber) {
    OMX_AUDIO_PARAM_PORTFORMATTYPE sAudioPortFormat;

    setHeader(&sAudioPortFormat, sizeof(OMX_AUDIO_PARAM_PORTFORMATTYPE));
    sAudioPortFormat.nIndex = 0;
    sAudioPortFormat.nPortIndex = portNumber;

    printf("Supported audio formats are:\n");
    for(;;) {
        err = OMX_GetParameter(handle,
                               OMX_IndexParamAudioPortFormat,
                               &sAudioPortFormat);
        if (err == OMX_ErrorNoMore) {
            printf("No more formats supported\n");
            return;
        }

        /* This shouldn't occur, but does with Broadcom library */
        if (sAudioPortFormat.eEncoding == OMX_AUDIO_CodingUnused) {
             printf("No coding format returned\n");
             return;
        }
```

```
        switch (sAudioPortFormat.eEncoding) {
        case OMX_AUDIO_CodingPCM:
            printf("Supported encoding is PCM\n");
            getPCMInformation(portNumber);
            break;
        case OMX_AUDIO_CodingVORBIS:
            printf("Supported encoding is Ogg Vorbis\n");
            break;
        case OMX_AUDIO_CodingMP3:
            printf("Supported encoding is MP3\n");
            getMP3Information(portNumber);
            break;
#ifdef RASPBERRY_PI
        case OMX_AUDIO_CodingFLAC:
            printf("Supported encoding is FLAC\n");
            break;
        case OMX_AUDIO_CodingDDP:
            printf("Supported encoding is DDP\n");
            break;
        case OMX_AUDIO_CodingDTS:
            printf("Supported encoding is DTS\n");
            break;
        case OMX_AUDIO_CodingWMAPRO:
            printf("Supported encoding is WMAPRO\n");
            break;
        case OMX_AUDIO_CodingATRAC3:
            printf("Supported encoding is ATRAC3\n");
            break;
        case OMX_AUDIO_CodingATRACX:
            printf("Supported encoding is ATRACX\n");
            break;
        case OMX_AUDIO_CodingATRACAAL:
            printf("Supported encoding is ATRACAAL\n");
            break;
#endif
        case OMX_AUDIO_CodingAAC:
            printf("Supported encoding is AAC\n");
            break;
        case OMX_AUDIO_CodingWMA:
            printf("Supported encoding is WMA\n");
            break;
        case OMX_AUDIO_CodingRA:
            printf("Supported encoding is RA\n");
            break;
        case OMX_AUDIO_CodingAMR:
            printf("Supported encoding is AMR\n");
            break;
        case OMX_AUDIO_CodingEVRC:
            printf("Supported encoding is EVRC\n");
            break;
```

```
        case OMX_AUDIO_CodingG726:
            printf("Supported encoding is G726\n");
            break;
        case OMX_AUDIO_CodingMIDI:
            printf("Supported encoding is MIDI\n");
            break;
        default:
            printf("Supported encoding is not PCM or MP3 or Vorbis, is 0x%X\n",
                  sAudioPortFormat.eEncoding);
        }
        sAudioPortFormat.nIndex++;
    }
}
```

So, to finally get down to the detail level, you need to look at the functions you define such as `getPCMInformation`, `getMP3Information`, and so on.

You will just look at PCM. Other types can be looked at in a similar way. Section 4.1.7 of the 1.1.2 specification defines the struct `OMX_AUDIO_PARAM_PCMMODETYPE`.

```
typedef struct OMX_AUDIO_PARAM_PCMMODETYPE {
    OMX_U32 nSize;
    OMX_VERSIONTYPE nVersion;
    OMX_U32 nPortIndex;
    OMX_U32 nChannels;
    OMX_NUMERICALDATATYPE eNumData;
    OMX_ENDIANTYPE eEndian;
    OMX_BOOL bInterleaved;
    OMX_U32 nBitPerSample;
    OMX_U32 nSamplingRate;
    OMX_AUDIO_PCMMODETYPE ePCMMode;
    OMX_AUDIO_CHANNELTYPE eChannelMapping[OMX_AUDIO_MAXCHANNELS];
} OMX_AUDIO_PARAM_PCMMODETYPE;
```

The index to look up this struct is `OMX_IndexParamAudioPcm`. (The full list of indices and data types for audio is in Table 4.2, for video is in Table 4-45, and for images is in Table 4-65.)

The following is the code to get this information:

```
void getPCMInformation(int portNumber) {
    /* assert: PCM is a supported mode */
    OMX_AUDIO_PARAM_PCMMODETYPE sPCMMode;

    /* set it into PCM format before asking for PCM info */
    if (setEncoding(portNumber, OMX_AUDIO_CodingPCM) != OMX_ErrorNone) {
        fprintf(stderr, "Error in setting coding to PCM\n");
        return;
    }

    setHeader(&sPCMMode, sizeof(OMX_AUDIO_PARAM_PCMMODETYPE));
    sPCMMode.nPortIndex = portNumber;
    err = OMX_GetParameter(handle, OMX_IndexParamAudioPcm, &sPCMMode);
```

```
    if(err != OMX_ErrorNone){
        printf("PCM mode unsupported\n");
    } else {
        printf("  PCM default sampling rate %d\n", sPCMMode.nSamplingRate);
        printf("  PCM default bits per sample %d\n", sPCMMode.nBitPerSample);
        printf("  PCM default number of channels %d\n", sPCMMode.nChannels);
    }
}
```

A program to dump all of this type of information is `info.c`.

```
/**
   Based on code
   Copyright (C) 2007-2009 STMicroelectronics
   Copyright (C) 2007-2009 Nokia Corporation and/or its subsidiary(-ies).
   under the LGPL
*/

#include <stdio.h>
#include <stdlib.h>
#include <fcntl.h>
#include <string.h>
#include <pthread.h>
#include <unistd.h>
#include <sys/stat.h>

#include <OMX_Core.h>
#include <OMX_Component.h>
#include <OMX_Types.h>
#include <OMX_Audio.h>

#ifdef RASPBERRY_PI
#include <bcm_host.h>
#endif

OMX_ERRORTYPE err;
OMX_HANDLETYPE handle;
OMX_VERSIONTYPE specVersion, compVersion;

OMX_CALLBACKTYPE callbacks;

#define indent {int n = 0; while (n++ < indentLevel*2) putchar(' ');}

static void setHeader(OMX_PTR header, OMX_U32 size) {
    /* header->nVersion */
    OMX_VERSIONTYPE* ver = (OMX_VERSIONTYPE*)(header + sizeof(OMX_U32));
    /* header->nSize */
    *((OMX_U32*)header) = size;

    /* for 1.2
       ver->s.nVersionMajor = OMX_VERSION_MAJOR;
       ver->s.nVersionMinor = OMX_VERSION_MINOR;
```

```
        ver->s.nRevision = OMX_VERSION_REVISION;
        ver->s.nStep = OMX_VERSION_STEP;
    */
    ver->s.nVersionMajor = specVersion.s.nVersionMajor;
    ver->s.nVersionMinor = specVersion.s.nVersionMinor;
    ver->s.nRevision = specVersion.s.nRevision;
    ver->s.nStep = specVersion.s.nStep;
}

void printState() {
    OMX_STATETYPE state;
    err = OMX_GetState(handle, &state);
    if (err != OMX_ErrorNone) {
        fprintf(stderr, "Error on getting state\n");
        exit(1);
    }
    switch (state) {
    case OMX_StateLoaded: fprintf(stderr, "StateLoaded\n"); break;
    case OMX_StateIdle: fprintf(stderr, "StateIdle\n"); break;
    case OMX_StateExecuting: fprintf(stderr, "StateExecuting\n"); break;
    case OMX_StatePause: fprintf(stderr, "StatePause\n"); break;
    case OMX_StateWaitForResources: fprintf(stderr, "StateWiat\n"); break;
    default:  fprintf(stderr, "State unknown\n"); break;
    }
}

OMX_ERRORTYPE setEncoding(int portNumber, OMX_AUDIO_CODINGTYPE encoding) {
    OMX_PARAM_PORTDEFINITIONTYPE sPortDef;

    setHeader(&sPortDef, sizeof(OMX_PARAM_PORTDEFINITIONTYPE));
    sPortDef.nPortIndex = portNumber;
    sPortDef.nPortIndex = portNumber;
    err = OMX_GetParameter(handle, OMX_IndexParamPortDefinition, &sPortDef);
    if(err != OMX_ErrorNone){
        fprintf(stderr, "Error in getting OMX_PORT_DEFINITION_TYPE parameter\n",
 0);
        exit(1);
    }

    sPortDef.format.audio.eEncoding = encoding;
    sPortDef.nBufferCountActual = sPortDef.nBufferCountMin;

    err = OMX_SetParameter(handle, OMX_IndexParamPortDefinition, &sPortDef);
    return err;
}

void getPCMInformation(int indentLevel, int portNumber) {
    /* assert: PCM is a supported mode */
    OMX_AUDIO_PARAM_PCMMODETYPE sPCMMode;
```

```
    /* set it into PCM format before asking for PCM info */
    if (setEncoding(portNumber, OMX_AUDIO_CodingPCM) != OMX_ErrorNone) {
        fprintf(stderr, "Error in setting coding to PCM\n");
        return;
    }

    setHeader(&sPCMMode, sizeof(OMX_AUDIO_PARAM_PCMMODETYPE));
    sPCMMode.nPortIndex = portNumber;
    err = OMX_GetParameter(handle, OMX_IndexParamAudioPcm, &sPCMMode);
    if(err != OMX_ErrorNone){
        indent printf("PCM mode unsupported\n");
    } else {
        indent printf("  PCM default sampling rate %d\n", sPCMMode.nSamplingRate);
        indent printf("  PCM default bits per sample %d\n", sPCMMode.nBitPerSample);
        indent printf("  PCM default number of channels %d\n", sPCMMode.nChannels);
    }

    /*
    setHeader(&sAudioPortFormat, sizeof(OMX_AUDIO_PARAM_PORTFORMATTYPE));
    sAudioPortFormat.nIndex = 0;
    sAudioPortFormat.nPortIndex = portNumber;
    */

}
void getMP3Information(int indentLevel, int portNumber) {
    /* assert: MP3 is a supported mode */
    OMX_AUDIO_PARAM_MP3TYPE sMP3Mode;

    /* set it into MP3 format before asking for MP3 info */
    if (setEncoding(portNumber, OMX_AUDIO_CodingMP3) != OMX_ErrorNone) {
        fprintf(stderr, "Error in setting coding to MP3\n");
        return;
    }

    setHeader(&sMP3Mode, sizeof(OMX_AUDIO_PARAM_MP3TYPE));
    sMP3Mode.nPortIndex = portNumber;
    err = OMX_GetParameter(handle, OMX_IndexParamAudioMp3, &sMP3Mode);
    if(err != OMX_ErrorNone){
        indent printf("MP3 mode unsupported\n");
    } else {
        indent printf("  MP3 default sampling rate %d\n", sMP3Mode.nSampleRate);
        indent printf("  MP3 default bits per sample %d\n", sMP3Mode.nBitRate);
        indent printf("  MP3 default number of channels %d\n", sMP3Mode.nChannels);
    }
}

void getSupportedImageFormats(int indentLevel, int portNumber) {
    OMX_IMAGE_PARAM_PORTFORMATTYPE sImagePortFormat;

    setHeader(&sImagePortFormat, sizeof(OMX_IMAGE_PARAM_PORTFORMATTYPE));
    sImagePortFormat.nIndex = 0;
    sImagePortFormat.nPortIndex = portNumber;
```

```
#ifdef LIM
    printf("LIM doesn't set image formats properly\n");
    return;
#endif

    indent printf("Supported image formats are:\n");
    indentLevel++;
    for(;;) {
        err = OMX_GetParameter(handle, OMX_IndexParamImagePortFormat, &sImagePortFormat);
        if (err == OMX_ErrorNoMore) {
            indent printf("No more formats supported\n");
            return;
        }

        /* This shouldn't occur, but does with Broadcom library */
        if (sImagePortFormat.eColorFormat == OMX_IMAGE_CodingUnused) {
             indent printf("No coding format returned\n");
             return;
        }

        indent printf("Image format compression format 0x%X\n",
                      sImagePortFormat.eCompressionFormat);
        indent printf("Image format color encoding 0x%X\n",
                      sImagePortFormat.eColorFormat);
        sImagePortFormat.nIndex++;
    }
}

void getSupportedVideoFormats(int indentLevel, int portNumber) {
    OMX_VIDEO_PARAM_PORTFORMATTYPE sVideoPortFormat;

    setHeader(&sVideoPortFormat, sizeof(OMX_VIDEO_PARAM_PORTFORMATTYPE));
    sVideoPortFormat.nIndex = 0;
    sVideoPortFormat.nPortIndex = portNumber;

#ifdef LIM
    printf("LIM doesn't set video formats properly\n");
    return;
#endif

    indent printf("Supported video formats are:\n");
    for(;;) {
        err = OMX_GetParameter(handle, OMX_IndexParamVideoPortFormat, &sVideoPortFormat);
        if (err == OMX_ErrorNoMore) {
            indent printf("No more formats supported\n");
            return;
        }

        /* This shouldn't occur, but does with Broadcom library */
        if (sVideoPortFormat.eColorFormat == OMX_VIDEO_CodingUnused) {
             indent printf("No coding format returned\n");
             return;
        }
```

```
        indent printf("Video format encoding 0x%X\n",
                      sVideoPortFormat.eColorFormat);
        sVideoPortFormat.nIndex++;
    }
}

void getSupportedAudioFormats(int indentLevel, int portNumber) {
    OMX_AUDIO_PARAM_PORTFORMATTYPE sAudioPortFormat;

    setHeader(&sAudioPortFormat, sizeof(OMX_AUDIO_PARAM_PORTFORMATTYPE));
    sAudioPortFormat.nIndex = 0;
    sAudioPortFormat.nPortIndex = portNumber;

#ifdef LIM
    printf("LIM doesn't set audio formats properly\n");
    return;
#endif

    indent printf("Supported audio formats are:\n");
    for(;;) {
        err = OMX_GetParameter(handle, OMX_IndexParamAudioPortFormat, &sAudioPortFormat);
        if (err == OMX_ErrorNoMore) {
            indent printf("No more formats supported\n");
            return;
        }

        /* This shouldn't occur, but does with Broadcom library */
        if (sAudioPortFormat.eEncoding == OMX_AUDIO_CodingUnused) {
             indent printf("No coding format returned\n");
             return;
        }

        switch (sAudioPortFormat.eEncoding) {
        case OMX_AUDIO_CodingPCM:
            indent printf("Supported encoding is PCM\n");
            getPCMInformation(indentLevel+1, portNumber);
            break;
        case OMX_AUDIO_CodingVORBIS:
            indent printf("Supported encoding is Ogg Vorbis\n");
            break;
        case OMX_AUDIO_CodingMP3:
            indent printf("Supported encoding is MP3\n");
            getMP3Information(indentLevel+1, portNumber);
            break;
#ifdef RASPBERRY_PI
        case OMX_AUDIO_CodingFLAC:
            indent printf("Supported encoding is FLAC\n");
            break;
        case OMX_AUDIO_CodingDDP:
            indent printf("Supported encoding is DDP\n");
            break;
```

```
        case OMX_AUDIO_CodingDTS:
            indent printf("Supported encoding is DTS\n");
            break;
        case OMX_AUDIO_CodingWMAPRO:
            indent printf("Supported encoding is WMAPRO\n");
            break;
        case OMX_AUDIO_CodingATRAC3:
            indent printf("Supported encoding is ATRAC3\n");
            break;
        case OMX_AUDIO_CodingATRACX:
            indent printf("Supported encoding is ATRACX\n");
            break;
        case OMX_AUDIO_CodingATRACAAL:
            indent printf("Supported encoding is ATRACAAL\n");
            break;
#endif
        case OMX_AUDIO_CodingAAC:
            indent printf("Supported encoding is AAC\n");
            break;
        case OMX_AUDIO_CodingWMA:
            indent printf("Supported encoding is WMA\n");
            break;
        case OMX_AUDIO_CodingRA:
            indent printf("Supported encoding is RA\n");
            break;
        case OMX_AUDIO_CodingAMR:
            indent printf("Supported encoding is AMR\n");
            break;
        case OMX_AUDIO_CodingEVRC:
            indent printf("Supported encoding is EVRC\n");
            break;
        case OMX_AUDIO_CodingG726:
            indent printf("Supported encoding is G726\n");
            break;
        case OMX_AUDIO_CodingMIDI:
            indent printf("Supported encoding is MIDI\n");
            break;

            /*
        case OMX_AUDIO_Coding:
            indent printf("Supported encoding is \n");
            break;
            */
        default:
            indent printf("Supported encoding is not PCM or MP3 or Vorbis, is 0x%X\n",
                  sAudioPortFormat.eEncoding);
        }
        sAudioPortFormat.nIndex++;
    }
}
```

```
void getAudioPortInformation(int indentLevel, int nPort, OMX_PARAM_PORTDEFINITIONTYPE
sPortDef) {
    indent printf("Port %d requires %d buffers\n", nPort, sPortDef.nBufferCountMin);
    indent printf("Port %d has min buffer size %d bytes\n", nPort, sPortDef.nBufferSize);

    if (sPortDef.eDir == OMX_DirInput) {
        indent printf("Port %d is an input port\n", nPort);
    } else {
        indent printf("Port %d is an output port\n",  nPort);
    }
    switch (sPortDef.eDomain) {
    case OMX_PortDomainAudio:
        indent printf("Port %d is an audio port\n", nPort);
        indent printf("Port mimetype %s\n",
               sPortDef.format.audio.cMIMEType);

        switch (sPortDef.format.audio.eEncoding) {
        case OMX_AUDIO_CodingPCM:
            indent printf("Port encoding is PCM\n");
            break;
        case OMX_AUDIO_CodingVORBIS:
            indent printf("Port encoding is Ogg Vorbis\n");
            break;
        case OMX_AUDIO_CodingMP3:
            indent printf("Port encoding is MP3\n");
            break;
        default:
            indent printf("Port encoding is not PCM or MP3 or Vorbis, is %d\n",
                   sPortDef.format.audio.eEncoding);
        }
        getSupportedAudioFormats(indentLevel+1, nPort);

        break;
        /* could put other port types here */
    default:
        indent printf("Port %d is not an audio port\n",  nPort);
    }
}

void getAllAudioPortsInformation(int indentLevel) {
    OMX_PORT_PARAM_TYPE param;
    OMX_PARAM_PORTDEFINITIONTYPE sPortDef;

    int startPortNumber;
    int nPorts;
    int n;

    setHeader(&param, sizeof(OMX_PORT_PARAM_TYPE));
```

```
    err = OMX_GetParameter(handle, OMX_IndexParamAudioInit, &param);
    if(err != OMX_ErrorNone){
        fprintf(stderr, "Error in getting audio OMX_PORT_PARAM_TYPE parameter\n", 0);
        return;
    }
    indent printf("Audio ports:\n");
    indentLevel++;

    startPortNumber = param.nStartPortNumber;
    nPorts = param.nPorts;
    if (nPorts == 0) {
        indent printf("No ports of this type\n");
        return;
    }

    indent printf("Ports start on %d\n", startPortNumber);
    indent printf("There are %d open ports\n", nPorts);

    for (n = 0; n < nPorts; n++) {
        setHeader(&sPortDef, sizeof(OMX_PARAM_PORTDEFINITIONTYPE));
        sPortDef.nPortIndex = startPortNumber + n;
        err = OMX_GetParameter(handle, OMX_IndexParamPortDefinition, &sPortDef);
        if(err != OMX_ErrorNone){
            fprintf(stderr, "Error in getting OMX_PORT_DEFINITION_TYPE parameter\n", 0);
            exit(1);
        }
        indent printf("Port %d has %d buffers of size %d\n",
                      sPortDef.nPortIndex,
                      sPortDef.nBufferCountActual,
                      sPortDef.nBufferSize);
        indent printf("Direction is %s\n",
                      (sPortDef.eDir == OMX_DirInput ? "input" : "output"));
        getAudioPortInformation(indentLevel+1, startPortNumber + n, sPortDef);
    }
}

void getAllVideoPortsInformation(int indentLevel) {
    OMX_PORT_PARAM_TYPE param;
    OMX_PARAM_PORTDEFINITIONTYPE sPortDef;
    int startPortNumber;
    int nPorts;
    int n;

    setHeader(&param, sizeof(OMX_PORT_PARAM_TYPE));

    err = OMX_GetParameter(handle, OMX_IndexParamVideoInit, &param);
    if(err != OMX_ErrorNone){
        fprintf(stderr, "Error in getting video OMX_PORT_PARAM_TYPE parameter\n", 0);
        return;
    }
```

```
    printf("Video ports:\n");
    indentLevel++;

    startPortNumber = param.nStartPortNumber;
    nPorts = param.nPorts;
    if (nPorts == 0) {
        indent printf("No ports of this type\n");
        return;
    }

    indent printf("Ports start on %d\n", startPortNumber);
    indent printf("There are %d open ports\n", nPorts);

    for (n = 0; n < nPorts; n++) {
        setHeader(&sPortDef, sizeof(OMX_PARAM_PORTDEFINITIONTYPE));
        sPortDef.nPortIndex = startPortNumber + n;
        err = OMX_GetParameter(handle, OMX_IndexParamPortDefinition, &sPortDef);
        if(err != OMX_ErrorNone){
            fprintf(stderr, "Error in getting OMX_PORT_DEFINITION_TYPE parameter\n", 0);
            exit(1);
        }
        //getVideoPortInformation(indentLevel+1, startPortNumber + n, sPortDef);
        indent printf("Port %d has %d buffers (minimum %d) of size %d\n",
                      sPortDef.nPortIndex,
                      sPortDef.nBufferCountActual,
                      sPortDef.nBufferCountMin,
                      sPortDef.nBufferSize);
        indent printf("Direction is %s\n",
                      (sPortDef.eDir == OMX_DirInput ? "input" : "output"));

        getSupportedVideoFormats(indentLevel+1,  startPortNumber + n);
    }
}

void getAllImagePortsInformation(int indentLevel) {
    OMX_PORT_PARAM_TYPE param;
    OMX_PARAM_PORTDEFINITIONTYPE sPortDef;
    int startPortNumber;
    int nPorts;
    int n;

    setHeader(&param, sizeof(OMX_PORT_PARAM_TYPE));

    err = OMX_GetParameter(handle, OMX_IndexParamImageInit, &param);
    if(err != OMX_ErrorNone){
        fprintf(stderr, "Error in getting image OMX_PORT_PARAM_TYPE parameter\n", 0);
        return;
    }
    printf("Image ports:\n");
    indentLevel++;
```

```
    startPortNumber = param.nStartPortNumber;
    nPorts = param.nPorts;
    if (nPorts == 0) {
        indent printf("No ports of this type\n");
        return;
    }

    indent printf("Ports start on %d\n", startPortNumber);
    indent printf("There are %d open ports\n", nPorts);

    for (n = 0; n < nPorts; n++) {
        setHeader(&sPortDef, sizeof(OMX_PARAM_PORTDEFINITIONTYPE));
        sPortDef.nPortIndex = startPortNumber + n;
        err = OMX_GetParameter(handle, OMX_IndexParamPortDefinition, &sPortDef);
        if(err != OMX_ErrorNone){
            fprintf(stderr, "Error in getting OMX_PORT_DEFINITION_TYPE parameter\n", 0);
            exit(1);
        }

        indent printf("Port %d has %d buffers (minimum %d) of size %d\n",
                      sPortDef.nPortIndex,
                      sPortDef.nBufferCountActual,
                      sPortDef.nBufferCountMin,
                      sPortDef.nBufferSize);
        indent printf("Direction is %s\n",
                      (sPortDef.eDir == OMX_DirInput ? "input" : "output"));

        //getImagePortInformation(indentLevel+1, startPortNumber + n, sPortDef);
        getSupportedImageFormats(indentLevel+1,  startPortNumber + n);
    }
}

void getAllOtherPortsInformation(int indentLevel) {
    OMX_PORT_PARAM_TYPE param;
    OMX_PARAM_PORTDEFINITIONTYPE sPortDef;
    int startPortNumber;
    int nPorts;
    int n;

    setHeader(&param, sizeof(OMX_PORT_PARAM_TYPE));

    err = OMX_GetParameter(handle, OMX_IndexParamOtherInit, &param);
    if(err != OMX_ErrorNone){
        fprintf(stderr, "Error in getting other OMX_PORT_PARAM_TYPE parameter\n", 0);
        exit(1);
    }
    printf("Other ports:\n");
    indentLevel++;

    startPortNumber = param.nStartPortNumber;
    nPorts = param.nPorts;
```

```
    if (nPorts == 0) {
        indent printf("No ports of this type\n");
        return;
    }

    indent printf("Ports start on %d\n", startPortNumber);
    indent printf("There are %d open ports\n", nPorts);

    indent printf("Port %d has %d buffers of size %d\n",
                  sPortDef.nPortIndex,
                  sPortDef.nBufferCountActual,
                  sPortDef.nBufferSize);
    indent printf("Direction is %s\n",
                  (sPortDef.eDir == OMX_DirInput ? "input" : "output"));
}

void getAllPortsInformation(int indentLevel) {
    OMX_PORT_PARAM_TYPE param;
    OMX_PARAM_PORTDEFINITIONTYPE sPortDef;
    int startPortNumber;
    int nPorts;
    int n;

    setHeader(&param, sizeof(OMX_PORT_PARAM_TYPE));

    err = OMX_GetParameter(handle, OMX_IndexParamVideoInit, &param);
    if(err != OMX_ErrorNone){
        fprintf(stderr, "Error in getting video OMX_PORT_PARAM_TYPE parameter\n", 0);
        return;
    }

    printf("Video ports:\n");
    indentLevel++;

    startPortNumber = param.nStartPortNumber;
    nPorts = param.nPorts;
    if (nPorts == 0) {
        indent printf("No ports of this type\n");
        return;
    }

    indent printf("Ports start on %d\n", startPortNumber);
    indent printf("There are %d open ports\n", nPorts);

    for (n = 0; n < nPorts; n++) {
        setHeader(&sPortDef, sizeof(OMX_PARAM_PORTDEFINITIONTYPE));
        sPortDef.nPortIndex = startPortNumber + n;
        err = OMX_GetParameter(handle, OMX_IndexParamPortDefinition, &sPortDef);
        if(err != OMX_ErrorNone){
            fprintf(stderr, "Error in getting OMX_PORT_DEFINITION_TYPE parameter\n", 0);
            exit(1);
        }
```

```
        indent printf("Port %d has %d buffers of size %d\n",
                      sPortDef.nPortIndex,
                      sPortDef.nBufferCountActual,
                      sPortDef.nBufferSize);
        indent printf("Direction is %s\n",
                      (sPortDef.eDir == OMX_DirInput ? "input" : "output"));
        switch (sPortDef.eDomain) {
        case  OMX_PortDomainVideo:
            indent printf("Domain is video\n");
            getSupportedVideoFormats(indentLevel+1,  startPortNumber + n);
            break;
        case  OMX_PortDomainImage:
            indent printf("Domain is image\n");
            getSupportedImageFormats(indentLevel+1,  startPortNumber + n);
            break;
        case  OMX_PortDomainAudio:
            indent printf("Domain is audio\n");
            getSupportedAudioFormats(indentLevel+1,  startPortNumber + n);
            break;
        case  OMX_PortDomainOther:
            indent printf("Domain is other\n");
            // getSupportedOtherFormats(indentLevel+1,  startPortNumber + n);
            break;
        }
        //getVideoPortInformation(indentLevel+1, startPortNumber + n, sPortDef);
        /*
        if (sPortDef.eDomain == OMX_PortDomainVideo)
            getSupportedVideoFormats(indentLevel+1,  startPortNumber + n);
        else
            indent printf("Not a video port\n");
        */
    }
}

int main(int argc, char** argv) {

    OMX_PORT_PARAM_TYPE param;
    OMX_PARAM_PORTDEFINITIONTYPE sPortDef;
    OMX_AUDIO_PORTDEFINITIONTYPE sAudioPortDef;
    OMX_AUDIO_PARAM_PORTFORMATTYPE sAudioPortFormat;
    OMX_AUDIO_PARAM_PCMMODETYPE sPCMMode;

#ifdef RASPBERRY_PI
    //char *componentName = "OMX.broadcom.audio_mixer";
    //char *componentName = "OMX.broadcom.audio_mixer";
    char *componentName = "OMX.broadcom.video_render";
#else
#ifdef LIM
    char *componentName = "OMX.limoi.alsa_sink";
```

```
#else
    char *componentName = "OMX.st.volume.component";
#endif
#endif
    unsigned char name[128]; /* spec says 128 is max name length */
    OMX_UUIDTYPE uid;
    int startPortNumber;
    int nPorts;
    int n;

    /* ovveride component name by command line argument */
    if (argc == 2) {
        componentName = argv[1];
    }

# ifdef RASPBERRY_PI
    bcm_host_init();
# endif

    err = OMX_Init();
    if(err != OMX_ErrorNone) {
        fprintf(stderr, "OMX_Init() failed\n", 0);
        exit(1);
    }
    /** Ask the core for a handle to the volume control component
     */
    err = OMX_GetHandle(&handle, componentName, NULL /*app private data */, &callbacks);
    if (err != OMX_ErrorNone) {
        fprintf(stderr, "OMX_GetHandle failed\n", 0);
        exit(1);
    }
    err = OMX_GetComponentVersion(handle, name, &compVersion, &specVersion, &uid);
    if (err != OMX_ErrorNone) {
        fprintf(stderr, "OMX_GetComponentVersion failed\n", 0);
        exit(1);
    }
    printf("Component name: %s version %d.%d, Spec version %d.%d\n",
           name, compVersion.s.nVersionMajor,
           compVersion.s.nVersionMinor,
           specVersion.s.nVersionMajor,
           specVersion.s.nVersionMinor);

    /** Get  ports information */
    //getAllPortsInformation(0);

    getAllAudioPortsInformation(0);
    getAllVideoPortsInformation(0);
    getAllImagePortsInformation(0);
    getAllOtherPortsInformation(0);

    exit(0);
}
```

The output for the video_render component is as follows:

```
Component name: OMX.broadcom.video_render:0 version 0.0, Spec version 1.1
Audio ports:
  No ports of this type
Video ports:
  Ports start on 90
  There are 1 open ports
  Port 90 has 3 buffers (minimum 2) of size 15360
  Direction is input
    Supported video formats are:
    Video format encoding 0x14
    Video format encoding 0x27
    Video format encoding 0x17
    Video format encoding 0x17
    Video format encoding 0x7F000003
    Video format encoding 0x6
    Video format encoding 0xC
    Video format encoding 0xB
    Video format encoding 0x7F000001
    No more formats supported
Image ports:
  No ports of this type
Other ports:
  No ports of this type
```

Conclusion

In this chapter, you looked at how to get information about OpenMAX components. So far it isn't hard; the information is just extremely voluminous because of the complexities and variations among all the components supported.

Resources

- *OpenMAX IL: The Standard for Media Library Portability*: `https://www.khronos.org/openmax/il/`
- *OpenMAX IL 1.1.2 Specification*: `https://www.khronos.org/registry/omxil/specs/OpenMAX_IL_1_1_2_Specification.pdf`
- *Raspberry Pi OpenMAX forum*: `www.raspberrypi.org/forums/viewforum.php?f=70`
- *VMCS-X OpenMAX IL Components* : `www.jvcref.com/files/PI/documentation/ilcomponents/`
- *Source code for ARM side libraries for interfacing to Raspberry Pi GPU*: `https://github.com/raspberrypi/userland`

CHAPTER 9

OpenMAX on the Raspberry Pi State

OpenMAX IL uses a state model for each component. Within each state certain operations can be performed, while others must be done before the next state transition can occur.

Building Programs

The programs in this chapter can be built using the following Makefile:

```
DMX_INC =  -I/opt/vc/include/ -I /opt/vc/include/interface/vmcs_host/ -I/opt/vc/include/
interface/vcos/pthreads -I/opt/vc/include/interface/vmcs_host/linux
EGL_INC =
OMX_INC =  -I /opt/vc/include/IL
INCLUDES = $(DMX_INC) $(EGL_INC) $(OMX_INC)

CFLAGS=-g -DRASPBERRY_PI -DOMX_SKIP64BIT $(INCLUDES)
CPPFLAGS =

DMX_LIBS =  -L/opt/vc/lib/ -lbcm_host -lvcos -lvchiq_arm -lpthread
EGL_LIBS = -L/opt/vc/lib/ -lEGL -lGLESv2
OMX_LIBS = -lopenmaxil

LDLIBS =  $(DMX_LIBS) $(EGL_LIBS) $(OMX_LIBS)

EXE = wontwork working event

all: $(EXE)
```

Component States

Every component has a state and can go through a series of state transitions. There are three “desirable” states: Loaded, Idle, and Executing. There are some less desirable states such as Wait for Resources and Invalid, and if a component is in one of those states except transiently, then your program probably won’t be working properly.

The state transition diagram is Figure 2-3 of the 1.1.2 specification.

J. Newmarch, *Raspberry Pi GPU Audio Video Programming*, DOI 10.1007/978-1-4842-2472-4_9

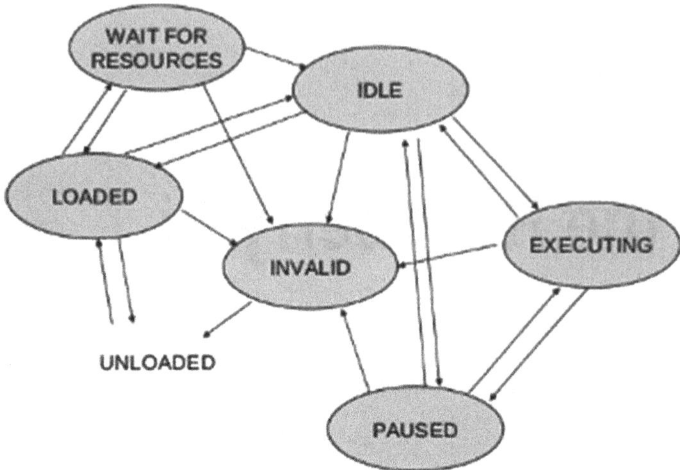

Figure 9-1. *OpenMAX state transitions*

Whenever a component changes state, it will invoke the event callback handler established for that component. I ignored this in earlier chapters and will cover it in more detail later. For now, you have to give a non-`NULL` function to the `EventHandler` callback struct, which is passed when getting a handle to a component. Minimally, it will look like this:

```
OMX_ERRORTYPE cEventHandler(
                           OMX_HANDLETYPE hComponent,
                           OMX_PTR pAppData,
                           OMX_EVENTTYPE eEvent,
                           OMX_U32 Data1,
                           OMX_U32 Data2,
                           OMX_PTR pEventData) {
    // do nothing
    return OMX_ErrorNone;
}

OMX_CALLBACKTYPE callbacks  = { .EventHandler = cEventHandler,
                                .EmptyBufferDone = NULL,
                                .FillBufferDone = NULL
};

err = OMX_GetHandle(&handle, componentName,
                    NULL, &callbacks);
```

A created component moves into the Loaded state. Requests to change to another state are done using the following call (section 3.2.2.2 of the 1.1.2 specification):

```
OMX_SendCommand(handle,
                OMX_CommandStateSet,
                <next_state>,
                <user_data>);
```

Loaded to Idle

What you want to do for any component is to move it from Loaded to Idle with the following:

```
OMX_SendCommand(handle,
                OMX_CommandStateSet,
                OMX_StateIdle,
                NULL);
```

A Nonworking Program

This is where you now hit hurdle #1 in OpenMAX programming: a request to change state has an effect only if all the prerequisites are met for the new state. To draw an analogy, if you want your car to change from a stopped state to a moving state, then in preparation for this you have to insert the ignition key, switch on the engine, release the parking brake, engage gears, and so on. If any one of these steps is missing, the state change just won't occur.

The first major change that you want to happen will be from `OMX_StateLoaded` to `OMX_StateIdle`. The following program creates a program and requests the state change. If you run it, you will see that it just sits in the Loaded state and never makes the transition. The program is `wontwork.c`. I've included some convenience functions: `err2str` and `printState`.

```
/*
 * WARNING: THIS PROGRAM DOESN'T WORK
 */

#include <stdio.h>
#include <stdlib.h>

#include <OMX_Core.h>
#include <OMX_Component.h>

#include <bcm_host.h>

char *err2str(int err) {
    switch (err) {
    case OMX_ErrorInsufficientResources: return "OMX_ErrorInsufficientResources";
    case OMX_ErrorUndefined: return "OMX_ErrorUndefined";
    case OMX_ErrorInvalidComponentName: return "OMX_ErrorInvalidComponentName";
    case OMX_ErrorComponentNotFound: return "OMX_ErrorComponentNotFound";
    case OMX_ErrorInvalidComponent: return "OMX_ErrorInvalidComponent";
    case OMX_ErrorBadParameter: return "OMX_ErrorBadParameter";
    case OMX_ErrorNotImplemented: return "OMX_ErrorNotImplemented";
    case OMX_ErrorUnderflow: return "OMX_ErrorUnderflow";
    case OMX_ErrorOverflow: return "OMX_ErrorOverflow";
    case OMX_ErrorHardware: return "OMX_ErrorHardware";
    case OMX_ErrorInvalidState: return "OMX_ErrorInvalidState";
    case OMX_ErrorStreamCorrupt: return "OMX_ErrorStreamCorrupt";
    case OMX_ErrorPortsNotCompatible: return "OMX_ErrorPortsNotCompatible";
    case OMX_ErrorResourcesLost: return "OMX_ErrorResourcesLost";
    case OMX_ErrorNoMore: return "OMX_ErrorNoMore";
```

```
    case OMX_ErrorVersionMismatch: return "OMX_ErrorVersionMismatch";
    case OMX_ErrorNotReady: return "OMX_ErrorNotReady";
    case OMX_ErrorTimeout: return "OMX_ErrorTimeout";
    case OMX_ErrorSameState: return "OMX_ErrorSameState";
    case OMX_ErrorResourcesPreempted: return "OMX_ErrorResourcesPreempted";
    case OMX_ErrorPortUnresponsiveDuringAllocation: return "OMX_
ErrorPortUnresponsiveDuringAllocation";
    case OMX_ErrorPortUnresponsiveDuringDeallocation: return "OMX_
ErrorPortUnresponsiveDuringDeallocation";
    case OMX_ErrorPortUnresponsiveDuringStop: return "OMX_ErrorPortUnresponsiveDuringStop";
    case OMX_ErrorIncorrectStateTransition: return "OMX_ErrorIncorrectStateTransition";
    case OMX_ErrorIncorrectStateOperation: return "OMX_ErrorIncorrectStateOperation";
    case OMX_ErrorUnsupportedSetting: return "OMX_ErrorUnsupportedSetting";
    case OMX_ErrorUnsupportedIndex: return "OMX_ErrorUnsupportedIndex";
    case OMX_ErrorBadPortIndex: return "OMX_ErrorBadPortIndex";
    case OMX_ErrorPortUnpopulated: return "OMX_ErrorPortUnpopulated";
    case OMX_ErrorComponentSuspended: return "OMX_ErrorComponentSuspended";
    case OMX_ErrorDynamicResourcesUnavailable: return "OMX_
ErrorDynamicResourcesUnavailable";
    case OMX_ErrorMbErrorsInFrame: return "OMX_ErrorMbErrorsInFrame";
    case OMX_ErrorFormatNotDetected: return "OMX_ErrorFormatNotDetected";
    case OMX_ErrorContentPipeOpenFailed: return "OMX_ErrorContentPipeOpenFailed";
    case OMX_ErrorContentPipeCreationFailed: return "OMX_ErrorContentPipeCreationFailed";
    case OMX_ErrorSeperateTablesUsed: return "OMX_ErrorSeperateTablesUsed";
    case OMX_ErrorTunnelingUnsupported: return "OMX_ErrorTunnelingUnsupported";
    default: return "unknown error";
    }
}

void printState(OMX_HANDLETYPE handle) {
    OMX_STATETYPE state;
    OMX_ERRORTYPE err;

    err = OMX_GetState(handle, &state);
    if (err != OMX_ErrorNone) {
        fprintf(stderr, "Error on getting state\n");
        exit(1);
    }
    switch (state) {
    case OMX_StateLoaded: printf("StateLoaded\n"); break;
    case OMX_StateIdle: printf("StateIdle\n"); break;
    case OMX_StateExecuting: printf("StateExecuting\n"); break;
    case OMX_StatePause: printf("StatePause\n"); break;
    case OMX_StateWaitForResources: printf("StateWait\n"); break;
    case OMX_StateInvalid: printf("StateInvalid\n"); break;
    default:  printf("State unknown\n"); break;
    }
}

OMX_ERRORTYPE cEventHandler(
                           OMX_HANDLETYPE hComponent,
                           OMX_PTR pAppData,
```

```
                          OMX_EVENTTYPE eEvent,
                          OMX_U32 Data1,
                          OMX_U32 Data2,
                          OMX_PTR pEventData) {

    printf("Hi there, I am in the %s callback\n", __func__);
    printf("Event is %i\n", (int)eEvent);
    printf("Param1 is %i\n", (int)Data1);
    printf("Param2 is %i\n", (int)Data2);

    return OMX_ErrorNone;
}

OMX_CALLBACKTYPE callbacks  = { .EventHandler = cEventHandler,
                                .EmptyBufferDone = NULL,
                                .FillBufferDone = NULL
};

int main(int argc, char** argv) {

    int i;
    char *componentName;
    OMX_ERRORTYPE err;
    OMX_HANDLETYPE handle;

    if (argc < 2) {
        fprintf(stderr, "Usage: %s component-name\n", argv[0]);
        exit(1);
    }
    componentName = argv[1];

    bcm_host_init();

    err = OMX_Init();
    if(err != OMX_ErrorNone) {
        fprintf(stderr, "OMX_Init() failed %s\n", err2str(err));
        exit(1);
    }
    /** Ask the core for a handle to the component
     */
    err = OMX_GetHandle(&handle, componentName,
                        // the next two fields are discussed later
                        NULL, &callbacks);
    if (err != OMX_ErrorNone) {
        fprintf(stderr, "OMX_GetHandle failed %s\n", err2str(err));
        exit(1);
    }

    // check our current state - should be Loaded
    printState(handle);
```

```
    // request a move to idle
    OMX_SendCommand(handle,
                    OMX_CommandStateSet,
                    OMX_StateIdle,
                    NULL);

    int n = 0;
    while (n++ < 10) {
        sleep(1);
        // are we there yet?
        printState(handle);
    }

    exit(0);
}
```

The output from

```
../wontwork OMX.broadcom.audio_capture
```

is similar to the following:

```
StateLoaded
Hi there, I am in the cEventHandler callback

Event is 1
Param1 is -2147479552
Param2 is 1
StateLoaded
StateLoaded
StateLoaded
...
```

What's Wrong with It?

The OpenMAX 1.1.2 specification in section 3.1.1.2.1.1, "OMX_StateLoaded to OMX_StateIdle," says the following:

> *If the IL client requests a state transition from OMX_StateLoaded to OMX_StateIdle, the component shall acquire all of its static resources, including buffers for all enabled ports, before completing the transition. The component does not acquire buffers for any disabled ports. Furthermore, before the transition can complete, the buffer supplier, which is always the IL client when not tunneling, shall ensure that the non-supplier possesses all of its buffers.*

Clear? Well, maybe not. What it is saying is that the component must have everything right, including allocating all of the input and output buffers. Alternatively, you can do the equivalent of leaving the car on a hill and just releasing the parking brake, which is a much simpler way of moving! For a component, this means disabling all of the ports. This can get you into an Idle state, and then you can worry about the next transition.

To disable all of the ports, you have to use the techniques in the previous chapter: find all of the video ports, all of the audio ports, all of the image ports, and all of the other ports. For each one, call the following:

```
OMX_SendCommand(handle, OMX_CommandPortDisable,
                <port_index>, NULL);
```

The progam is `working.c`.

```
#include <stdio.h>
#include <stdlib.h>

#include <OMX_Core.h>
#include <OMX_Component.h>

#include <bcm_host.h>

char *err2str(int err) {
    return "omitted";
}

void printState(OMX_HANDLETYPE handle) {
    OMX_STATETYPE state;
    OMX_ERRORTYPE err;

    err = OMX_GetState(handle, &state);
    if (err != OMX_ErrorNone) {
        fprintf(stderr, "Error on getting state\n");
        exit(1);
    }
    switch (state) {
    case OMX_StateLoaded: printf("StateLoaded\n"); break;
    case OMX_StateIdle: printf("StateIdle\n"); break;
    case OMX_StateExecuting: printf("StateExecuting\n"); break;
    case OMX_StatePause: printf("StatePause\n"); break;
    case OMX_StateWaitForResources: printf("StateWait\n"); break;
    case OMX_StateInvalid: printf("StateInvalid\n"); break;
    default:  printf("State unknown\n"); break;
    }
}

OMX_ERRORTYPE cEventHandler(
                            OMX_HANDLETYPE hComponent,
                            OMX_PTR pAppData,
                            OMX_EVENTTYPE eEvent,
                            OMX_U32 Data1,
                            OMX_U32 Data2,
                            OMX_PTR pEventData) {

    printf("Hi there, I am in the %s callback\n", __func__);
    printf("Event is %i\n", (int)eEvent);
    printf("Param1 is %i\n", (int)Data1);
```

```
    printf("Param2 is %i\n", (int)Data2);

    return OMX_ErrorNone;
}

OMX_CALLBACKTYPE callbacks  = { .EventHandler = cEventHandler,
                                .EmptyBufferDone = NULL,
                                .FillBufferDone = NULL
};

void disableSomePorts(OMX_HANDLETYPE handle, OMX_INDEXTYPE indexType) {
    OMX_PORT_PARAM_TYPE param;
    int startPortNumber, endPortNumber;
    int nPorts;
    int n;
    OMX_ERRORTYPE err;

    memset(&param, 0, sizeof(OMX_PORT_PARAM_TYPE));
    param.nSize = sizeof(OMX_PORT_PARAM_TYPE);
    param.nVersion.nVersion = OMX_VERSION;

    err = OMX_GetParameter(handle, indexType, &param);
    if(err != OMX_ErrorNone){
        fprintf(stderr, "Error in getting image OMX_PORT_PARAM_TYPE parameter\n"
);
        return;
    }

    startPortNumber = param.nStartPortNumber;
    nPorts = param.nPorts;
    endPortNumber = startPortNumber + nPorts;

    for (n = startPortNumber; n < endPortNumber; n++) {
        OMX_SendCommand(handle, OMX_CommandPortDisable,
                        n, NULL);
    }
}

void disableAllPorts(OMX_HANDLETYPE handle) {
    disableSomePorts(handle, OMX_IndexParamVideoInit);
    disableSomePorts(handle, OMX_IndexParamImageInit);
    disableSomePorts(handle, OMX_IndexParamAudioInit);
    disableSomePorts(handle, OMX_IndexParamOtherInit);
}

int main(int argc, char** argv) {

    int i;
    char *componentName;
    OMX_ERRORTYPE err;
    OMX_HANDLETYPE handle;
```

```
    if (argc < 2) {
        fprintf(stderr, "Usage: %s component-name\n", argv[0]);
        exit(1);
    }
    componentName = argv[1];

    bcm_host_init();

    err = OMX_Init();
    if(err != OMX_ErrorNone) {
        fprintf(stderr, "OMX_Init() failed %s\n", err2str(err));
        exit(1);
    }
    /** Ask the core for a handle to the component
     */
    err = OMX_GetHandle(&handle, componentName,
                        // the next two fields are discussed later
                        NULL, &callbacks);
    if (err != OMX_ErrorNone) {
        fprintf(stderr, "OMX_GetHandle failed %s\n", err2str(err));
        exit(1);
    }

    sleep(1);
    // check our current state - should be Loaded
    printState(handle);

    disableAllPorts(handle);

    // request a move to idle
    OMX_SendCommand(handle,
                    OMX_CommandStateSet,
                    OMX_StateIdle,
                    NULL);

    int n = 0;
    while (n++ < 10) {
        sleep(1);
        // are we there yet?
        printState(handle);
    }

    exit(0);
}
```

When run with the following:

```
./working OMX.broadcom.video_render
```

it successfully makes the transition to Idle state, as shown here:

```
./working OMX.broadcom.video_render
StateLoaded
Hi there, I am in the cEventHandler callback
Event is 0
Param1 is 2
Param2 is 90
Hi there, I am in the cEventHandler callback
Event is 0
Param1 is 0
Param2 is 2
StateIdle
StateIdle
StateIdle
...
```

Idle to Executing

Once a component is in Idle state, a request can be made to move it to Executing. This is actually pretty easy and generally succeeds. Once it is in Executing state, it can start filling and emptying buffers (you haven't looked at buffers yet).

When Do State Changes Occur?

In the "working" program of an earlier section, you did a loop with a sleep to print the current state. If you want to detect when a state change has occurred, you could do a similar sort of "busy wait."

That isn't really satisfactory: OpenMAX should be able to able to tell you when a state change occurs. And it does: when a state change occurs in a component, it generates an event that is given to the event handler callback.

The event handler callback is discussed in section 3.1.2.9.1 of the 1.1.2 specification. It is defined as follows:

```
OMX_ERRORTYPE (*EventHandler)(
    OMX_IN OMX_HANDLETYPE hComponent,
    OMX_IN OMX_PTR pAppData,
    OMX_IN OMX_EVENTTYPE eEvent,
    OMX_IN OMX_U32 nData1,
    OMX_IN OMX_U32 nData2,
    OMX_IN OMX_PTR pEventData);
```

The `hComponent` is the handle to the component. The `eEvent` is an enumerated data type giving possible event types. The data values contain values appropriate to the event. These are described in Table 3-7 of the 1.1.2 specification.

Table 9-1. *Parameters in events*

eEvent	nData1	nData2	pEventData
OMX_EventCmdComplete	OMX_CommandStateSet	State reached	Null
	OMX_CommandFlush	Port index	Null
	OMX_CommandPort Disable	Port index	Null
	OMX_CommandPort Enable	Port index	Null
	OMX_CommandMark Buffer	Port index	Null
OMX_EventError	Error code	0	Null
OMX_EventMark	0	0	Data linked to the mark, if any
OMX_EventPortSettings Changed	port index	0	Null
OMX_EventBufferFlag	port index	nFlags unaltered	Null
OMX_EventResources Acquired	0	0	Null
OMX_EventDynamic ResourcesAvailable	0	0	Null

For now, the important event is OMX_EventCmdComplete with nData1 set to OMX_CommandStateSet. The value in nData2 is then the state reached: OMX_StateIdle, and so on.

The program event.c monitors events. When run on the OMX.broadcom.clock component, it reports disabling all five ports, transitioning to Idle, and then transitioning to Executing.

```
#include <stdio.h>
#include <stdlib.h>

#include <OMX_Core.h>
#include <OMX_Component.h>

#include <bcm_host.h>

char *err2str(int err) {
    return "omitted";
}

void printState(OMX_HANDLETYPE handle) {
    OMX_STATETYPE state;
    OMX_ERRORTYPE err;

    err = OMX_GetState(handle, &state);
    if (err != OMX_ErrorNone) {
        fprintf(stderr, "Error on getting state\n");
        exit(1);
    }
    switch (state) {
    case OMX_StateLoaded: printf("StateLoaded\n"); break;
    case OMX_StateIdle: printf("StateIdle\n"); break;
    case OMX_StateExecuting: printf("StateExecuting\n"); break;
```

```
    case OMX_StatePause: printf("StatePause\n"); break;
    case OMX_StateWaitForResources: printf("StateWait\n"); break;
    case OMX_StateInvalid: printf("StateInvalid\n"); break;
    default:  printf("State unknown\n"); break;
    }
}

OMX_ERRORTYPE cEventHandler(
                            OMX_HANDLETYPE hComponent,
                            OMX_PTR pAppData,
                            OMX_EVENTTYPE eEvent,
                            OMX_U32 Data1,
                            OMX_U32 Data2,
                            OMX_PTR pEventData) {

    if(eEvent == OMX_EventCmdComplete) {
        if (Data1 == OMX_CommandStateSet) {
            printf("Component State changed to ");
            switch ((int)Data2) {
            case OMX_StateInvalid:
                printf("OMX_StateInvalid\n");
                break;
            case OMX_StateLoaded:
                printf("OMX_StateLoaded\n");
                break;
            case OMX_StateIdle:
                printf("OMX_StateIdle\n");
                break;
            case OMX_StateExecuting:
                printf("OMX_StateExecuting\n");
                break;
            case OMX_StatePause:
                printf("OMX_StatePause\n");
                break;
            case OMX_StateWaitForResources:
                printf("OMX_StateWaitForResources\n");
               break;
            }
        } else  if (Data1 == OMX_CommandPortEnable){
            printf("OMX State Port enabled %d\n", (int) Data2);
        } else if (Data1 == OMX_CommandPortDisable){
            printf("OMX State Port disabled %d\n", (int) Data2);
        }
    } else if(eEvent == OMX_EventBufferFlag) {
        if((int)Data2 == OMX_BUFFERFLAG_EOS) {
            printf("Event is buffer end of stream\n");
        }
    } else if(eEvent == OMX_EventError) {
      if (Data1 == OMX_ErrorSameState) {
        printf("Already in requested state\n");
```

```
        } else {
          printf("Event is Error %X\n", Data1);
        }
    } else  if(eEvent == OMX_EventMark) {
          printf("Event is Buffer Mark\n");
    } else  if(eEvent == OMX_EventPortSettingsChanged) {
          printf("Event is PortSettingsChanged\n");
    }

    return OMX_ErrorNone;
}

OMX_CALLBACKTYPE callbacks  = { .EventHandler = cEventHandler,
                                .EmptyBufferDone = NULL,
                                .FillBufferDone = NULL
};

void disableSomePorts(OMX_HANDLETYPE handle, OMX_INDEXTYPE indexType) {
    OMX_PORT_PARAM_TYPE param;
    int startPortNumber, endPortNumber;
    int nPorts;
    int n;
    OMX_ERRORTYPE err;

    //setHeader(&param, sizeof(OMX_PORT_PARAM_TYPE));

    memset(&param, 0, sizeof(OMX_PORT_PARAM_TYPE));
    param.nSize = sizeof(OMX_PORT_PARAM_TYPE);
    param.nVersion.nVersion = OMX_VERSION;

    err = OMX_GetParameter(handle, indexType, &param);
    if(err != OMX_ErrorNone){
        fprintf(stderr, "Error in getting image OMX_PORT_PARAM_TYPE parameter\n"
, 0);
        return;
    }

    startPortNumber = param.nStartPortNumber;
    nPorts = param.nPorts;
    endPortNumber = startPortNumber + nPorts;

    for (n = startPortNumber; n < endPortNumber; n++) {
        OMX_SendCommand(handle, OMX_CommandPortDisable,
                        n, NULL);
    }
}
```

```
void disableAllPorts(OMX_HANDLETYPE handle) {
    disableSomePorts(handle, OMX_IndexParamVideoInit);
    disableSomePorts(handle, OMX_IndexParamImageInit);
    disableSomePorts(handle, OMX_IndexParamAudioInit);
    disableSomePorts(handle, OMX_IndexParamOtherInit);
}

int main(int argc, char** argv) {

    int i;
    char *componentName;
    OMX_ERRORTYPE err;
    OMX_HANDLETYPE handle;

    if (argc < 2) {
        fprintf(stderr, "Usage: %s component-name\n", argv[0]);
        exit(1);
    }
    componentName = argv[1];

    bcm_host_init();

    err = OMX_Init();
    if(err != OMX_ErrorNone) {
        fprintf(stderr, "OMX_Init() failed %s\n", err2str(err));
        exit(1);
    }
    /** Ask the core for a handle to the component
     */
    err = OMX_GetHandle(&handle, componentName,
                        // the next two fields are discussed later
                        NULL, &callbacks);
    if (err != OMX_ErrorNone) {
        fprintf(stderr, "OMX_GetHandle failed %s\n", err2str(err));
        exit(1);
    }

    sleep(1);
    // check our current state - should be Loaded
    printState(handle);

    disableAllPorts(handle);

    // request a move to idle
    OMX_SendCommand(handle,
                    OMX_CommandStateSet,
                    OMX_StateIdle,
                    NULL);

    sleep(2);
    printState(handle);
```

```
    // and to executing
    OMX_SendCommand(handle,
                    OMX_CommandStateSet,
                    OMX_StateExecuting,
                    NULL);
    sleep(2);
    printState(handle);

    exit(0);
}
```

Here's the output:

```
$ ./event OMX.broadcom.clock
StateLoaded
OMX State Port disabled 80
OMX State Port disabled 81
OMX State Port disabled 82
OMX State Port disabled 83
OMX State Port disabled 84
OMX State Port disabled 85
Component State changed to OMX_StateIdle
StateIdle
Component State changed to OMX_StateExecuting
StateExecuting
```

What to Do When a State Change Occurs?

OpenMAX is inherently an asynchronous system. Each component is a "black box," and the client using it has no idea nor control over how a component does its work. It may be fast, it may be slow, or it may never complete the requested activity (as in the wontwork program).

Events do not occur synchronously in the client's thread. They occur in a separate thread. Calls to the event handler are, however, blocking synchronous calls to the component. That is, the component cannot do any more work until the event is handled by the client's code. OpenMAX puts a time limit on this (section 3.1.2.9.1) of five milliseconds.

That isn't much time. What the client needs to do in this component's thread is to signal a wakeup call in its *own* thread. That is, the client should block in its own thread, waiting on a wakeup event from the component's thread.

The client can do this in a portable way using Posix threads. Posix threads are well documented. You would need code such as the following:

```
pthread_mutex_t mutex;
OMX_STATETYPE currentState = OMX_StateLoaded;
pthread_cond_t stateCond;

void waitFor(OMX_STATETYPE state) {
    printf("Waiting for %p\n", state);
    pthread_mutex_lock(&mutex);
```

```
    while (currentState != state)
        pthread_cond_wait(&stateCond, &mutex);
    printf("Wait successfully completed\n");
    pthread_mutex_unlock(&mutex);
}

void wakeUp(OMX_STATETYPE newState) {
    printf("Waking up %p\n", newState);
    pthread_mutex_lock(&mutex);

    currentState = newState;
    pthread_cond_signal(&stateCond);
    pthread_mutex_unlock(&mutex);
}
```

`waitFor` is called in the client thread, and `wakeUp` is called in the event handler.

I won't give any examples of using this here, but it is easy to add to the event example. Examples of this are in my Linux sound book.[1] The RPi's GPU has its own VCOS thread system and has built a library, the `ilclient` library, around these. So, it's off to this topic in the next chapter.

Conclusion

A key issue in OpenMAX programming is handling state changes for each component. This chapter has discussed one of the critical issues using general OpenMAX calls: setting parameters on components in order that they change state in an appropriate manner.

Resources

- *OpenMAX IL: The Standard for Media Library Portability*: `https://www.khronos.org/openmax/il/`
- *OpenMAX IL 1.1.2 Specification*: `https://www.khronos.org/registry/omxil/specs/OpenMAX_IL_1_1_2_Specification.pdf`
- *Raspberry Pi OpenMAX forum*: `www.raspberrypi.org/forums/viewforum.php?f=70`
- *VMCS-X OpenMAX IL Components*: `www.jvcref.com/files/PI/documentation/ilcomponents/`
- *Source code for ARM side libraries for interfacing to Raspberry Pi GPU*: `https://github.com/raspberrypi/userland`

[1]To be published by Apress

CHAPTER 10

OpenMAX IL Client Library on the Raspberry Pi

The IL Client library is designed to make it easier to use OpenMAX on the RPi. It is not portable, but if you want to build only RPi applications (or for other Broadcom GPU systems), it can reduce the effort and frustration of using OpenMAX.

Building Programs

The IL Client library is not a standard library on the RPi. It is actually in the `hello_pi` directory, in `/opt/vc/src/hello_pi/libs/ilclient/`. You may have to build this yourself using the `Makefile` in that directory.

You can build the program in this chapter using the following `Makefile`:

```
DMX_INC =  -I/opt/vc/include/ -I /opt/vc/include/interface/vmcs_host/ -I/opt/vc/include/
interface/vcos/pthreads -I/opt/vc/include/interface/vmcs_host/linux
EGL_INC =
OMX_INC =  -I /opt/vc/include/IL
OMX_ILCLIENT_INC = -I/opt/vc/src/hello_pi/libs/ilclient
INCLUDES = $(DMX_INC) $(EGL_INC) $(OMX_INC) $(OMX_ILCLIENT_INC)

CFLAGS=-g -DRASPBERRY_PI -DOMX_SKIP64BIT $(INCLUDES)
CPPFLAGS =

DMX_LIBS =  -L/opt/vc/lib/ -lbcm_host -lvcos -lvchiq_arm -lpthread
EGL_LIBS = -L/opt/vc/lib/ -lEGL -lGLESv2
OMX_LIBS = -lopenmaxil
OMX_ILCLIENT_LIBS = -L/opt/vc/src/hello_pi/libs/ilclient -lilclient

LDLIBS =  $(DMX_LIBS) $(EGL_LIBS) $(OMX_LIBS) $(OMX_ILCLIENT_LIBS)

all: il_working
```


J. Newmarch, *Raspberry Pi GPU Audio Video Programming*, DOI 10.1007/978-1-4842-2472-4_10

Public Functions

The Il Client functions are documented in the file /opt/vc/src/hello_pi/libs/ilclient/ilclient.h. The types ILCLIENT_T and COMPONENT_T are also defined in this file, but you don't really need to look at them; just use them as though they were opaque types.

```
ILCLIENT_T *ilclient_init(void);

int ilclient_create_component(ILCLIENT_T *handle,
                              COMPONENT_T **comp,
                              char *name,
                              ILCLIENT_CREATE_FLAGS_T flags)

void ilclient_cleanup_components(COMPONENT_T *list[])

            il_working.c
int ilclient_change_component_state(COMPONENT_T *comp,
                                    OMX_STATETYPE state);

void ilclient_disable_port(COMPONENT_T *comp,
                           int portIndex)

void ilclient_enable_port(COMPONENT_T *comp,
                          int portIndex);

int ilclient_enable_port_buffers(COMPONENT_T *comp,
                                 int portIndex,
                                 ILCLIENT_MALLOC_T ilclient_malloc,
                                 ILCLIENT_FREE_T ilclient_free,
                                 void *userdata);

void ilclient_disable_port_buffers(COMPONENT_T *comp,
                                   int portIndex,
                                   OMX_BUFFERHEADERTYPE *bufferList,
                                   ILCLIENT_FREE_T ilclient_free,
                                   void *userdata);

int ilclient_setup_tunnel(TUNNEL_T *tunnel,
                          unsigned int portStream,
                          int timeout);

void ilclient_disable_tunnel(TUNNEL_T *tunnel);

void ilclient_flush_tunnels(TUNNEL_T *tunnel,
                                         int max);

void ilclient_teardown_tunnels(TUNNEL_T *tunnels)

OMX_BUFFERHEADERTYPE* ilclient_get_output_buffer(COMPONENT_T *comp,
                                                 int portIndex,
                                                 int block);
```

```
OMX_BUFFERHEADERTYPE* ilclient_get_input_buffer(COMPONENT_T *comp,
                                                int portIndex,
                                                int block);

int ilclient_remove_event(COMPONENT_T *comp,
                          OMX_EVENTTYPE event,
                          OMX_U32 nData1,
                          int ignore1,
                          OMX_U32 nData2,
                          int ignore2);

void ilclient_return_events(COMPONENT_T *comp);

 int ilclient_wait_for_event(COMPONENT_T *comp,
                             OMX_EVENTTYPE event,
                             OMX_U32 nData1,
                             int ignore1,
                             OMX_U32 nData2,
                             int ignore2,
                             int event_flag,
                             int timeout);

int ilclient_wait_for_command_complete(COMPONENT_T *comp,
                                       OMX_COMMANDTYPE command,
                                       OMX_U32 nData2);

int ilclient_wait_for_command_complete_dual(COMPONENT_T *comp,
                                            OMX_COMMANDTYPE command,
                                            OMX_U32 nData2,
                                            COMPONENT_T *related);

void ilclient_debug_output(char *format, ...);

int ilclient_get_port_index(COMPONENT_T *comp,
                            OMX_DIRTYPE dir,
                            OMX_PORTDOMAINTYPE type,
                            int index);

OMX_HANDLETYPE ilclient_get_handle(COMPONENT_T *comp);

 int ilclient_suggest_bufsize(COMPONENT_T *comp,
                              OMX_U32 nBufSizeHint);
```

Creating a Client

In the previous chapter, you created a handle to a component using OMX_GetHandle and then promptly disabled all ports. The function ilclient_create_component will do this for you with the following:

```
ILCLIENT_T  *handle;
```

```
COMPONENT_T *component;
int         ret;

char *component_name = "image_decode"; // for example
handle = ilclient_init();

ret = ilclient_create_component(handle,
                                &component,
                                component_name,
                                ILCLIENT_DISABLE_ALL_PORTS);
if (ret < 0) {
    fprintf(stderr, "Error initialising componentn");
    exit(1);
}
```

Please note that the names used for the component are *not* the full names; this library prefixes OMX.broadcom. to the component name, so you just specify the last part of the name: clock, image_decode, and so on.

You don't need this yet, but this is the right point to mention it: components use buffers, and this function can also be used to set which types of buffers are allowed. The default is no buffers. Input and output buffers can be allowed by extending the flags of the last argument.

```
ret = ilclient_create_component(handle,
                                &component,
                                component_name,
                                ILCLIENT_DISABLE_ALL_PORTS
                                |
                                ILCLIENT_ENABLE_INPUT_BUFFERS
                                |
                                ILCLIENT_ENABLE_OUTPUT_BUFFERS);
```

Changing Component State

The function ilclient_change_component_state should now be used to ask for a change of state rather than OMX_SendCommand. Not only does it call for a state change, but it also blocks, waiting for the change to occur. For example, to change a component to Idle state, you call the following:

```
ret = ilclient_change_component_state(component,
                                      OMX_StateIdle);
if (ret < 0) {
    // error
}
```

In the previous chapter, you monitored state changes by installing an event handler on the component. This library installs its own handlers, so you don't need to do that. (If you do try, you will probably mess up the library.) The IL Client library looks after event handling.

The equivalent of the event.c program to change the state of the previous chapter is much simpler now. It is il_working.c.

```
#include <stdio.h>
#include <stdlib.h>

#include <OMX_Core.h>
#include <OMX_Component.h>

#include <bcm_host.h>
#include <ilclient.h>

void printState(OMX_HANDLETYPE handle) {
    OMX_STATETYPE state;
    OMX_ERRORTYPE err;

    err = OMX_GetState(handle, &state);
    if (err != OMX_ErrorNone) {
        fprintf(stderr, "Error on getting state\n");
        exit(1);
    }
    switch (state) {
    case OMX_StateLoaded:           printf("StateLoaded\n"); break;
    case OMX_StateIdle:             printf("StateIdle\n"); break;
    case OMX_StateExecuting:        printf("StateExecuting\n"); break;
    case OMX_StatePause:            printf("StatePause\n"); break;
    case OMX_StateWaitForResources: printf("StateWait\n"); break;
    case OMX_StateInvalid:          printf("StateInvalid\n"); break;
    default:                        printf("State unknown\n"); break;
    }
}

int main(int argc, char** argv) {

    int i;
    char *componentName;
    int err;
    ILCLIENT_T  *handle;
    COMPONENT_T *component;

    if (argc < 2) {
        fprintf(stderr, "Usage: %s component-name\n", argv[0]);
        exit(1);
    }
    componentName = argv[1];

    bcm_host_init();

    handle = ilclient_init();
    if (handle == NULL) {
        fprintf(stderr, "IL client init failed\n");
        exit(1);
    }
```

```
   if (OMX_Init() != OMX_ErrorNone) {
        ilclient_destroy(handle);
        fprintf(stderr, "OMX init failed\n");
        exit(1);
    }

    err = ilclient_create_component(handle,
                                    &component,
                                    componentName,
                                    ILCLIENT_DISABLE_ALL_PORTS
                                    );
    if (err == -1) {
        fprintf(stderr, "Component create failed\n");
        exit(1);
    }
    printState(ilclient_get_handle(component));

    err = ilclient_change_component_state(component,
                                          OMX_StateIdle);
    if (err < 0) {
        fprintf(stderr, "Couldn't change state to Idle\n");
        exit(1);
    }
    printState(ilclient_get_handle(component));

    err = ilclient_change_component_state(component,
                                          OMX_StateExecuting);
    if (err < 0) {
        fprintf(stderr, "Couldn't change state to Executing\n");
        exit(1);
    }
    printState(ilclient_get_handle(component));

    exit(0);
}
```

This can be run with the following, for example (note the IL Client version of the component name):

```
./il_working video_render
```

It has the following output:

```
StateLoaded
StateIdle
StateExecuting
```

Waiting for Events

Although the library generally takes care of events, there are times when your application needs to ensure that certain events have occurred. The function to do this is `ilclient_wait_for_event`.

```
int ilclient_wait_for_event(COMPONENT_T *comp, OMX_EVENTTYPE event,
                            OMX_U32 nData1, int ignore1,
                            OMX_IN OMX_U32 nData2, int ignore2,
                            int event_flag, int suspend)
```

You will see uses of this later.

Debugging Clients

OpenMAX is not a friendly environment when things go wrong. If you are lucky, OpenMAX will generate an error. More often, expected state changes won't occur, buffers won't be consumed or produced, and so on, and you don't get any indication apart from a hung application. The IL Client library can help isolate where these problems occur.

OpenMAX errors are generally handled silently by the IL Client library. In other words, you aren't told about them. However, you can set an error callback function with `ilclient_set_error_callback`, and in this you can try to handle the error.

The simplest handler is just to print the error.

```
ilclient_set_error_callback(handle,
                            error_callback,
                            NULL);

void error_callback(void *userdata, COMPONENT_T *comp, OMX_U32 data) {
    fprintf(stderr, "OMX error %s\n", err2str(data));
}

char *err2str(int err) {
    switch (err) {
    case OMX_ErrorInsufficientResources: return "OMX_ErrorInsufficientResources";
    case OMX_ErrorUndefined: return "OMX_ErrorUndefined";
    case OMX_ErrorInvalidComponentName: return "OMX_ErrorInvalidComponentName";
    case OMX_ErrorComponentNotFound: return "OMX_ErrorComponentNotFound";
    case OMX_ErrorInvalidComponent: return "OMX_ErrorInvalidComponent";
    case OMX_ErrorBadParameter: return "OMX_ErrorBadParameter";
    ...
    }
}
```

Many IL Client functions return zero on success or a negative value on failure. Sometimes this will tell you what the failure is, but anyway it tells you where it occurs. Combined with the error callback, this can help pin down problems. For example, the command `ilclient_wait_for_command_complete` returns 0 if the command successfully completed (or returns `OMX_ErrorSameState`, which doesn't really mean an error) or -1 if a different error occurred.

Some functions have a timeout parameter. These functions return 0 on success while the failure value will show that the timeout has occurred. This is a strong indication that, for example, an expected change has not occurred. For example, the command `ilclient_setup_tunnel` returns -1 on timeout and other negative values for other errors.

Conclusion

I have briefly introduced the IL Client library. You will use it extensively in the following chapters.

Resources

- *OpenMAX IL: The Standard for Media Library Portability*: `https://www.khronos.org/openmax/il/`
- *OpenMAX IL 1.1.2 Specification*: `https://www.khronos.org/registry/omxil/specs/OpenMAX_IL_1_1_2_Specification.pdf`
- *Raspberry Pi OpenMAX forum*: `www.raspberrypi.org/forums/viewforum.php?f=70`
- *VMCS-X OpenMAX IL Components*: `www.jvcref.com/files/PI/documentation/ilcomponents/`
- *Source code for ARM side libraries for interfacing to Raspberry Pi GPU*: `https://github.com/raspberrypi/userland`

CHAPTER 11

OpenMAX Buffers on the Raspberry Pi

OpenMAX components have input buffers to give data to the component and output buffers to return data either to the application or to the next component in a pipeline. This chapter looks at buffer management both using OpenMAX and using the IL Client library on the Raspberry Pi.

Building Programs

You can build the programs in this chapter using the following `Makefile`:

```
DMX_INC =  -I/opt/vc/include/ -I /opt/vc/include/interface/vmcs_host/ -I/opt/vc/include/
interface/vcos/pthreads -I/opt/vc/include/interface/vmcs_host/linux
EGL_INC =
OMX_INC =  -I /opt/vc/include/IL
OMX_ILCLIENT_INC = -I/opt/vc/src/hello_pi/libs/ilclient
INCLUDES = $(DMX_INC) $(EGL_INC) $(OMX_INC) $(OMX_ILCLIENT_INC)

CFLAGS=-g -DRASPBERRY_PI -DOMX_SKIP64BIT $(INCLUDES)
CPPFLAGS =

DMX_LIBS =  -L/opt/vc/lib/ -lbcm_host -lvcos -lvchiq_arm -lpthread
EGL_LIBS = -L/opt/vc/lib/ -lEGL -lGLESv2
OMX_LIBS = -lopenmaxil
OMX_ILCLIENT_LIBS = -L/opt/vc/src/hello_pi/libs/ilclient -lilclient

LDLIBS =  $(DMX_LIBS) $(EGL_LIBS) $(OMX_LIBS) $(OMX_ILCLIENT_LIBS)

all: il_buffer il_decode_image
```

Buffers

Data is transferred into and out of components using buffers, which are just byte arrays. Each component has ports, and the ports each have a number of buffers. The program `portinfo` of Chapter 8 can be used to list how many buffers a port has, how many it must minimally have, and what the recommended buffer size is.

J. Newmarch, *Raspberry Pi GPU Audio Video Programming*, DOI 10.1007/978-1-4842-2472-4_11

For example, running this on the Broadcom video_decode component's input port gives the following:

```
$./portinfo OMX.broadcom.video_decode 130
Component name: OMX.broadcom.video_decode:46 version 0.0, Spec version 1.1
Port 130
  is input port
  Domain is Video
  Buffer count 20
  Buffer minimum count 1
  Buffer size 81920 bytes
```

When a component is created, it has an array of pointers to buffer headers (a placeholder for buffers and information about them) but doesn't actually have any memory allocated either for the headers or for the buffers. These have to be allocated. In the next section, you'll look at the OpenMAX calls to do this, and then in the following section you'll see how the IL Client library manages this.

OpenMAX Buffer Allocation

There are two OpenMAX calls to get buffers for a port. The first is OMX_AllocateBuffer where the IL client asks the component to do all the work for them (section 3.2.2.15 of the specification). Typical code is as follows:

```
for (i = 0; i < pClient-> nBufferCount; i++) {
    OMX_AllocateBuffer(hComp,
                       &pClient->pBufferHdr[i],
                       pClient->nPortIndex,
                       pClient,
                       pClient->nBufferSize);
}
```

The component must be in the Loaded state or the port must be disabled. Typically, an application will disable the port and then allocate the buffers.

The other call is OMX_UseBuffer (section 3.2.2.14). This takes a buffer allocated either by the client or by another component and uses that as its own buffer.

The IL Client Library

You don't need to worry about buffer allocation. The library does it for you. The function ilclient_enable_port_buffers does the relevant buffer allocation for that port. The following is a typical call:

```
ilclient_enable_port_buffers(component, <port_index>,
                             NULL, NULL, NULL);
```

(You can give your own malloc function, but it is easier to let the library use its own.)

Once buffers have been allocated, the port can be moved to an enabled state as follows:

```
ilclient_enable_port(component, <port_index>)
```

This is a blocking call and will complete when the port has changed state.

You can modify the state change program of the previous chapter to create buffers for its ports. For simplicity, you hard-code a component (in this example, the Broadcom image_encode component), which has one input port 340 and one output port 341. Note that you have to set the component create flags to the following:

```
ILCLIENT_DISABLE_ALL_PORTS
 |
ILCLIENT_ENABLE_INPUT_BUFFERS
 |
ILCLIENT_ENABLE_OUTPUT_BUFFERS
```

If you don't, you may get obscure errors such as the following:

```
assertion failure:ilclient.c:747:ilclient_change_component_state():error == OMX_ErrorNone

Program received signal SIGABRT, Aborted.
0xb6e41bfc in raise () from /lib/arm-linux-gnueabihf/libc.so.6
```

If this occurs, it's best to use a debugger such as gdb to step into ilclient_change_component_state to find out what the actual error is.

The program is il_buffer.c.

```
#include <stdio.h>
#include <stdlib.h>

#include <OMX_Core.h>
#include <OMX_Component.h>

#include <bcm_host.h>
#include <ilclient.h>

void printState(OMX_HANDLETYPE handle) {
    OMX_STATETYPE state;
    OMX_ERRORTYPE err;

    err = OMX_GetState(handle, &state);
    if (err != OMX_ErrorNone) {
        fprintf(stderr, "Error on getting state\n");
        exit(1);
    }
    switch (state) {
    case OMX_StateLoaded:           printf("StateLoaded\n"); break;
    case OMX_StateIdle:             printf("StateIdle\n"); break;
    case OMX_StateExecuting:        printf("StateExecuting\n"); break;
    case OMX_StatePause:            printf("StatePause\n"); break;
    case OMX_StateWaitForResources: printf("StateWait\n"); break;
    case OMX_StateInvalid:          printf("StateInvalid\n"); break;
    default:                        printf("State unknown\n"); break;
    }
}
```

```
char *err2str(int err) {
    switch (err) {
    case OMX_ErrorInsufficientResources: return "OMX_ErrorInsufficientResources";
    case OMX_ErrorUndefined: return "OMX_ErrorUndefined";
    case OMX_ErrorInvalidComponentName: return "OMX_ErrorInvalidComponentName";
    case OMX_ErrorComponentNotFound: return "OMX_ErrorComponentNotFound";
    case OMX_ErrorInvalidComponent: return "OMX_ErrorInvalidComponent";
    case OMX_ErrorBadParameter: return "OMX_ErrorBadParameter";
    case OMX_ErrorNotImplemented: return "OMX_ErrorNotImplemented";
    case OMX_ErrorUnderflow: return "OMX_ErrorUnderflow";
    case OMX_ErrorOverflow: return "OMX_ErrorOverflow";
    case OMX_ErrorHardware: return "OMX_ErrorHardware";
    case OMX_ErrorInvalidState: return "OMX_ErrorInvalidState";
    case OMX_ErrorStreamCorrupt: return "OMX_ErrorStreamCorrupt";
    case OMX_ErrorPortsNotCompatible: return "OMX_ErrorPortsNotCompatible";
    case OMX_ErrorResourcesLost: return "OMX_ErrorResourcesLost";
    case OMX_ErrorNoMore: return "OMX_ErrorNoMore";
    case OMX_ErrorVersionMismatch: return "OMX_ErrorVersionMismatch";
    case OMX_ErrorNotReady: return "OMX_ErrorNotReady";
    case OMX_ErrorTimeout: return "OMX_ErrorTimeout";
    case OMX_ErrorSameState: return "OMX_ErrorSameState";
    case OMX_ErrorResourcesPreempted: return "OMX_ErrorResourcesPreempted";
    case OMX_ErrorPortUnresponsiveDuringAllocation: return
"OMX_ErrorPortUnresponsiveDuringAllocation";
    case OMX_ErrorPortUnresponsiveDuringDeallocation: return
"OMX_ErrorPortUnresponsiveDuringDeallocation";
    case OMX_ErrorPortUnresponsiveDuringStop: return "OMX_ErrorPortUnresponsiveDuringStop";
    case OMX_ErrorIncorrectStateTransition: return "OMX_ErrorIncorrectStateTransition";
    case OMX_ErrorIncorrectStateOperation: return "OMX_ErrorIncorrectStateOperation";
    case OMX_ErrorUnsupportedSetting: return "OMX_ErrorUnsupportedSetting";
    case OMX_ErrorUnsupportedIndex: return "OMX_ErrorUnsupportedIndex";
    case OMX_ErrorBadPortIndex: return "OMX_ErrorBadPortIndex";
    case OMX_ErrorPortUnpopulated: return "OMX_ErrorPortUnpopulated";
    case OMX_ErrorComponentSuspended: return "OMX_ErrorComponentSuspended";
    case OMX_ErrorDynamicResourcesUnavailable: return
"OMX_ErrorDynamicResourcesUnavailable";
    case OMX_ErrorMbErrorsInFrame: return "OMX_ErrorMbErrorsInFrame";
    case OMX_ErrorFormatNotDetected: return "OMX_ErrorFormatNotDetected";
    case OMX_ErrorContentPipeOpenFailed: return "OMX_ErrorContentPipeOpenFailed";
    case OMX_ErrorContentPipeCreationFailed: return "OMX_ErrorContentPipeCreationFailed";
    case OMX_ErrorSeperateTablesUsed: return "OMX_ErrorSeperateTablesUsed";
    case OMX_ErrorTunnelingUnsupported: return "OMX_ErrorTunnelingUnsupported";
    default: return "unknown error";
    }
}

void error_callback(void *userdata, COMPONENT_T *comp, OMX_U32 data) {
    fprintf(stderr, "OMX error %s\n", err2str(data));
}

int main(int argc, char** argv) {
```

```
    int i;
    char *componentName;
    int err;
    ILCLIENT_T  *handle;
    COMPONENT_T *component;

    componentName = "image_encode";

    bcm_host_init();

    handle = ilclient_init();
    if (handle == NULL) {
        fprintf(stderr, "IL client init failed\n");
        exit(1);
    }

   if (OMX_Init() != OMX_ErrorNone) {
        ilclient_destroy(handle);
        fprintf(stderr, "OMX init failed\n");
        exit(1);
    }

   ilclient_set_error_callback(handle,
                               error_callback,
                               NULL);

    err = ilclient_create_component(handle,
                                &component,
                                componentName,
                                ILCLIENT_DISABLE_ALL_PORTS
                                    |
                                    ILCLIENT_ENABLE_INPUT_BUFFERS
                                    |
                                    ILCLIENT_ENABLE_OUTPUT_BUFFERS
                                );
    if (err == -1) {
        fprintf(stderr, "Component create failed\n");
        exit(1);
    }
    printState(ilclient_get_handle(component));

    err = ilclient_change_component_state(component,
                                          OMX_StateIdle);
    if (err < 0) {
        fprintf(stderr, "Couldn't change state to Idle\n");
        exit(1);
    }
    printState(ilclient_get_handle(component));

    // input port
    ilclient_enable_port_buffers(component, 340,
                                 NULL, NULL, NULL);
```

```
    ilclient_enable_port(component, 340);
    // the input port is enabled and has input buffers allocated

    // output port
    ilclient_enable_port_buffers(component, 341,
                                 NULL, NULL, NULL);
    ilclient_enable_port(component, 341);
    // the output port is enabled and has output buffers allocated

    err = ilclient_change_component_state(component,
                                          OMX_StateExecuting);
    if (err < 0) {
        fprintf(stderr, "Couldn't change state to Executing\n");
        exit(1);
    }
    printState(ilclient_get_handle(component));

    exit(0);
}
```

The output from running `./il_buffer video_decode` is as follows:

```
StateLoaded
StateIdle
OMX error OMX_ErrorSameState
OMX error OMX_ErrorSameState
StateExecuting
```

The “errors” called `OMX_ErrorSameState` aren’t errors at all and can be ignored.

Writing to and Reading from Buffers

The whole point of using OpenMAX on the RPi is to get data into and out of the Broadcom GPU using the OpenMAX components. You do this by loading up the input buffers of a component with data and asking the component to empty them with a call `OMX_EmptyThisBuffer`. When the component has processed the data, it signals an `OMX_FillBufferDone` event for an output buffer, at which point your application will take the processed data and do something with it.

When a component has finished processing its input buffer data, it signals back to the client with an `OMX_EmptyBufferDone` event. At this point, the client can put more data into the buffer for processing by the component.

Similarly, when a client has finished doing something with the output data, it returns the buffer to the component by calling the function `OMX_FillThisBuffer`.

One input buffer may produce none or more buffers of output data. An output buffer can consume data from one or more input buffers. There is no direct correlation between emptying an input buffer and filling an output buffer. Essentially these processes should be done concurrently by the client.

```
whenever there is an empty input buffer        Whenever an output buffer is full
    put data into the input buffer                 process its data
    call EmptyThis Buffer                          call FillThisBuffer
```

Now of course things are not that simple. Unless you want to get into Video Core or Posix threads, a sequential version of this is easier. It looks like this:

```
while there is more input data
    if there is an empty input buffer
         put data into the input buffer
         call EmptyThis Buffer
    if there is a full output buffer
         process its data
         call FillThisBuffer

// cleanup any remaining output data
while there is another full output buffer
     process its data
     call FillThisBuffer
```

Of course, this begs the question of how you know that there are empty input buffers and full output buffers. There are IL Client calls to do this.

```
ilclient_get_input_buffer
ilclient_get_output_buffer
```

These look after the background event processing for `OMX_EmptyBufferDone` and `OMX_FillBufferDone`. These calls can be blocking or nonblocking. Block if you know that there must be a buffer available before you can continue, but don't block otherwise (busy wait).

Sequence of Actions

The pseudo-code for a simple application would ideally look like this:

```
create component in loaded state with all ports disabled
move the component to idle state
enable input ports (creating input port buffers)
enable output ports (creating output port buffers)
move the component to executing state

while there is more input data
    if there is an empty input buffer
         put data into the input buffer
         call EmptyThis Buffer
    if there is a full output buffer
         process its data
         call FillThisBuffer

while there is another full output buffer
     process its data
     call FillThisBuffer
```

The EOS Flag

The client will generally know when there is no more data. It will have finished reading from a file, a socket connection will have closed, and so on. It will need to pass this information into the component so that it in turn can pass it back out to the client to signal that there are no more output buffers.

The client signals this by setting the flag OMX_BUFFERFLAG_EOS in the flags field of the buffer header. Typical code to read data from a file into input buffers, setting EOS on completion, looks like this:

```
OMX_ERRORTYPE read_into_buffer_and_empty(FILE *fp,
                                          COMPONENT_T *component,
                                          OMX_BUFFERHEADERTYPE *buff_header,
                                          int *toread) {
    OMX_ERRORTYPE r;

    int buff_size = buff_header->nAllocLen;
    int nread = fread(buff_header->pBuffer, 1, buff_size, fp);

    printf("Read %d\n", nread);

    buff_header->nFilledLen = nread;
    *toread -= nread;
    if (*toread <= 0) {
        printf("Setting EOS on input\n");
        buff_header->nFlags |= OMX_BUFFERFLAG_EOS;
    }
    r = OMX_EmptyThisBuffer(ilclient_get_handle(component),
                            buff_header);
    if (r != OMX_ErrorNone) {
        fprintf(stderr, "Empty buffer error %s\n",
                err2str(r));
    }
    return r;
}
```

When getting information from the output buffers, the client will be watching for that flag. Minimal code to just print the size of output data and exit on end-of-stream is as follows:

```
OMX_ERRORTYPE save_info_from_filled_buffer(COMPONENT_T *component,
                                           OMX_BUFFERHEADERTYPE * buff_header) {
    OMX_ERRORTYPE r;

    printf("Got a filled buffer with %d, allocated %d\n",
           buff_header->nFilledLen,
           buff_header->nAllocLen);

    // do something here, like save the data - do nothing this time

    // quit if no more data coming
    if (buff_header->nFlags & OMX_BUFFERFLAG_EOS) {
        printf("Got EOS on output\n");
        exit(0);
    }
```

```
    // otherwise refill it
    r = OMX_FillThisBuffer(ilclient_get_handle(component),
                           buff_header);
    if (r != OMX_ErrorNone) {
        fprintf(stderr, "Fill buffer error %s\n",
                err2str(r));
    }
    return r;
}
```

A Hiccup: Port Settings Changed

The little flies in this ointment are these:

- Most components need to know at least something about their input data before they can begin to handle it. For example, to handle image data, the `image_decode` component must be told that the data is in JPEG format (for example).
- Even if the component knows the input data format, it won't know up front how much space to allocate for output buffers. It can't know this until it has processed at least some of the input data.

The first point will be examined in detail in later chapters. In the following program, just ignore `set_image_decoder_input_format` for now.

The second problem is dealt with by OpenMAX generating an `OMX_PortSettingsChanged` event. When this occurs is problematic: it may occur after (or during) the processing of the first block. This seems to occur with JPEG images, for example. But there may need to be a number of blocks read. This seems to occur with H.264 video files.

The IL Client library has two relevant functions to look for `OMX_PortSettingsChanged` events.

- The call `ilclient_wait_for_event` is a blocking call with an error timeout. This can be used if you know there is (or should be) an event coming within the timeout period. The return value will tell you whether it succeeded.
- The call `ilclient_remove_event` checks the previous event list and returns success if it is able to remove an event from this list. This mechanism waits until the event has already occurred and then detects it.

The blocking `ilclient_wait_for_event` works on a JPEG image. The second method works on an H.264 movie. I haven't found the middle ground yet.

To handle the case of a blocking wait, the following changes need to be made to the pseudo-code:

- Enable the input ports but not the output ones
- Send a single block of data into the component's input port
- Wait for a `PortSettingsChanged` event on the component's output port
- Enable the output port
- Continue emptying and filling buffers

The pseudo-code now looks like this:

```
create component in loaded state with all ports disabled
move the component to idle state
```

```
enable input ports (creating input port buffers)
move the component to executing state

put data into an input buffer
call EmptyThis Buffer

wait for port settings changed on the output port
enable output ports (creating output port buffers)

while there is more input data
    if there is an empty input buffer
         put data into the input buffer
         call EmptyThis Buffer
    if there is a full output buffer
         process its data
         call FillThisBuffer

while there is another full output buffer
     process its data
     call FillThisBuffer
```

Example Image Decoding

To make this more concrete, the following program decodes an image using the decode_image component. You aren't (at this point) interested in decoding images, so ignore the function void set_image_decoder_input_format, which just lets the decoder know the input format (JPEG data).

The program almost follows the pseudo-code of the previous section. The divergence is caused by what seems to be a bug to me. When the client has finished its data, it adds OMX_BUFFERFLAG_EOS. However, if the total data is less than the size of the image_decode's input buffers, the component doesn't pass this through. The program hangs. The workaround is to reset the data and send it again.

The program is il_decode_image.c.

The output on a typical JPEG file from, for example, ./il_decode_image jan-us-2006.jpg looks like this:

```
StateLoaded
StateIdle
Setting image decoder format
OMX error OMX_ErrorSameState
StateExecuting
Read 81920
OMX error OMX_ErrorSameState
Read 81920
Got a filled buffer with 0, allocated 10616832
Read 81920
Read 81920
...
Read 81920
Read 3901
Setting EOS on input
Getting last output buffers
```

```
Got eos event
Got a filled buffer with 10616832, allocated 10616832
Got EOS on output
```

The program is

```
#include <stdio.h>
#include <stdlib.h>
#include <sys/stat.h>

#include <OMX_Core.h>
#include <OMX_Component.h>

#include <bcm_host.h>
#include <ilclient.h>

void printState(OMX_HANDLETYPE handle) {
    OMX_STATETYPE state;
    OMX_ERRORTYPE err;

    err = OMX_GetState(handle, &state);
    if (err != OMX_ErrorNone) {
        fprintf(stderr, "Error on getting state\n");
        exit(1);
    }
    switch (state) {
    case OMX_StateLoaded:            printf("StateLoaded\n"); break;
    case OMX_StateIdle:              printf("StateIdle\n"); break;
    case OMX_StateExecuting:         printf("StateExecuting\n"); break;
    case OMX_StatePause:             printf("StatePause\n"); break;
    case OMX_StateWaitForResources: printf("StateWait\n"); break;
    case OMX_StateInvalid:           printf("StateInvalid\n"); break;
    default:                         printf("State unknown\n"); break;
    }
}

char *err2str(int err) {
    return "error deleted";
}

void eos_callback(void *userdata, COMPONENT_T *comp, OMX_U32 data) {
    fprintf(stderr, "Got eos event\n");
}

void error_callback(void *userdata, COMPONENT_T *comp, OMX_U32 data) {
    fprintf(stderr, "OMX error %s\n", err2str(data));
}

int get_file_size(char *fname) {
    struct stat st;

    if (stat(fname, &st) == -1) {
        perror("Stat'ing img file");
```

```
        return -1;
    }
    return(st.st_size);
}

static void set_image_decoder_input_format(COMPONENT_T *component) {
   // set input image format
    OMX_IMAGE_PARAM_PORTFORMATTYPE imagePortFormat;

    memset(&imagePortFormat, 0, sizeof(OMX_IMAGE_PARAM_PORTFORMATTYPE));
    imagePortFormat.nSize = sizeof(OMX_IMAGE_PARAM_PORTFORMATTYPE);
    imagePortFormat.nVersion.nVersion = OMX_VERSION;

    imagePortFormat.nPortIndex = 320;
    imagePortFormat.eCompressionFormat = OMX_IMAGE_CodingJPEG;
    OMX_SetParameter(ilclient_get_handle(component),
                     OMX_IndexParamImagePortFormat, &imagePortFormat);

}

OMX_ERRORTYPE read_into_buffer_and_empty(FILE *fp,
                                         COMPONENT_T *component,
                                         OMX_BUFFERHEADERTYPE *buff_header,
                                         int *toread) {
    OMX_ERRORTYPE r;

    int buff_size = buff_header->nAllocLen;
    int nread = fread(buff_header->pBuffer, 1, buff_size, fp);

    printf("Read %d\n", nread);

    buff_header->nFilledLen = nread;
    *toread -= nread;
    if (*toread <= 0) {
        printf("Setting EOS on input\n");
        buff_header->nFlags |= OMX_BUFFERFLAG_EOS;
    }
    r = OMX_EmptyThisBuffer(ilclient_get_handle(component),
                        buff_header);
    if (r != OMX_ErrorNone) {
        fprintf(stderr, "Empty buffer error %s\n",
                err2str(r));
    }
    return r;
}

OMX_ERRORTYPE save_info_from_filled_buffer(COMPONENT_T *component,
                                           OMX_BUFFERHEADERTYPE * buff_header) {
    OMX_ERRORTYPE r;

    printf("Got a filled buffer with %d, allocated %d\n",
```

```
        buff_header->nFilledLen,
        buff_header->nAllocLen);
    if (buff_header->nFlags & OMX_BUFFERFLAG_EOS) {
        printf("Got EOS on output\n");
        exit(0);
    }

    // do something here, like save the data - do nothing this time

    // and then refill it
    r = OMX_FillThisBuffer(ilclient_get_handle(component),
                           buff_header);
    if (r != OMX_ErrorNone) {
        fprintf(stderr, "Fill buffer error %s\n",
                err2str(r));
    }
    return r;
}

int main(int argc, char** argv) {

    int i;
    char *componentName;
    int err;
    char *img;
    ILCLIENT_T  *handle;
    COMPONENT_T *component;
    FILE *fp;
    int toread;
    OMX_BUFFERHEADERTYPE *buff_header;

    if (argc < 2) {
        fprintf(stderr, "Usage: %s jpeg-image\n", argv[0]);
        exit(1);
    }
    img = argv[1];
    if ((fp = fopen(img, "r")) == NULL)  {
        fprintf(stderr, "Can't open: %s\n", img);
        exit(2);
    }
    if (( toread = get_file_size(img)) == -1) {
        fprintf(stderr, "Can't stat: %s\n", img);
        exit(2);
    }

    componentName = "image_decode";

    bcm_host_init();

    handle = ilclient_init();
    if (handle == NULL) {
```

```
        fprintf(stderr, "IL client init failed\n");
        exit(1);
    }

   if (OMX_Init() != OMX_ErrorNone) {
        ilclient_destroy(handle);
        fprintf(stderr, "OMX init failed\n");
        exit(1);
    }

   ilclient_set_error_callback(handle,
                               error_callback,
                               NULL);
   ilclient_set_eos_callback(handle,
                             eos_callback,
                             NULL);

    err = ilclient_create_component(handle,
                                    &component,
                                    componentName,
                                    ILCLIENT_DISABLE_ALL_PORTS
                                        |
                                        ILCLIENT_ENABLE_INPUT_BUFFERS
                                        |
                                        ILCLIENT_ENABLE_OUTPUT_BUFFERS
                                    );
    if (err == -1) {
        fprintf(stderr, "Component create failed\n");
        exit(1);
    }
    printState(ilclient_get_handle(component));

    err = ilclient_change_component_state(component,
                                          OMX_StateIdle);
    if (err < 0) {
        fprintf(stderr, "Couldn't change state to Idle\n");
        exit(1);
    }
    printState(ilclient_get_handle(component));

    // must be before we enable buffers
    set_image_decoder_input_format(component);

    // input port
    ilclient_enable_port_buffers(component, 320,
                                 NULL, NULL, NULL);
    ilclient_enable_port(component, 320);

    err = ilclient_change_component_state(component,
                                          OMX_StateExecuting);
    if (err < 0) {
```

```
    fprintf(stderr, "Couldn't change state to Executing\n");
    exit(1);
}
printState(ilclient_get_handle(component));

// Read the first block so that the component can get
// the dimensions of the image and call port settings
// changed on the output port to configure it
buff_header =
    ilclient_get_input_buffer(component,
                              320,
                              1 /* block */);
if (buff_header != NULL) {
    read_into_buffer_and_empty(fp,
                               component,
                               buff_header,
                               &toread);

    // If all the file has been read in, then
    // we have to re-read this first block.
    // Broadcom bug?
    if (toread <= 0) {
        printf("Rewinding\n");
        // wind back to start and repeat
        fp = freopen(IMG, "r", fp);
        toread = get_file_size(IMG);
    }
}

// wait for first input block to set params for output port
ilclient_wait_for_event(component,
                        OMX_EventPortSettingsChanged,
                        321, 0, 0, 1,
                        ILCLIENT_EVENT_ERROR | ILCLIENT_PARAMETER_CHANGED,
                        10000);

// now enable output port since port params have been set
ilclient_enable_port_buffers(component, 321,
                             NULL, NULL, NULL);
ilclient_enable_port(component, 321);

// now work through the file
while (toread > 0) {
    OMX_ERRORTYPE r;

    // do we have an input buffer we can fill and empty?
    buff_header =
        ilclient_get_input_buffer(component,
                                  320,
                                  1 /* block */);
    if (buff_header != NULL) {
```

```
            read_into_buffer_and_empty(fp,
                                       component,
                                       buff_header,
                                       &toread);
        }

        // do we have an output buffer that has been filled?
        buff_header =
            ilclient_get_output_buffer(component,
                                          321,
                                          0 /* no block */);
        if (buff_header != NULL) {
            save_info_from_filled_buffer(component,
                                             buff_header);
        }
    }

    while (1) {
        printf("Getting last output buffers\n");
        buff_header =
            ilclient_get_output_buffer(component,
                                           321,
                                           1 /* block */);
        if (buff_header != NULL) {
            save_info_from_filled_buffer(component,
                                             buff_header);
        }
    }
    exit(0);
}
```

Conclusion

Buffers are the means of communicating data to and from components. This chapter has looked at how to manage buffers. You briefly looked at the OpenMAX calls but spent the most time using the Broadcom IL library.

Resources

- *OpenMAX IL: The Standard for Media Library Portability*: https://www.khronos.org/openmax/il/
- *OpenMAX IL 1.1.2 Specification*: https://www.khronos.org/registry/omxil/specs/OpenMAX_IL_1_1_2_Specification.pdf
- *Raspberry Pi OpenMAX forum*: www.raspberrypi.org/forums/viewforum.php?f=70
- *VMCS-X OpenMAX IL Components*: www.jvcref.com/files/PI/documentation/ilcomponents/
- *Source code for ARM side libraries for interfacing to Raspberry Pi GPU*: https://github.com/raspberrypi/userland

CHAPTER 12

Image Processing on the Raspberry Pi

This chapter looks at handling single images using OpenMAX on the Raspberry Pi. It also introduces a pipeline of components and looks at alternative ways of handling communication between them.

Building Programs

You can build the programs in this chapter using the following Makefile:

```
DMX_INC =  -I/opt/vc/include/ -I /opt/vc/include/interface/vmcs_host/ -I/opt/vc/include/
interface/vcos/pthreads -I/opt/vc/include/interface/vmcs_host/linux
EGL_INC =
OMX_INC =  -I /opt/vc/include/IL
OMX_ILCLIENT_INC = -I/opt/vc/src/hello_pi/libs/ilclient
INCLUDES = $(DMX_INC) $(EGL_INC) $(OMX_INC) $(OMX_ILCLIENT_INC)

CFLAGS=-g -DRASPBERRY_PI -DOMX_SKIP64BIT $(INCLUDES)
CPPFLAGS =

DMX_LIBS =  -L/opt/vc/lib/ -lbcm_host -lvcos -lvchiq_arm -lpthread
EGL_LIBS = -L/opt/vc/lib/ -lEGL -lGLESv2
OMX_LIBS = -lopenmaxil
OMX_ILCLIENT_LIBS = -L/opt/vc/src/hello_pi/libs/ilclient -lilclient

LDLIBS =  $(DMX_LIBS) $(EGL_LIBS) $(OMX_LIBS) $(OMX_ILCLIENT_LIBS)

all: il_decode_image_redone  il_render_image_notunnel il_render_image
```

Image Components

The Raspberry Pi has a number of OpenMAX components specifically for image processing.

- image_decode
- image_encode
- image_read

J. Newmarch, *Raspberry Pi GPU Audio Video Programming*, DOI 10.1007/978-1-4842-2472-4_12

- image_write
- image_fx
- resize (also does format changes)
- source (creates raw bitmap images)
- transition (transitions between two images)
- write_still (writes an image from pixel data)
- video_render (renders both images and movies)

Image Formats

OpenMAX has a number of data structures used to get and set information about components. You have seen some of these before.

- OMX_PORT_PARAM_TYPE: This is used with the index parameter OMX_IndexParamImageInit in a call to OMX_GetParameter. This gives the number of image ports and the port number of the first port.
- OMX_PARAM_PORTDEFINITIONTYPE: This is used with the index parameter OMX_PARAM_PORTDEFINITIONTYPE to give, for each port, the number of buffers, the size of each buffer, and the direction (input or output) of the port.
- OMX_IMAGE_PORTDEFINITIONTYPE: This is a field of the OMX_PARAM_PORTDEFINITIONTYPE for images.
- OMX_IMAGE_PARAM_PORTFORMATTYPE: This is used to get information about the different formats supported by each port.

Some of these were discussed in Chapter 8.

You haven't looked at the field OMX_IMAGE_PORTDEFINITIONTYPE, which is part of the port definition information. It contains the following relevant fields:

```
typedef struct OMX_IMAGE_PORTDEFINITIONTYPE {
    OMX_STRING cMIMEType;
    OMX_U32 nFrameWidth;
    OMX_U32 nFrameHeight;
    OMX_S32 nStride;
    OMX_U32 nSliceHeight;
    OMX_IMAGE_CODINGTYPE eCompressionFormat;
    OMX_COLOR_FORMATTYPE eColorFormat;
} OMX_IMAGE_PORTDEFINITIONTYPE;
```

The last two fields are the current values set for the port. The *possible* values are obtained from the next structure, OMX_IMAGE_PARAM_PORTFORMATTYPE, so I will discuss it in the next paragraph. The major fields you get here, though, are parameters about the image size: nFrameWidth, nFrameHeight, nStride, and nSliceHeight. These are parameters you often need if you are, say, saving a decoded image in a different format such as TGA.

OMX_IMAGE_PARAM_PORTFORMATTYPE is defined (in section 4.4.4 in the 1.1.2 specification) as follows:

```
typedef struct OMX_IMAGE_PARAM_PORTFORMATTYPE {
    OMX_U32 nSize;
    OMX_VERSIONTYPE nVersion;
    OMX_U32 nPortIndex;
    OMX_U32 nIndex;
    OMX_IMAGE_CODINGTYPE eCompressionFormat;
    OMX_COLOR_FORMATTYPE eColorFormat;
} OMX_IMAGE_PARAM_PORTFORMATTYPE;
```

The first two fields are common to all OpenMAX structures. The nPortIndex field is the port you are looking at. The nIndex field is to distinguish between all of the different format types supported by this port. The eCompressionFormat and eColorFormat fields give information about the format.

The values for OMX_IMAGE_CODINGTYPE are given in Table 4-66 of the 1.1.2 specification and on the RPi are given in the file /opt/vc/include/IL/OMX_Image.h, as follows:

```
typedef enum OMX_IMAGE_CODINGTYPE {
    OMX_IMAGE_CodingUnused,      /** Value when format is N/A */
    OMX_IMAGE_CodingAutoDetect,  /** Auto detection of image format */
    OMX_IMAGE_CodingJPEG,        /** JPEG/JFIF image format */
    OMX_IMAGE_CodingJPEG2K,      /** JPEG 2000 image format */
    OMX_IMAGE_CodingEXIF,        /** EXIF image format */
    OMX_IMAGE_CodingTIFF,        /** TIFF image format */
    OMX_IMAGE_CodingGIF,         /** Graphics image format */
    OMX_IMAGE_CodingPNG,         /** PNG image format */
    OMX_IMAGE_CodingLZW,         /** LZW image format */
    OMX_IMAGE_CodingBMP,         /** Windows Bitmap format */
    OMX_IMAGE_CodingKhronosExtensions = 0x6F000000, /** Reserved region for introducing
Khronos Standard Extensions */
    OMX_IMAGE_CodingVendorStartUnused = 0x7F000000, /** Reserved region for introducing
Vendor Extensions */

    OMX_IMAGE_CodingTGA,
    OMX_IMAGE_CodingPPM,

    OMX_IMAGE_CodingMax = 0x7FFFFFFF
} OMX_IMAGE_CODINGTYPE;
```

Unfortunately, on the RPi, the only value given is OMX_IMAGE_CodingUnused, so this field does not give useful information about queries. However, the values here may need to be set in the OMX_IMAGE_PORTDEFINITIONTYPE field to set the data compression type.

The type OMX_COLOR_FORMATTYPE is more informative. It is defined in the file /opt/vc/include/IL/OMX_IVCommon.h as follows:

```
typedef enum OMX_COLOR_FORMATTYPE {
    OMX_COLOR_FormatUnused,
    OMX_COLOR_FormatMonochrome,
    OMX_COLOR_Format8bitRGB332,
    OMX_COLOR_Format12bitRGB444,
    OMX_COLOR_Format16bitARGB4444,
    OMX_COLOR_Format16bitARGB1555,
    OMX_COLOR_Format16bitRGB565,
    OMX_COLOR_Format16bitBGR565,
    OMX_COLOR_Format18bitRGB666,
```

```
    OMX_COLOR_Format18bitARGB1665,
    OMX_COLOR_Format19bitARGB1666,
    OMX_COLOR_Format24bitRGB888,
    OMX_COLOR_Format24bitBGR888,
    OMX_COLOR_Format24bitARGB1887,
    OMX_COLOR_Format25bitARGB1888,
    OMX_COLOR_Format32bitBGRA8888,
    OMX_COLOR_Format32bitARGB8888,
    OMX_COLOR_FormatYUV411Planar,
    OMX_COLOR_FormatYUV411PackedPlanar,
    OMX_COLOR_FormatYUV420Planar,
    OMX_COLOR_FormatYUV420PackedPlanar,
    OMX_COLOR_FormatYUV420SemiPlanar,
    OMX_COLOR_FormatYUV422Planar,
    OMX_COLOR_FormatYUV422PackedPlanar,
    OMX_COLOR_FormatYUV422SemiPlanar,
    OMX_COLOR_FormatYCbYCr,
    OMX_COLOR_FormatYCrYCb,
    OMX_COLOR_FormatCbYCrY,
    OMX_COLOR_FormatCrYCbY,
    OMX_COLOR_FormatYUV444Interleaved,
    OMX_COLOR_FormatRawBayer8bit,
    OMX_COLOR_FormatRawBayer10bit,
    OMX_COLOR_FormatRawBayer8bitcompressed,
    OMX_COLOR_FormatL2,
    OMX_COLOR_FormatL4,
    OMX_COLOR_FormatL8,
    OMX_COLOR_FormatL16,
    OMX_COLOR_FormatL24,
    OMX_COLOR_FormatL32,
    OMX_COLOR_FormatYUV420PackedSemiPlanar,
    OMX_COLOR_FormatYUV422PackedSemiPlanar,
    OMX_COLOR_Format18BitBGR666,
    OMX_COLOR_Format24BitARGB6666,
    OMX_COLOR_Format24BitABGR6666,
    OMX_COLOR_FormatKhronosExtensions = 0x6F000000, /** Reserved region for introducing
    Khronos Standard Extensions */
    OMX_COLOR_FormatVendorStartUnused = 0x7F000000, /** Reserved region for introducing
    Vendor Extensions */
    OMX_COLOR_Format32bitABGR8888,
    OMX_COLOR_Format8bitPalette,
    OMX_COLOR_FormatYUVUV128,
    OMX_COLOR_FormatRawBayer12bit,
    OMX_COLOR_FormatBRCMEGL,
    OMX_COLOR_FormatBRCMOpaque,
    OMX_COLOR_FormatYVU420PackedPlanar,
    OMX_COLOR_FormatYVU420PackedSemiPlanar,
    OMX_COLOR_FormatMax = 0x7FFFFFFF
} OMX_COLOR_FORMATTYPE;
```

That's a rather large number of formats!

Running the program info from Chapter 8 shows the following for the image_decode component:

```
Image ports:
  Ports start on 320
  There are 2 open ports
  Port 320 has 3 buffers (minimum 2) of size 81920
  Direction is input
    Supported image formats are:
      No coding format returned
  Port 321 has 1 buffers (minimum 1) of size 541696
  Direction is output
    Supported image formats are:
      Image format compression format 0
      Image format color encoding 0x6
      Image format compression format 0
      Image format color encoding 0x14
      Image format compression format 0
      Image format color encoding 0x17
      Image format compression format 0
      Image format color encoding 0x7F000001
      Image format compression format 0
      Image format color encoding 0x10
      No more formats supported
```

The color codings are as follows:

- *0x6*: OMX_COLOR_Format16bitRGB565
- *0x14*: OMX_COLOR_FormatYUV420PackedPlanar
- *0x17*: OMX_COLOR_FormatYUV422PackedPlanar
- *0x7F000001*: OMX_COLOR_Format32bitABGR8888
- *0x10*: OMX_COLOR_Format32bitARGB8888

This looks good: both YUV and RGB formats are supported. But if you look into the Broadcom specification for the image_decode component, you get this statement for the OMX_IndexParamImagePortFormat:

> *The input port supports several compressed formats, the output port supports several color formats. However, the component doesn't support color format changing, so the color format will be set to match that emitted by the encountered image format.*

In other words, you only get out the format that you put in. If the image input is RGB, then so is the output, while if the image input is YUV, then so is the output. Format changing is supported by the Broadcom component resize and will be shown in a later example.

Decoding a JPEG Image

In Chapter 11, I gave an example that decoded a JPEG image from a file. There the emphasis was on buffer management; now you will redo it with the emphasis on decoding images.

Well, actually you won't do much more. You just get the output image parameters with the following call:

```
void get_output_port_settings(COMPONENT_T *component) {
    OMX_PARAM_PORTDEFINITIONTYPE portdef;

    printf("Port settings changed\n");
    // need to setup the input for the resizer with the output of the
    // decoder
    portdef.nSize = sizeof(OMX_PARAM_PORTDEFINITIONTYPE);
    portdef.nVersion.nVersion = OMX_VERSION;
    portdef.nPortIndex = 321;
    OMX_GetParameter(ilclient_get_handle(component),
                     OMX_IndexParamPortDefinition, &portdef);

    unsigned int uWidth =
        (unsigned int) portdef.format.image.nFrameWidth;
    unsigned int uHeight =
        (unsigned int) portdef.format.image.nFrameHeight;
    unsigned int uStride =
        (unsigned int) portdef.format.image.nStride;
    unsigned int uSliceHeight =
        (unsigned int) portdef.format.image.nSliceHeight;
    printf("Frame width %d, frame height %d, stride %d, slice height %d\n",
           uWidth,
           uHeight,
           uStride,
           uSliceHeight);
    printf("Getting format Compression 0x%x Color Format: 0x%x\n",
           (unsigned int) portdef.format.image.eCompressionFormat,
           (unsigned int) portdef.format.image.eColorFormat);
}
```

This call should appear immediately after the port setting's event has been received. The code is in `il_decode_image_redone.c` (not shown).

The new output from one JPEG file run by `./il_decode_image_redone cimg0135.jpg` is as follows:

```
Port settings changed
Frame width 3072, frame height 2304, stride 3072, slice height 2304
Getting format Compression 0x0 Color Format: 0x14
```

So, the output color format is `OMX_COLOR_FormatYUV420PackedPlanar`.

Tunneling

The previous example used just a single component. This won't be the norm. To display an image, you will need a decode component and a render component, for example. The IL client can look after all of the input and output buffers, copying data or pointers between them. Alternatively, the components may be able to exchange information directly between themselves, a technique called *tunneling*. (There is a third mechanism called *proprietary communication* that I won't spend time on.)

These communication models are shown in Figure 2-2 of the 1.1.2 specification, reproduced here:

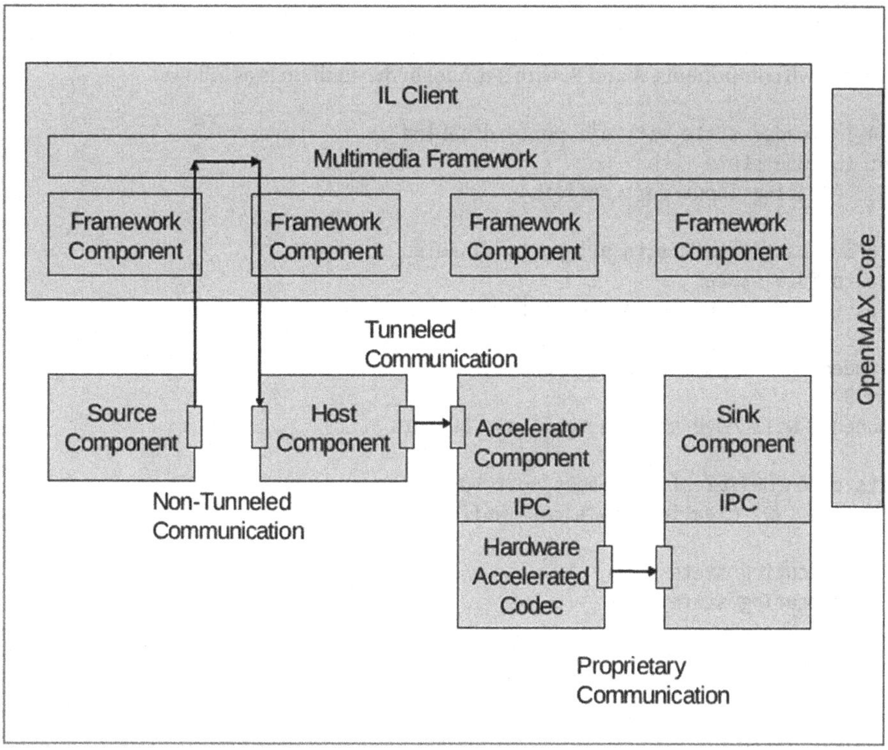

Figure 12-1. *Possible communication paths between components*

The Broadcom components support tunneling. This can make it much easier to write an application using multiple components (but not that easy, of course!). The basic simplification is as follows: you don't have to set up buffers between tunneled ports, you don't have to negotiate data formats between them, and you don't have to move data between the buffers; the components look after this themselves.

You still have to feed data into the first component's input buffers and take data out of the last component's output buffers. You also have to look after the state management of the components and their ports, even the tunneled ones.

The pseudo-code for a single component was given in the previous chapter as follows:

```
create component in loaded state with all ports disabled
move the component to idle state
enable input ports (creating input port buffers)
enable output ports (creating output port buffers)
move the component to executing state

while there is more input data
    if there is an empty input buffer
         put data into the input buffer
         call EmptyThis Buffer
    if there is a full output buffer
         process its data
         call FillThisBuffer
```

```
while there is another full output buffer
     process its data
     call FillThisBuffer
```

The modification for two components, A and B, with a tunnel between them is as follows:

```
create component A in loaded state with all ports disabled
move the component to idle state
enable input ports (creating input port buffers)

create component B in loaded state with all ports disabled
move the component to idle state
enable output ports (creating output port buffers)

setup a tunnel between the output ports of A and the
    input ports of B
// note: the components will have created their own buffers

enable output ports of A (NOT creating output port buffers)
enable input ports of B (NOT creating input port buffers)

move component A to executing state
move component B to executing state

while there is more input data
    if there is an empty input buffer on A
         put data into the input buffer
         call EmptyThis Buffer
    if there is a full output buffer on B
         process its data
         call FillThisBuffer

while there is another full output buffer on B
     process its data
     call FillThisBuffer
```

Essentially you just split the input and output sides, with OpenMAX doing the tunnel communication between the components.

The previous pseudo-code is fine if the sizes of buffers are fixed and known from the beginning. This isn't the case with components such as `image_decode`, where the size of the output buffers isn't known until at least one input buffer has been processed. This complicates the code as the tunnel now cannot be established until a port setting's changed event has been received on the first component's output port. A tunnel can be established only if the components are in the Loaded state or if their relevant ports are disabled.

There is a convenience function called `ilclient_setup_tunnel` in the IL Client library that can handle the state issues of the components and their ports. It also has a timeout parameter, which if nonzero will manage waiting for a `PortSettingsEvent` option. You will ignore this option for now, as it hides too much of what is going on.

The IL Client library tunneling calls use a struct.

```
typedef struct {
   COMPONENT_T *source;  /** The source component */
   int source_port;      /** The output port index on the source component */
```

```
    COMPONENT_T *sink;      /** The sink component */
    int sink_port;          /** The input port index on the sink component */
} TUNNEL_T;
```

The calls also use the convenience macro set_tunnel to set parameters, as follows:

```
TUNNEL_T tunnel;
set_tunnel(&tunnel, decodeComponent, 321, renderComponent, 90);
```

The function ilclient_setup_tunnel takes three parameters.

```
int ilclient_setup_tunnel(TUNNEL_T *tunnel,
                          unsigned int portStream,
                          int timeout);
```

You will set the third parameter to 0. The timeout will be 0 if you want to handle PortSettingsChanged events yourself.

The revised pseudo-code for this case is as follows:

```
create component A in loaded state with all ports disabled
move the component to idle state
enable input ports (creating input port buffers)

create component B in loaded state with all ports disabled
move the component to idle state
enable output ports (creating output port buffers)

move component A to executing state

put data into an input buffer of component A
call EmptyThis Buffer

wait for port settings changed on the output port of component A

// ilclient_setup_tunnel will handle these steps
// move component A back to idle state
// disable output ports of component A

setup the tunnel between the ports of components A and B
// note: the components will have created their own buffers

enable output ports of component A (NOT creating output port buffers)
enable output ports of component B (NOT creating input port buffers)

while there is more input data
    if there is an empty input buffer on A
         put data into the input buffer
         call EmptyThis Buffer
    if there is a full output buffer on B
         process its data
         call FillThisBuffer
```

```
while there is another full output buffer on B
     process its data
     call FillThisBuffer
```

Rendering an Image Using Tunnelling

In the first program in this chapter, you used the image_decode component to decode a JPEG image into YUV format. To render this, you need a render component. The OpenMAX specification surprisingly does not specify any image render components. However, the Broadcom implementation has the component video_render, which will render both single images and videos.

The video_render component has one input port, 60, and no output ports (its output goes to the render device). So, you can simplify the pseudo-code because you won't have any data output buffers and can just concentrate on the input side.

The program to decode and render an image is il_render_image.c.

```
#include <stdio.h>
#include <stdlib.h>
#include <sys/stat.h>

#include <OMX_Core.h>
#include <OMX_Component.h>

#include <bcm_host.h>
#include <ilclient.h>

void printState(OMX_HANDLETYPE handle) {
    // code elided
}

char *err2str(int err) {
    return "error elided";
}

void eos_callback(void *userdata, COMPONENT_T *comp, OMX_U32 data) {
    fprintf(stderr, "Got eos event\n");
}

void error_callback(void *userdata, COMPONENT_T *comp, OMX_U32 data) {
    fprintf(stderr, "OMX error %s\n", err2str(data));
}

int get_file_size(char *fname) {
    struct stat st;

    if (stat(fname, &st) == -1) {
            perror("Stat'ing img file");
            return -1;
        }
    return(st.st_size);
}
```

```
unsigned int uWidth;
unsigned int uHeight;

OMX_ERRORTYPE read_into_buffer_and_empty(FILE *fp,
                                         COMPONENT_T *component,
                                         OMX_BUFFERHEADERTYPE *buff_header,
                                         int *toread) {
    OMX_ERRORTYPE r;

    int buff_size = buff_header->nAllocLen;
    int nread = fread(buff_header->pBuffer, 1, buff_size, fp);

    printf("Read %d\n", nread);

    buff_header->nFilledLen = nread;
    *toread -= nread;
    if (*toread <= 0) {
        printf("Setting EOS on input\n");
        buff_header->nFlags |= OMX_BUFFERFLAG_EOS;
    }
    r = OMX_EmptyThisBuffer(ilclient_get_handle(component),
                            buff_header);
    if (r != OMX_ErrorNone) {
        fprintf(stderr, "Empty buffer error %s\n",
                err2str(r));
    }
    return r;
}

static void set_image_decoder_input_format(COMPONENT_T *component) {
   // set input image format
    printf("Setting image decoder format\n");
    OMX_IMAGE_PARAM_PORTFORMATTYPE imagePortFormat;
    //setHeader(&imagePortFormat,  sizeof(OMX_IMAGE_PARAM_PORTFORMATTYPE));
    memset(&imagePortFormat, 0, sizeof(OMX_IMAGE_PARAM_PORTFORMATTYPE));
    imagePortFormat.nSize = sizeof(OMX_IMAGE_PARAM_PORTFORMATTYPE);
    imagePortFormat.nVersion.nVersion = OMX_VERSION;

    imagePortFormat.nPortIndex = 320;
    imagePortFormat.eCompressionFormat = OMX_IMAGE_CodingJPEG;
    OMX_SetParameter(ilclient_get_handle(component),
                     OMX_IndexParamImagePortFormat, &imagePortFormat);

}

void setup_decodeComponent(ILCLIENT_T  *handle,
                           char *decodeComponentName,
                           COMPONENT_T **decodeComponent) {
    int err;
```

```
    err = ilclient_create_component(handle,
                                    decodeComponent,
                                    decodeComponentName,
                                    ILCLIENT_DISABLE_ALL_PORTS
                                        |
                                        ILCLIENT_ENABLE_INPUT_BUFFERS
                                        /* |
                                        ILCLIENT_ENABLE_OUTPUT_BUFFERS
                                        */
                                    );
    if (err == -1) {
        fprintf(stderr, "DecodeComponent create failed\n");
        exit(1);
    }
    printState(ilclient_get_handle(*decodeComponent));

    err = ilclient_change_component_state(*decodeComponent,
                                          OMX_StateIdle);
    if (err < 0) {
        fprintf(stderr, "Couldn't change state to Idle\n");
        exit(1);
    }
    printState(ilclient_get_handle(*decodeComponent));

    // must be before we enable buffers
    set_image_decoder_input_format(*decodeComponent);
}

void setup_renderComponent(ILCLIENT_T  *handle,
                           char *renderComponentName,
                           COMPONENT_T **renderComponent) {
    int err;

   err = ilclient_create_component(handle,
                                    renderComponent,
                                    renderComponentName,
                                    ILCLIENT_DISABLE_ALL_PORTS
                                       /* |
                                        ILCLIENT_ENABLE_INPUT_BUFFERS
                                       */
                                    );
    if (err == -1) {
        fprintf(stderr, "RenderComponent create failed\n");
        exit(1);
    }
    printState(ilclient_get_handle(*renderComponent));

    err = ilclient_change_component_state(*renderComponent,
                                          OMX_StateIdle);
```

```
    if (err < 0) {
        fprintf(stderr, "Couldn't change state to Idle\n");
        exit(1);
    }
    printState(ilclient_get_handle(*renderComponent));
}

int main(int argc, char** argv) {

    int i;
    char *decodeComponentName;
    char *renderComponentName;
    int err;
    char *img;
    ILCLIENT_T  *handle;
    COMPONENT_T *decodeComponent;
    COMPONENT_T *renderComponent;
    FILE *fp;
    int toread;
    OMX_BUFFERHEADERTYPE *buff_header;

    if (argc < 2) {
        fprintf(stderr, "Usage: %s jpeg-image\n", argv[0]);
        exit(1);
    }
    img = argv[1];
    if ((fp = fopen(img, "r")) == NULL)  {
          fprintf(stderr, "Can't open: %s\n", img);
          exit(2);
    }
    if (( toread = get_file_size(img)) == -1) {
          fprintf(stderr, "Can't stat: %s\n", img);
          exit(2);
    }
    decodeComponentName = "image_decode";
    renderComponentName = "video_render";

    bcm_host_init();

    handle = ilclient_init();
    if (handle == NULL) {
        fprintf(stderr, "IL client init failed\n");
        exit(1);
    }

   if (OMX_Init() != OMX_ErrorNone) {
        ilclient_destroy(handle);
        fprintf(stderr, "OMX init failed\n");
        exit(1);
    }
```

```
    ilclient_set_error_callback(handle,
                                error_callback,
                                NULL);
    ilclient_set_eos_callback(handle,
                              eos_callback,
                              NULL);

    setup_decodeComponent(handle, decodeComponentName, &decodeComponent);
    setup_renderComponent(handle, renderComponentName, &renderComponent);
    // both components now in Idle state, no buffers, ports disabled

     // input port
     ilclient_enable_port_buffers(decodeComponent, 320,
                                  NULL, NULL, NULL);
     ilclient_enable_port(decodeComponent, 320);

     err = ilclient_change_component_state(decodeComponent,
                                           OMX_StateExecuting);
     if (err < 0) {
         fprintf(stderr, "Couldn't change state to Executing\n");
         exit(1);
     }
     printState(ilclient_get_handle(decodeComponent));

     // Read the first block so that the decodeComponent can get
     // the dimensions of the image and call port settings
     // changed on the output port to configure it
     buff_header =
         ilclient_get_input_buffer(decodeComponent,
                                   320,
                                   1 /* block */);
     if (buff_header != NULL) {
         read_into_buffer_and_empty(fp,
                                    decodeComponent,
                                    buff_header,
                                    &toread);

         // If all the file has been read in, then
         // we have to re-read this first block.
         // Broadcom bug?
         if (toread <= 0) {
             printf("Rewinding\n");
             // wind back to start and repeat
             fp = freopen(img, "r", fp);
             toread = get_file_size(img);
         }
     }

     // wait for first input block to set params for output port
     ilclient_wait_for_event(decodeComponent,
                             OMX_EventPortSettingsChanged,
                             321, 0, 0, 1,
                             ILCLIENT_EVENT_ERROR | ILCLIENT_PARAMETER_CHANGED,
                             5);
```

```
    printf("Port settings changed\n");

    TUNNEL_T tunnel;
    set_tunnel(&tunnel, decodeComponent, 321, renderComponent, 90);
    if ((err = ilclient_setup_tunnel(&tunnel, 0, 0)) < 0) {
        fprintf(stderr, "Error setting up tunnel %X\n", err);
        exit(1);
    } else {
        printf("Tunnel set up ok\n");
    }

   // Okay to go back to processing data
   // enable the decode output ports

   OMX_SendCommand(ilclient_get_handle(decodeComponent),
                   OMX_CommandPortEnable, 321, NULL);

   ilclient_enable_port(decodeComponent, 321);

   // enable the render output ports
   /*
   OMX_SendCommand(ilclient_get_handle(renderComponent),
                   OMX_CommandPortEnable, 90, NULL);
   */
   ilclient_enable_port(renderComponent, 90);

   // set both components to executing state
    err = ilclient_change_component_state(decodeComponent,
                                          OMX_StateExecuting);
    if (err < 0) {
        fprintf(stderr, "Couldn't change state to Idle\n");
        exit(1);
    }
    err = ilclient_change_component_state(renderComponent,
                                          OMX_StateExecuting);
    if (err < 0) {
        fprintf(stderr, "Couldn't change state to Idle\n");
        exit(1);
    }

    // now work through the file
    while (toread > 0) {
        OMX_ERRORTYPE r;

        // do we have a decode input buffer we can fill and empty?
        buff_header =
            ilclient_get_input_buffer(decodeComponent,
                                      320,
                                      1 /* block */);
```

```
        if (buff_header != NULL) {
            read_into_buffer_and_empty(fp,
                                       decodeComponent,
                                       buff_header,
                                       &toread);
        }
    }

    ilclient_wait_for_event(renderComponent,
                            OMX_EventBufferFlag,
                            90, 0, OMX_BUFFERFLAG_EOS, 0,
                            ILCLIENT_BUFFER_FLAG_EOS, 10000);
    printf("EOS on render\n");

    sleep(100);

    exit(0);
}
```

Rendering an Image Without Tunneling

If you choose not to use tunneling, then you have to handle all of the intermediate steps in managing the transfer of information from the decoder to the renderer. This adds a measure of complexity but may be needed if, say, you want to overlay text onto a decoded image. The basic mechanism is to create buffers for one component using OMX_AllocateBuffer or ilclient_enable_port_buffers and to share the resultant buffers using OMX_UseBuffer.

In the tunneling case, you had to wait until one input buffer had been processed before you knew enough about the input image and could establish the tunnel. Similarly, you have to wait for the first buffer before you can set up your nontunneled equivalent.

The first major difference lies in structure formats: the decoder uses the image type OMX_IMAGE_PORTDEFINITIONTYPE, while the renderer uses the video type OMX_VIDEO_PORTDEFINITIONTYPE. While these share many common fields, they are not all aligned in the same place in memory. So, some fields have to be copied across from the image format to the video format.

This involves getting the OMX_IMAGE_PORTDEFINITIONTYPE structure for each component, changing some fields in the video component, and setting them back into the video component. The code to do this is as follows:

```
int setup_shared_buffer_format(COMPONENT_T *decodeComponent,
                               COMPONENT_T *renderComponent) {
    OMX_PARAM_PORTDEFINITIONTYPE portdef,  rportdef;;
    int ret;
    OMX_ERRORTYPE err;

    // need to setup the input for the render with the output of the
    // decoder
    portdef.nSize = sizeof(OMX_PARAM_PORTDEFINITIONTYPE);
    portdef.nVersion.nVersion = OMX_VERSION;
    portdef.nPortIndex = 321;
    OMX_GetParameter(ilclient_get_handle(decodeComponent),
                     OMX_IndexParamPortDefinition, &portdef);
```

```
    // Get default values of render
    rportdef.nSize = sizeof(OMX_PARAM_PORTDEFINITIONTYPE);
    rportdef.nVersion.nVersion = OMX_VERSION;
    rportdef.nPortIndex = 90;
    rportdef.nBufferSize = portdef.nBufferSize;
    nBufferSize = portdef.nBufferSize;

    err = OMX_GetParameter(ilclient_get_handle(renderComponent),
                           OMX_IndexParamPortDefinition, &rportdef);
    if (err != OMX_ErrorNone) {
        fprintf(stderr, "Error getting render port params %s\n", err2str(err));
        return err;
    }

    // tell render input what the decoder output will be providing
    //Copy some
    rportdef.format.video.nFrameWidth = portdef.format.image.nFrameWidth;
    rportdef.format.video.nFrameHeight = portdef.format.image.nFrameHeight;
    rportdef.format.video.nStride = portdef.format.image.nStride;
    rportdef.format.video.nSliceHeight = portdef.format.image.nSliceHeight;

    err = OMX_SetParameter(ilclient_get_handle(renderComponent),
                           OMX_IndexParamPortDefinition, &rportdef);
    if (err != OMX_ErrorNone) {
        fprintf(stderr, "Error setting render port params %s\n", err2str(err));
        return err;
    } else {
        printf("Render port params set up ok, buf size %d\n",
               portdef.nBufferSize);
    }

    return  OMX_ErrorNone;
}
```

Once the first input buffer has been processed, you can call the previous function to synchronize formats and also create output buffers (only one!) for the decoder. You can use the standard IL Client library call to create the buffers `ilclient_enable_port_buffers` and enable this port with the following:

```
    ilclient_enable_port_buffers(decodeComponent, 321,
                                 NULL, NULL, NULL);
    ilclient_enable_port(decodeComponent, 321);
```

There is no need at this point to share the buffer. Anyway, you don't yet have access to the decoder's buffers, not until a nonblocking call to `ilclient_get_output_buffer` has succeeded in returning an output buffer from the decoder. Once that has been done, a buffer header of type `OMX_BUFFERHEADERTYPE *` is made available to the client. You can then reuse the buffer in this header.

The following function is a bit messy: you have only one output buffer header from the decoder but three input buffer headers for the renderer. You disable two of them by setting their buffers to `NULL`. For the shared one, you can share only the *buffer*; the buffer *header* must be distinct. The call `OMX_UseBuffer` creates a distinct header but shares the buffer. Unfortunately, you must also copy across information from one buffer header to the other. This type of behavior isn't documented, just something you discover painfully after an hour or so with the debugger and guessing a lot. What has to be copied is the length of the filled buffer.

```
ppRenderInputBufferHeader[0]->nFilledLen =
                        buff_header->nFilledLen;
```

There is one documented behavior in the 1.1.2 specification's section 3.2.2.14.

> *OMX_UseBuffer macro shall be executed under the following conditions:*
>
> - ...
> - *On a disabled port when the component is in the OMX_StateExecuting, the OMX_StatePause, or the OMX_StateIdle state*

The problem is that this is either wrong or misleading or the Broadcom implementation is broken. Even when the state is Executing, you have to *enable* the port, not *disable* it! Ho-hum, another two days wasted figuring that out.

The relevant code is as follows:

```
OMX_BUFFERHEADERTYPE *pOutputBufferHeader = NULL;
OMX_BUFFERHEADERTYPE **ppRenderInputBufferHeader;

int use_buffer(COMPONENT_T *renderComponent,
               OMX_BUFFERHEADERTYPE *buff_header) {
    int ret;
    OMX_PARAM_PORTDEFINITIONTYPE portdef;

    ppRenderInputBufferHeader =
        (OMX_BUFFERHEADERTYPE **) malloc(sizeof(void) *
                                         3);

    OMX_SendCommand(ilclient_get_handle(renderComponent),
                    OMX_CommandPortEnable, 90, NULL);

    ilclient_wait_for_event(renderComponent,
                            OMX_EventCmdComplete,
                            OMX_CommandPortEnable, 1,
                            90, 1, 0,
                            5000);

    printState(ilclient_get_handle(renderComponent));

    ret = OMX_UseBuffer(ilclient_get_handle(renderComponent),
                        &ppRenderInputBufferHeader[0],
                        90,
                        NULL,
                        nBufferSize,
                        buff_header->pBuffer);
```

```
    if (ret != OMX_ErrorNone) {
        fprintf(stderr, "Eror sharing buffer %s\n", err2str(ret));
        return ret;
    } else {
        printf("Sharing buffer ok\n");
    }

    ppRenderInputBufferHeader[0]->nAllocLen =
        buff_header->nAllocLen;

    int n;
    for (n = 1; n < 3; n++) {
        printState(ilclient_get_handle(renderComponent));
        ret = OMX_UseBuffer(ilclient_get_handle(renderComponent),
                            &ppRenderInputBufferHeader[n],
                            90,
                            NULL,
                            0,
                            NULL);
        if (ret != OMX_ErrorNone) {
            fprintf(stderr, "Eror sharing null buffer %s\n", err2str(ret));
            return ret;
        }
    }

    ilclient_enable_port(renderComponent, 90);

    ret = ilclient_change_component_state(renderComponent,
                                          OMX_StateExecuting);
    if (ret < 0) {
        fprintf(stderr, "Couldn't change render state to Executing\n");
        exit(1);
    }
    return 0;
```

Those are the major issues. Playing games with component state and port state until it works is standard. The program is `il_render_image_notunnel.c`.

```
#include <stdio.h>
#include <stdlib.h>
#include <sys/stat.h>

#include <OMX_Component.h>

#include <bcm_host.h>
#include <ilclient.h>

#define IMG "hype.jpg"
char *img_file = IMG;
```

```
void printState(OMX_HANDLETYPE handle) {
    // elided
}

char *err2str(int err) {
    return "elided";
}

void eos_callback(void *userdata, COMPONENT_T *comp, OMX_U32 data) {
    fprintf(stderr, "Got eos event\n");
}

void error_callback(void *userdata, COMPONENT_T *comp, OMX_U32 data) {
    fprintf(stderr, "OMX error %s\n", err2str(data));
}

int get_file_size(char *fname) {
    struct stat st;

    if (stat(fname, &st) == -1) {
        perror("Stat'ing img file");
        return -1;
    }
    return(st.st_size);
}

unsigned int uWidth;
unsigned int uHeight;

unsigned int nBufferSize;

OMX_ERRORTYPE read_into_buffer_and_empty(FILE *fp,
                                         COMPONENT_T *component,
                                         OMX_BUFFERHEADERTYPE *buff_header,
                                         int *toread) {
    OMX_ERRORTYPE r;

    int buff_size = buff_header->nAllocLen;
    int nread = fread(buff_header->pBuffer, 1, buff_size, fp);

    printf("Read %d\n", nread);

    buff_header->nFilledLen = nread;
    *toread -= nread;
    if (*toread <= 0) {
        printf("Setting EOS on input\n");
        buff_header->nFlags |= OMX_BUFFERFLAG_EOS;
    }
    r = OMX_EmptyThisBuffer(ilclient_get_handle(component),
                            buff_header);
```

```
    if (r != OMX_ErrorNone) {
        fprintf(stderr, "Empty buffer error %s\n",
                err2str(r));
    }
    return r;
}

static void set_image_decoder_input_format(COMPONENT_T *component) {

    printf("Setting image decoder format\n");
    OMX_IMAGE_PARAM_PORTFORMATTYPE imagePortFormat;

    memset(&imagePortFormat, 0, sizeof(OMX_IMAGE_PARAM_PORTFORMATTYPE));
    imagePortFormat.nSize = sizeof(OMX_IMAGE_PARAM_PORTFORMATTYPE);
    imagePortFormat.nVersion.nVersion = OMX_VERSION;

    imagePortFormat.nPortIndex = 320;
    imagePortFormat.eCompressionFormat = OMX_IMAGE_CodingJPEG;
    OMX_SetParameter(ilclient_get_handle(component),
                     OMX_IndexParamImagePortFormat, &imagePortFormat);

}

void setup_decodeComponent(ILCLIENT_T  *handle,
                           char *decodeComponentName,
                           COMPONENT_T **decodeComponent) {
    int err;

    err = ilclient_create_component(handle,
                                    decodeComponent,
                                    decodeComponentName,
                                    ILCLIENT_DISABLE_ALL_PORTS
                                    |
                                    ILCLIENT_ENABLE_INPUT_BUFFERS
                                    |
                                    ILCLIENT_ENABLE_OUTPUT_BUFFERS
                                    );
    if (err == -1) {
        fprintf(stderr, "DecodeComponent create failed\n");
        exit(1);
    }
    printState(ilclient_get_handle(*decodeComponent));

    err = ilclient_change_component_state(*decodeComponent,
                                          OMX_StateIdle);
    if (err < 0) {
        fprintf(stderr, "Couldn't change state to Idle\n");
        exit(1);
    }
    printState(ilclient_get_handle(*decodeComponent));
```

```
    // must be before we enable buffers
    set_image_decoder_input_format(*decodeComponent);
}

void setup_renderComponent(ILCLIENT_T  *handle,
                           char *renderComponentName,
                           COMPONENT_T **renderComponent) {
    int err;

    err = ilclient_create_component(handle,
                                    renderComponent,
                                    renderComponentName,
                                    ILCLIENT_DISABLE_ALL_PORTS
                                    |
                                    ILCLIENT_ENABLE_INPUT_BUFFERS
                                    );
    if (err == -1) {
        fprintf(stderr, "RenderComponent create failed\n");
        exit(1);
    }
    printState(ilclient_get_handle(*renderComponent));

    err = ilclient_change_component_state(*renderComponent,
                                          OMX_StateIdle);
    if (err < 0) {
        fprintf(stderr, "Couldn't change state to Idle\n");
        exit(1);
    }
    printState(ilclient_get_handle(*renderComponent));
}

OMX_BUFFERHEADERTYPE *pOutputBufferHeader = NULL;
OMX_BUFFERHEADERTYPE **ppRenderInputBufferHeader;

int setup_shared_buffer_format(COMPONENT_T *decodeComponent,
                               COMPONENT_T *renderComponent) {
    OMX_PARAM_PORTDEFINITIONTYPE portdef,  rportdef;;
    int ret;
    OMX_ERRORTYPE err;

    // need to setup the input for the render with the output of the
    // decoder
    portdef.nSize = sizeof(OMX_PARAM_PORTDEFINITIONTYPE);
    portdef.nVersion.nVersion = OMX_VERSION;
    portdef.nPortIndex = 321;
    OMX_GetParameter(ilclient_get_handle(decodeComponent),
                     OMX_IndexParamPortDefinition, &portdef);

    // Get default values of render
    rportdef.nSize = sizeof(OMX_PARAM_PORTDEFINITIONTYPE);
    rportdef.nVersion.nVersion = OMX_VERSION;
    rportdef.nPortIndex = 90;
```

```
    rportdef.nBufferSize = portdef.nBufferSize;
    nBufferSize = portdef.nBufferSize;

    err = OMX_GetParameter(ilclient_get_handle(renderComponent),
                           OMX_IndexParamPortDefinition, &rportdef);
    if (err != OMX_ErrorNone) {
        fprintf(stderr, "Error getting render port params %s\n", err2str(err));
        return err;
    }

    // tell render input what the decoder output will be providing
    //Copy some
    rportdef.format.video.nFrameWidth = portdef.format.image.nFrameWidth;
    rportdef.format.video.nFrameHeight = portdef.format.image.nFrameHeight;
    rportdef.format.video.nStride = portdef.format.image.nStride;
    rportdef.format.video.nSliceHeight = portdef.format.image.nSliceHeight;

    err = OMX_SetParameter(ilclient_get_handle(renderComponent),
                           OMX_IndexParamPortDefinition, &rportdef);
    if (err != OMX_ErrorNone) {
        fprintf(stderr, "Error setting render port params %s\n", err2str(err));
        return err;
    } else {
        printf("Render port params set up ok, buf size %d\n",
               portdef.nBufferSize);
    }

    return  OMX_ErrorNone;
}

int use_buffer(COMPONENT_T *renderComponent,
               OMX_BUFFERHEADERTYPE *buff_header) {
    int ret;
    OMX_PARAM_PORTDEFINITIONTYPE portdef;

    ppRenderInputBufferHeader =
        (OMX_BUFFERHEADERTYPE **) malloc(sizeof(void) *
                                         3);

    OMX_SendCommand(ilclient_get_handle(renderComponent),
                    OMX_CommandPortEnable, 90, NULL);

    ilclient_wait_for_event(renderComponent,
                            OMX_EventCmdComplete,
                            OMX_CommandPortEnable, 1,
                            90, 1, 0,
                            5000);

    printState(ilclient_get_handle(renderComponent));
```

```
    ret = OMX_UseBuffer(ilclient_get_handle(renderComponent),
                        &ppRenderInputBufferHeader[0],
                        90,
                        NULL,
                        nBufferSize,
                        buff_header->pBuffer);
    if (ret != OMX_ErrorNone) {
        fprintf(stderr, "Eror sharing buffer %s\n", err2str(ret));
        return ret;
    } else {
        printf("Sharing buffer ok\n");
    }

    ppRenderInputBufferHeader[0]->nAllocLen =
        buff_header->nAllocLen;

    int n;
    for (n = 1; n < 3; n++) {
        printState(ilclient_get_handle(renderComponent));
        ret = OMX_UseBuffer(ilclient_get_handle(renderComponent),
                            &ppRenderInputBufferHeader[n],
                            90,
                            NULL,
                            0,
                            NULL);
        if (ret != OMX_ErrorNone) {
            fprintf(stderr, "Eror sharing null buffer %s\n", err2str(ret));
            return ret;
        }
    }

    ilclient_enable_port(renderComponent, 90);

    ret = ilclient_change_component_state(renderComponent,
                                          OMX_StateExecuting);
    if (ret < 0) {
        fprintf(stderr, "Couldn't change render state to Executing\n");
        exit(1);
    }
    return 0;
}

int main(int argc, char** argv) {

    int i;
    char *decodeComponentName;
    char *renderComponentName;
    int err;
    FILE *fp;
    int toread;
    ILCLIENT_T  *handle;
```

```
COMPONENT_T *decodeComponent;
COMPONENT_T *renderComponent;

if (argc == 2) {
    img_file = argv[1];
}
if ((fp = fopen(img_file, "r")) == NULL)  {
    fprintf(stderr, "Can't open: %s\n", img_file);
    exit(2);
}
if ((toread = get_file_size(img_file) == -1)  {
    fprintf(stderr, "Can't stat: %s\n", img_file);
    exit(2);
}

OMX_BUFFERHEADERTYPE *buff_header;

decodeComponentName = "image_decode";
renderComponentName = "video_render";

bcm_host_init();

handle = ilclient_init();
if (handle == NULL) {
    fprintf(stderr, "IL client init failed\n");
    exit(1);
}

if (OMX_Init() != OMX_ErrorNone) {
    ilclient_destroy(handle);
    fprintf(stderr, "OMX init failed\n");
    exit(1);
}

ilclient_set_error_callback(handle,
                            error_callback,
                            NULL);
ilclient_set_eos_callback(handle,
                          eos_callback,
                          NULL);

setup_decodeComponent(handle, decodeComponentName, &decodeComponent);
setup_renderComponent(handle, renderComponentName, &renderComponent);
// both components now in Idle state, no buffers, ports disabled

// input port
ilclient_enable_port_buffers(decodeComponent, 320,
                             NULL, NULL, NULL);
ilclient_enable_port(decodeComponent, 320);
```

```
    err = ilclient_change_component_state(decodeComponent,
                                          OMX_StateExecuting);
    if (err < 0) {
        fprintf(stderr, "Couldn't change decode state to Executing\n");
        exit(1);
    }
    printState(ilclient_get_handle(decodeComponent));

    // Read the first block so that the decodeComponent can get
    // the dimensions of the image and call port settings
    // changed on the output port to configure it
    buff_header =
        ilclient_get_input_buffer(decodeComponent,
                                  320,
                                  1 /* block */);
    if (buff_header != NULL) {
        read_into_buffer_and_empty(fp,
                                   decodeComponent,
                                   buff_header,
                                   &toread);

        // If all the file has been read in, then
        // we have to re-read this first block.
        // Broadcom bug?
        if (toread <= 0) {
            printf("Rewinding\n");
            // wind back to start and repeat
            fp = freopen(img_file, "r", fp);
            toread = get_file_size(img_file);
        }
    }

    // wait for first input block to set params for output port
    ilclient_wait_for_event(decodeComponent,
                            OMX_EventPortSettingsChanged,
                            321, 0, 0, 1,
                            ILCLIENT_EVENT_ERROR | ILCLIENT_PARAMETER_CHANGED,
                            10000);
    printf("Port settings changed\n");

    setup_shared_buffer_format(decodeComponent, renderComponent);

   ilclient_enable_port_buffers(decodeComponent, 321,
                                NULL, NULL, NULL);
    ilclient_enable_port(decodeComponent, 321);

    // set decoder only to executing state
    err = ilclient_change_component_state(decodeComponent,
                                          OMX_StateExecuting);
```

```
if (err < 0) {
    fprintf(stderr, "Couldn't change state to Executing\n");
    exit(1);
}

// now work through the file
while (toread > 0) {
    OMX_ERRORTYPE r;

    // do we have a decode input buffer we can fill and empty?
    buff_header =
        ilclient_get_input_buffer(decodeComponent,
                                  320,
                                  1 /* block */);
    if (buff_header != NULL) {
        read_into_buffer_and_empty(fp,
                                   decodeComponent,
                                   buff_header,
                                   &toread);
    }

    // do we have an output buffer that has been filled?
    buff_header =
        ilclient_get_output_buffer(decodeComponent,
                                   321,
                                   0 /* no block */);
    if (buff_header != NULL) {
        printf("Got an output buffer length %d\n",
               buff_header->nFilledLen);
        if (buff_header->nFlags & OMX_BUFFERFLAG_EOS) {
            printf("Got EOS\n");
        }
        if (pOutputBufferHeader == NULL) {
            use_buffer(renderComponent, buff_header);
            pOutputBufferHeader = buff_header;
        }

        if (buff_header->nFilledLen > 0) {
            OMX_EmptyThisBuffer(ilclient_get_handle(renderComponent),
                                buff_header);
        }

        OMX_FillThisBuffer(ilclient_get_handle(decodeComponent),
                           buff_header);
    }
}

int done = 0;
while ( !done ) {
    printf("Getting last output buffers\n");
    buff_header =
```

```
            ilclient_get_output_buffer(decodeComponent,
                                       321,
                                       1 /* block */);
        printf("Got a final output buffer length %d\n",
               buff_header->nFilledLen);
        if (buff_header->nFlags & OMX_BUFFERFLAG_EOS) {
            printf("Got EOS\n");
            done = 1;
        }

        if (pOutputBufferHeader == NULL) {
            use_buffer(renderComponent, buff_header);
            pOutputBufferHeader = buff_header;
        }

        ppRenderInputBufferHeader[0]->nFilledLen = buff_header->nFilledLen;
        OMX_EmptyThisBuffer(ilclient_get_handle(renderComponent),
                            ppRenderInputBufferHeader[0]);
    }

    sleep(100);

    exit(0);
}
```

Conclusion

This chapter has shown basic image handling using multiple components. There are more complex things that can be done, which will be addressed in later editions of this book.

Resources

- *OpenMAX IL: The Standard for Media Library Portability*: `https://www.khronos.org/openmax/il/`
- *OpenMAX IL 1.1.2 Specification*: `https://www.khronos.org/registry/omxil/specs/OpenMAX_IL_1_1_2_Specification.pdf`
- *Raspberry Pi OpenMAX forum*: `www.raspberrypi.org/forums/viewforum.php?f=70`
- *VMCS-X OpenMAX IL Components*: `www.jvcref.com/files/PI/documentation/ilcomponents/`
- *Source code for ARM side libraries for interfacing to Raspberry Pi GPU*: `https://github.com/raspberrypi/userland`

CHAPTER 13

OpenMAX Video Processing on the Raspberry Pi

This chapter looks at video processing using the Broadcom GPU on the Raspberry Pi.

Building Programs

You can build the programs in this chapter using the following Makefile:

```
DMX_INC =  -I/opt/vc/include/ -I /opt/vc/include/interface/vmcs_host/ -I/opt/vc/include/
interface/vcos/pthreads -I/opt/vc/include/interface/vmcs_host/linux
EGL_INC =
OMX_INC =  -I /opt/vc/include/IL
OMX_ILCLIENT_INC = -I/opt/vc/src/hello_pi/libs/ilclient
INCLUDES = $(DMX_INC) $(EGL_INC) $(OMX_INC) $(OMX_ILCLIENT_INC)

CFLAGS=-g -DRASPBERRY_PI -DOMX_SKIP64BIT $(INCLUDES)
CPPFLAGS =

DMX_LIBS =  -L/opt/vc/lib/ -lbcm_host -lvcos -lvchiq_arm -lpthread
EGL_LIBS = -L/opt/vc/lib/ -lEGL -lGLESv2
OMX_LIBS = -lopenmaxil
OMX_ILCLIENT_LIBS = -L/opt/vc/src/hello_pi/libs/ilclient -lilclient

LDLIBS =  $(DMX_LIBS) $(EGL_LIBS) $(OMX_LIBS) $(OMX_ILCLIENT_LIBS)

all: il_decode_video  il_render_video
```

Video Components

The Raspberry Pi has a number of OpenMAX components specifically for video processing, as shown here:

- camera
- egl_render
- video_decode

J. Newmarch, *Raspberry Pi GPU Audio Video Programming*, DOI 10.1007/978-1-4842-2472-4_13

- video_encode
- video_render
- video_scheduler
- video_splitter

Video Formats

OpenMAX has a number of data structures used to get and set information about components. You have seen some of these before.

- OMX_PORT_PARAM_TYPE: This is used with the index parameter OMX_IndexParamVideoInit in a call to OMX_GetParameter. This gives the number of video ports and the port number of the first port.
- OMX_PARAM_PORTDEFINITIONTYPE: This is used with the index parameter OMX_PARAM_PORTDEFINITIONTYPE to give, for each port, the number of buffers, the size of each buffer, and the direction (input or output) of the port.
- OMX_VIDEO_PORTDEFINITIONTYPE: This is a field of the OMX_PARAM_PORTDEFINITIONTYPE for videos.
- OMX_VIDEO_PARAM_PORTFORMATTYPE: This is used to get information about the different formats supported by each port.

Some of these were discussed in Chapter 8.

You haven't looked at the field OMX_VIDEO_PORTDEFINITIONTYPE, which is part of the port definition information. It contains the following relevant fields:

```
typedef struct OMX_VIDEO_PORTDEFINITIONTYPE {
    OMX_STRING cMIMEType;
    OMX_NATIVE_DEVICETYPE pNativeRender;
    OMX_U32 nFrameWidth;
    OMX_U32 nFrameHeight;
    OMX_S32 nStride;
    OMX_U32 nSliceHeight;
    OMX_U32 nBitrate;
    OMX_U32 xFramerate;
    OMX_BOOL bFlagErrorConcealment;
    OMX_VIDEO_CODINGTYPE eCompressionFormat;
    OMX_COLOR_FORMATTYPE eColorFormat;
    OMX_NATIVE_WINDOWTYPE pNativeWindow;
} OMX_VIDEO_PORTDEFINITIONTYPE;
```

The last two but one fields are the current values set for the port. The *possible* values are obtained from the next structure OMX_VIDEO_PARAM_PORTFORMATTYPE, so I will discuss it in the next paragraph. The major fields you get here, though, are parameters about the video size: nFrameWidth, nFrameHeight, nStride, and nSliceHeight. These are parameters you often need if you are, say, saving a decoded video in a different format such as MPEG4.

The OMX_VIDEO_PARAM_PORTFORMATTYPE is defined (in section 4.3.5 in the 1.1.2 specification) as follows:

```
typedef struct OMX_VIDEO_PARAM_PORTFORMATTYPE {
    OMX_U32 nSize;
    OMX_VERSIONTYPE nVersion;
    OMX_U32 nPortIndex;
    OMX_U32 nIndex;
    OMX_VIDEO_CODINGTYPE eCompressionFormat;
    OMX_COLOR_FORMATTYPE eColorFormat;
    OMX_U32 xFramerate;
} OMX_VIDEO_PARAM_PORTFORMATTYPE;
```

The first two fields are common to all OpenMAX structures. The `nPortIndex` field is the port you are looking at. The `nIndex` field is to distinguish between all the different format types supported by this port. The `eCompressionFormat` and `eColorFormat` fields give information about the format, while the last field gives the frame rate of the video in frames per second.

The values for `OMX_VIDEO_CODINGTYPE` are given in Table 4-66 of the 1.1.2 specification and on the RPi are given in the file `/opt/vc/include/IL/OMX_Video.h`, as shown here:

```
typedef enum OMX_VIDEO_CODINGTYPE {
    OMX_VIDEO_CodingUnused,     /** Value when coding is N/A */
    OMX_VIDEO_CodingAutoDetect, /** Autodetection of coding type */
    OMX_VIDEO_CodingMPEG2,      /** AKA: H.262 */
    OMX_VIDEO_CodingH263,       /** H.263 */
    OMX_VIDEO_CodingMPEG4,      /** MPEG-4 */
    OMX_VIDEO_CodingWMV,        /** all versions of Windows Media Video */
    OMX_VIDEO_CodingRV,         /** all versions of Real Video */
    OMX_VIDEO_CodingAVC,        /** H.264/AVC */
    OMX_VIDEO_CodingMJPEG,      /** Motion JPEG */
    OMX_VIDEO_CodingKhronosExtensions = 0x6F000000, /** Reserved region for introducing
Khronos Standard Extensions */
    OMX_VIDEO_CodingVendorStartUnused = 0x7F000000, /** Reserved region for introducing
Vendor Extensions */

    OMX_VIDEO_CodingVP6,        /** On2 VP6 */
    OMX_VIDEO_CodingVP7,        /** On2 VP7 */
    OMX_VIDEO_CodingVP8,        /** On2 VP8 */
    OMX_VIDEO_CodingYUV,        /* raw YUV video */
    OMX_VIDEO_CodingSorenson,   /** Sorenson */
    OMX_VIDEO_CodingTheora,     /** Theora */
    OMX_VIDEO_CodingMVC,        /** H.264/MVC */

    OMX_VIDEO_CodingMax = 0x7FFFFFFF
} OMX_VIDEO_CODINGTYPE;
```

The type `OMX_COLOR_FORMATTYPE` is more informative. It is defined in the file `/opt/vc/include/IL/OMX_IVCommon.h`, as follows:

```
typedef enum OMX_COLOR_FORMATTYPE {
    OMX_COLOR_FormatUnused,
    OMX_COLOR_FormatMonochrome,
    OMX_COLOR_Format8bitRGB332,
    OMX_COLOR_Format12bitRGB444,
    OMX_COLOR_Format16bitARGB4444,
    OMX_COLOR_Format16bitARGB1555,
```

```
    OMX_COLOR_Format16bitRGB565,
    OMX_COLOR_Format16bitBGR565,
    OMX_COLOR_Format18bitRGB666,
    OMX_COLOR_Format18bitARGB1665,
    OMX_COLOR_Format19bitARGB1666,
    OMX_COLOR_Format24bitRGB888,
    OMX_COLOR_Format24bitBGR888,
    OMX_COLOR_Format24bitARGB1887,
    OMX_COLOR_Format25bitARGB1888,
    OMX_COLOR_Format32bitBGRA8888,
    OMX_COLOR_Format32bitARGB8888,
    OMX_COLOR_FormatYUV411Planar,
    OMX_COLOR_FormatYUV411PackedPlanar,
    OMX_COLOR_FormatYUV420Planar,
    OMX_COLOR_FormatYUV420PackedPlanar,
    OMX_COLOR_FormatYUV420SemiPlanar,
    OMX_COLOR_FormatYUV422Planar,
    OMX_COLOR_FormatYUV422PackedPlanar,
    OMX_COLOR_FormatYUV422SemiPlanar,
    OMX_COLOR_FormatYCbYCr,
    OMX_COLOR_FormatYCrYCb,
    OMX_COLOR_FormatCbYCrY,
    OMX_COLOR_FormatCrYCbY,
    OMX_COLOR_FormatYUV444Interleaved,
    OMX_COLOR_FormatRawBayer8bit,
    OMX_COLOR_FormatRawBayer10bit,
    OMX_COLOR_FormatRawBayer8bitcompressed,
    OMX_COLOR_FormatL2,
    OMX_COLOR_FormatL4,
    OMX_COLOR_FormatL8,
    OMX_COLOR_FormatL16,
    OMX_COLOR_FormatL24,
    OMX_COLOR_FormatL32,
    OMX_COLOR_FormatYUV420PackedSemiPlanar,
    OMX_COLOR_FormatYUV422PackedSemiPlanar,
    OMX_COLOR_Format18BitBGR666,
    OMX_COLOR_Format24BitARGB6666,
    OMX_COLOR_Format24BitABGR6666,
    OMX_COLOR_FormatKhronosExtensions = 0x6F000000, /** Reserved region for introducing
Khronos Standard Extensions */
    OMX_COLOR_FormatVendorStartUnused = 0x7F000000, /** Reserved region for introducing
Vendor Extensions */
    OMX_COLOR_Format32bitABGR8888,
    OMX_COLOR_Format8bitPalette,
    OMX_COLOR_FormatYUVUV128,
    OMX_COLOR_FormatRawBayer12bit,
    OMX_COLOR_FormatBRCMEGL,
    OMX_COLOR_FormatBRCMOpaque,
    OMX_COLOR_FormatYVU420PackedPlanar,
    OMX_COLOR_FormatYVU420PackedSemiPlanar,
    OMX_COLOR_FormatMax = 0x7FFFFFFF
} OMX_COLOR_FORMATTYPE;
```

That's a rather large number of formats!

Running the program `info` from Chapter 8 shows the following for the `video_decode` component:

```
Video ports:
  Ports start on 130
  There are 2 open ports
  Port 130 has 20 buffers (minimum 1) of size 81920
  Direction is input
    Supported video formats are:
    Video format encoding 0x14
    Video compression format 0x4
    Video format encoding 0x14
    Video compression format 0x7
    Video format encoding 0x14
    Video compression format 0x7F000007
    Video format encoding 0x14
    Video compression format 0x8
    Video format encoding 0x14
    Video compression format 0x5
    Video format encoding 0x14
    Video compression format 0x3
    Video format encoding 0x14
    Video compression format 0x2
    Video format encoding 0x14
    Video compression format 0x7F000001
    Video format encoding 0x14
    Video compression format 0x7F000002
    Video format encoding 0x14
    Video compression format 0x7F000003
    Video format encoding 0x14
    Video compression format 0x6
    Video format encoding 0x14
    Video compression format 0x7F000004
    Video format encoding 0x14
    Video compression format 0x7F000005
    Video format encoding 0x14
    Video compression format 0x7F000006
    Video format encoding 0x14
    Video compression format 0x0
    No more formats supported
  Port 131 has 1 buffers (minimum 1) of size 115200
  Direction is output
    Supported video formats are:
    Video format encoding 0x14
    Video compression format 0x0
    No more formats supported
```

The compression formats are as follows:

- *0x2*: `OMX_VIDEO_CodingMPEG2`
- *0x3*: `OMX_VIDEO_CodingH263`

- *0x4*: OMX_VIDEO_CodingMPEG4
- *0x5*: OMX_VIDEO_CodingWMV
- *0x6*: OMX_VIDEO_CodingRV
- *0x7*: OMX_VIDEO_CodingAVC
- *0x8*: OMX_VIDEO_CodingMJPEG
- *0x7F000001*: OMX_VIDEO_CodingVP6
- *0x7F000002*: OMX_VIDEO_CodingVP7
- *0x7F000003*: OMX_VIDEO_CodingVP8
- *0x7F000004*: OMX_VIDEO_CodingYUV
- *0x7F000005*: OMX_VIDEO_CodingSorenson
- *0x7F000006*: OMX_VIDEO_CodingTheora

Currently I have not confirmed any of these except for H.264 (CodingAVC).

Decoding an H.264 File

The program to decode a video into an uncompressed format is essentially the same as the one for decoding an image. There is an essential difference: reading one block is not enough for the component to get information about the video file format, and multiple block reads generally have to occur.

In this example, you just slap a loop around reading the initial blocks until a `PortSettingsChanged` event occurs. The call to `ilclient_wait_for_event` will time out for the first set of calls, so you drop the timeout period from 10 seconds down to 2 seconds. This isn't any good for interactive viewing, of course, but here you are just decoding.

The significant change from the image-decoding program is this initial loop:

```
int port_settings_changed = 0;
while (!port_settings_changed) {
    buff_header =
        ilclient_get_input_buffer(component,
                                  130,
                                  1 /* block */);
    if (buff_header != NULL) {
        read_into_buffer_and_empty(fp,
                                   component,
                                   buff_header,
                                   &toread);

        // If all the file has been read in, then
        // we have to re-read this first block.
        // Broadcom bug?
        if (toread <= 0) {
            printf("Rewinding\n");
            // wind back to start and repeat
            fp = freopen(VIDEO, "r", fp);
            toread = get_file_size(VIDEO);
```

```
            }
        }

        // try if this block sets params for output port
        err = ilclient_wait_for_event(component,
                                      OMX_EventPortSettingsChanged,
                                      131, 0, 0, 1,
                                      ILCLIENT_EVENT_ERROR | ILCLIENT_PARAMETER_CHANGED,
                                      2000);
        if (err < 0) {
            printf("Wait for port settings changed timed out\n");
        } else {
            port_settings_changed = 1;
        }
    }
```

The full program is `il_decode_video.c`.

```
#include <stdio.h>
#include <stdlib.h>
#include <sys/stat.h>

#include <OMX_Core.h>
#include <OMX_Component.h>

#include <bcm_host.h>
#include <ilclient.h>

#define video  "/opt/vc/src/hello_pi/hello_video/test.h264"

void printState(OMX_HANDLETYPE handle) {
    // elided
}

char *err2str(int err) {
    return "elided";
}

void eos_callback(void *userdata, COMPONENT_T *comp, OMX_U32 data) {
    fprintf(stderr, "Got eos event\n");
}

void error_callback(void *userdata, COMPONENT_T *comp, OMX_U32 data) {
    fprintf(stderr, "OMX error %s\n", err2str(data));
}

int get_file_size(char *fname) {
    struct stat st;

    if (stat(fname, &st) == -1) {
            perror("Stat'ing video file");
            return -1;
```

```
        }
    return(st.st_size);
}

static void set_video_decoder_input_format(COMPONENT_T *component) {
   // set input video format
    printf("Setting video decoder format\n");
    OMX_VIDEO_PARAM_PORTFORMATTYPE videoPortFormat;
    //setHeader(&videoPortFormat,  sizeof(OMX_VIDEO_PARAM_PORTFORMATTYPE));
    memset(&videoPortFormat, 0, sizeof(OMX_VIDEO_PARAM_PORTFORMATTYPE));
    videoPortFormat.nSize = sizeof(OMX_VIDEO_PARAM_PORTFORMATTYPE);
    videoPortFormat.nVersion.nVersion = OMX_VERSION;

    videoPortFormat.nPortIndex = 130;
    videoPortFormat.eCompressionFormat = OMX_VIDEO_CodingAVC;
    OMX_SetParameter(ilclient_get_handle(component),
                     OMX_IndexParamVideoPortFormat, &videoPortFormat);

}

OMX_ERRORTYPE read_into_buffer_and_empty(FILE *fp,
                                         COMPONENT_T *component,
                                         OMX_BUFFERHEADERTYPE *buff_header,
                                         int *toread) {
    OMX_ERRORTYPE r;

    int buff_size = buff_header->nAllocLen;
    int nread = fread(buff_header->pBuffer, 1, buff_size, fp);

    buff_header->nFilledLen = nread;
    *toread -= nread;
    printf("Read %d, %d still left\n", nread, *toread);

    if (*toread <= 0) {
        printf("Setting EOS on input\n");
        buff_header->nFlags |= OMX_BUFFERFLAG_EOS;
    }
    r = OMX_EmptyThisBuffer(ilclient_get_handle(component),
                            buff_header);
    if (r != OMX_ErrorNone) {
        fprintf(stderr, "Empty buffer error %s\n",
                err2str(r));
    }
    return r;
}

OMX_ERRORTYPE save_info_from_filled_buffer(COMPONENT_T *component,
                                           OMX_BUFFERHEADERTYPE * buff_header) {
    OMX_ERRORTYPE r;

    printf("Got a filled buffer with %d, allocated %d\n",
           buff_header->nFilledLen,
```

```
            buff_header->nAllocLen);
    if (buff_header->nFlags & OMX_BUFFERFLAG_EOS) {
        printf("Got EOS on output\n");
        exit(0);
    }

    // and then refill it
    r = OMX_FillThisBuffer(ilclient_get_handle(component),
                           buff_header);
    if (r != OMX_ErrorNone) {
        fprintf(stderr, "Fill buffer error %s\n",
                err2str(r));
    }
    return r;
}

void get_output_port_settings(COMPONENT_T *component) {
    OMX_PARAM_PORTDEFINITIONTYPE portdef;

    printf("Port settings changed\n");
    // need to setup the input for the resizer with the output of the
    // decoder
    portdef.nSize = sizeof(OMX_PARAM_PORTDEFINITIONTYPE);
    portdef.nVersion.nVersion = OMX_VERSION;
    portdef.nPortIndex = 131;
    OMX_GetParameter(ilclient_get_handle(component),
                     OMX_IndexParamPortDefinition, &portdef);

    unsigned int uWidth =
        (unsigned int) portdef.format.video.nFrameWidth;
    unsigned int uHeight =
        (unsigned int) portdef.format.video.nFrameHeight;
    unsigned int uStride =
        (unsigned int) portdef.format.video.nStride;
    unsigned int uSliceHeight =
        (unsigned int) portdef.format.video.nSliceHeight;
    printf("Frame width %d, frame height %d, stride %d, slice height %d\n",
           uWidth,
           uHeight,
           uStride,
           uSliceHeight);
    printf("Getting format Compression 0x%x Color Format: 0x%x\n",
           (unsigned int) portdef.format.video.eCompressionFormat,
           (unsigned int) portdef.format.video.eColorFormat);
}

int main(int argc, char** argv) {

    int i;
    char *componentName;
    int err;
    ILCLIENT_T  *handle;
```

```
    COMPONENT_T *component;
    FILE *fp = fopen(video, "r");
    int toread = get_file_size(video);
    OMX_BUFFERHEADERTYPE *buff_header;

    componentName = "video_decode";

    bcm_host_init();

    handle = ilclient_init();
    if (handle == NULL) {
        fprintf(stderr, "IL client init failed\n");
        exit(1);
    }

   if (OMX_Init() != OMX_ErrorNone) {
        ilclient_destroy(handle);
        fprintf(stderr, "OMX init failed\n");
        exit(1);
    }

   ilclient_set_error_callback(handle,
                               error_callback,
                               NULL);
   ilclient_set_eos_callback(handle,
                             eos_callback,
                             NULL);

    err = ilclient_create_component(handle,
                                    &component,
                                    componentName,
                                    ILCLIENT_DISABLE_ALL_PORTS
                                        |
                                        ILCLIENT_ENABLE_INPUT_BUFFERS
                                        |
                                        ILCLIENT_ENABLE_OUTPUT_BUFFERS
                                    );
    if (err == -1) {
        fprintf(stderr, "Component create failed\n");
        exit(1);
    }
    printState(ilclient_get_handle(component));

    err = ilclient_change_component_state(component,
                                          OMX_StateIdle);
    if (err < 0) {
        fprintf(stderr, "Couldn't change state to Idle\n");
        exit(1);
    }
    printState(ilclient_get_handle(component));

    // must be before we enable buffers
    set_video_decoder_input_format(component);
```

```
// input port
ilclient_enable_port_buffers(component, 130,
                             NULL, NULL, NULL);
ilclient_enable_port(component, 130);

err = ilclient_change_component_state(component,
                                      OMX_StateExecuting);
if (err < 0) {
    fprintf(stderr, "Couldn't change state to Executing\n");
    exit(1);
}
printState(ilclient_get_handle(component));

// Read the first block so that the component can get
// the dimensions of the video and call port settings
// changed on the output port to configure it
int port_settings_changed = 0;
while (!port_settings_changed) {
buff_header =
    ilclient_get_input_buffer(component,
                              130,
                              1 /* block */);
if (buff_header != NULL) {
    read_into_buffer_and_empty(fp,
                               component,
                               buff_header,
                               &toread);

    // If all the file has been read in, then
    // we have to re-read this first block.
    // Broadcom bug?
    if (toread <= 0) {
        printf("Rewinding\n");
        // wind back to start and repeat
        fp = freopen(video, "r", fp);
        toread = get_file_size(video);
    }
}

// wait for first input block to set params for output port
err = ilclient_wait_for_event(component,
                              OMX_EventPortSettingsChanged,
                              131, 0, 0, 1,
                              ILCLIENT_EVENT_ERROR | ILCLIENT_PARAMETER_CHANGED,
                              2000);
if (err < 0) {
    printf("Wait for port settings changed timed out\n");
} else {
    port_settings_changed = 1;
}
err = ilclient_remove_event(component,
```

```
                                          OMX_EventPortSettingsChanged,
                                          131, 0, 0, 1);
    if (err < 0) {
        printf("Wait for remove port settings changed timed out\n");
    } else {
        port_settings_changed = 1;
    }
    }

    get_output_port_settings(component);

    // now enable output port since port params have been set
    ilclient_enable_port_buffers(component, 131,
                                 NULL, NULL, NULL);
    ilclient_enable_port(component, 131);

    // now work through the file
    while (toread > 0) {
        OMX_ERRORTYPE r;

        // do we have an input buffer we can fill and empty?
        buff_header =
            ilclient_get_input_buffer(component,
                                      130,
                                      1 /* block */);
        if (buff_header != NULL) {
            read_into_buffer_and_empty(fp,
                                       component,
                                       buff_header,
                                       &toread);
        }

        // do we have an output buffer that has been filled?
        buff_header =
            ilclient_get_output_buffer(component,
                                       131,
                                       0 /* no block */);
        if (buff_header != NULL) {
            save_info_from_filled_buffer(component,
                                         buff_header);
        } else {
            printf("No filled buffer\n");
        }
    }

    while (1) {
        printf("Getting last output buffers\n");
        buff_header =
            ilclient_get_output_buffer(component,
                                       131,
                                       1 /* block */);
```

```
        if (buff_header != NULL) {
            save_info_from_filled_buffer(component,
                                         buff_header);
        }
    }
    exit(0);
}
```

The program runs with no command-line parameters. The Videocore library comes with a standard H.264 example clip, /opt/vc/src/hello_pi/hello_video/test.h264, from Big Buck Bunny. It reads multiple frames before a PortSettngsChanged event occurs at which time the video format can be determined.

```
...
Read 81920, 30353369 still left
Empty buffer done
Wait for port settings changed timed out
Wait for remove port settings changed timed out
Read 81920, 30271449 still left
Empty buffer done
Wait for remove port settings changed timed out
Port settings changed
Frame width 1920, frame height 1080, stride 1920, slice height 1088
Getting format Compression 0x0 Color Format: 0x14
...
```

Rendering an H.264 Video

In a similar manner to decoding an image, rendering a video closely follows the code for rendering an image. However, there is now a major difference in how you wait for the initial PortSettingsChanged event. You have seen these two possibilities so far:

- Read the first part of the data and then block, waiting for the PortSettingsChanged event that you know will come.
- Keep reading data, each time block waiting for the PortSettingsChanged event and timing out if it doesn't arrive.

Neither of these two is acceptable for the immediate rendering of videos. It may take more than one read, and waiting on timeouts is too slow. Instead, you need to adopt an alternative approach based on the ilclient_remove_event function, which returns true if it can remove an event from the list of those events already seen.

The revised wait for the PortSettingsChanged event is as follows:

```
while there is more data
    read a new buffer
    empty the buffer

    if a PortSettingsChanged event has occurred
        break

    if there is no more data
```

```
        block waiting for a PortSettingsChanged event
        if failed on timeout
            error
        else
            break
```

(This algorithm should be generally applicable. But it breaks on the tunneled image rendering for unknown reasons.)

The video-rendering program using this is `il_render_video.c`.

```
#include <stdio.h>
#include <stdlib.h>
#include <sys/stat.h>

#include <OMX_Core.h>
#include <OMX_Component.h>

#include <bcm_host.h>
#include <ilclient.h>

#define VIDEO "/opt/vc/src/hello_pi/hello_video/test.h264"
char *video_file = VIDEO;

void printState(OMX_HANDLETYPE handle) {
    // elided
}

char *err2str(int err) {
    return "elided";
}

void eos_callback(void *userdata, COMPONENT_T *comp, OMX_U32 data) {
    fprintf(stderr, "Got eos event\n");
}

void error_callback(void *userdata, COMPONENT_T *comp, OMX_U32 data) {
    fprintf(stderr, "OMX error %s\n", err2str(data));
}

int get_file_size(char *fname) {
    struct stat st;

    if (stat(fname, &st) == -1) {
        perror("Stat'ing video file");
        return -1;
    }
    return(st.st_size);
}

unsigned int uWidth;
unsigned int uHeight;
```

```
OMX_ERRORTYPE read_into_buffer_and_empty(FILE *fp,
                                         COMPONENT_T *component,
                                         OMX_BUFFERHEADERTYPE *buff_header,
                                         int *toread) {
    OMX_ERRORTYPE r;

    int buff_size = buff_header->nAllocLen;
    int nread = fread(buff_header->pBuffer, 1, buff_size, fp);

    buff_header->nFilledLen = nread;
    *toread -= nread;
    printf("Read %d, %d still left\n", nread, *toread);

    if (*toread <= 0) {
        printf("Setting EOS on input\n");
        buff_header->nFlags |= OMX_BUFFERFLAG_EOS;
    }
    r = OMX_EmptyThisBuffer(ilclient_get_handle(component),
                            buff_header);
    if (r != OMX_ErrorNone) {
        fprintf(stderr, "Empty buffer error %s\n",
                err2str(r));
    }
    return r;
}

static void set_video_decoder_input_format(COMPONENT_T *component) {
    int err;

    // set input video format
    printf("Setting video decoder format\n");
    OMX_VIDEO_PARAM_PORTFORMATTYPE videoPortFormat;
    //setHeader(&videoPortFormat,  sizeof(OMX_VIDEO_PARAM_PORTFORMATTYPE));
    memset(&videoPortFormat, 0, sizeof(OMX_VIDEO_PARAM_PORTFORMATTYPE));
    videoPortFormat.nSize = sizeof(OMX_VIDEO_PARAM_PORTFORMATTYPE);
    videoPortFormat.nVersion.nVersion = OMX_VERSION;

    videoPortFormat.nPortIndex = 130;
    videoPortFormat.eCompressionFormat = OMX_VIDEO_CodingAVC;

    err = OMX_SetParameter(ilclient_get_handle(component),
                           OMX_IndexParamVideoPortFormat, &videoPortFormat);
    if (err != OMX_ErrorNone) {
        fprintf(stderr, "Error setting video decoder format %s\n", err2str(err));
        exit(1);
    } else {
        printf("Video decoder format set up ok\n");
    }

}
```

```
void setup_decodeComponent(ILCLIENT_T  *handle,
                           char *decodeComponentName,
                           COMPONENT_T **decodeComponent) {
    int err;

    err = ilclient_create_component(handle,
                                    decodeComponent,
                                    decodeComponentName,
                                    ILCLIENT_DISABLE_ALL_PORTS
                                    |
                                    ILCLIENT_ENABLE_INPUT_BUFFERS
                                    |
                                    ILCLIENT_ENABLE_OUTPUT_BUFFERS
                                    );
    if (err == -1) {
        fprintf(stderr, "DecodeComponent create failed\n");
        exit(1);
    }
    printState(ilclient_get_handle(*decodeComponent));

    err = ilclient_change_component_state(*decodeComponent,
                                          OMX_StateIdle);
    if (err < 0) {
        fprintf(stderr, "Couldn't change state to Idle\n");
        exit(1);
    }
    printState(ilclient_get_handle(*decodeComponent));

    // must be before we enable buffers
    set_video_decoder_input_format(*decodeComponent);
}

void setup_renderComponent(ILCLIENT_T  *handle,
                           char *renderComponentName,
                           COMPONENT_T **renderComponent) {
    int err;

    err = ilclient_create_component(handle,
                                    renderComponent,
                                    renderComponentName,
                                    ILCLIENT_DISABLE_ALL_PORTS
                                    |
                                    ILCLIENT_ENABLE_INPUT_BUFFERS
                                    );
    if (err == -1) {
        fprintf(stderr, "RenderComponent create failed\n");
        exit(1);
    }
    printState(ilclient_get_handle(*renderComponent));

    err = ilclient_change_component_state(*renderComponent,
                                          OMX_StateIdle);
```

```
    if (err < 0) {
        fprintf(stderr, "Couldn't change state to Idle\n");
        exit(1);
    }
    printState(ilclient_get_handle(*renderComponent));
}

int main(int argc, char** argv) {

    int i;
    char *decodeComponentName;
    char *renderComponentName;
    int err;
    ILCLIENT_T  *handle;
    COMPONENT_T *decodeComponent;
    COMPONENT_T *renderComponent;
    FILE *fp;
    int toread;
    OMX_BUFFERHEADERTYPE *buff_header;

    if (argc == 2) {
        video_file = argv[1];
    }
    if ((fp = fopen(video_file, "r")) == NULL)  {
        fprintf(stderr, "Can't open: %s\n", video_file);
        exit(2);
    }
    if ((toread = get_file_size(video_file)) == -1)  {
        fprintf(stderr, "Can't stat: %s\n", video_file);
        exit(2);
    }

    decodeComponentName = "video_decode";
    renderComponentName = "video_render";

    bcm_host_init();

    handle = ilclient_init();
    if (handle == NULL) {
        fprintf(stderr, "IL client init failed\n");
        exit(1);
    }

    if (OMX_Init() != OMX_ErrorNone) {
        ilclient_destroy(handle);
        fprintf(stderr, "OMX init failed\n");
        exit(1);
    }

    ilclient_set_error_callback(handle,
                                error_callback,
                                NULL);
```

```
    ilclient_set_eos_callback(handle,
                              eos_callback,
                              NULL);

    setup_decodeComponent(handle, decodeComponentName, &decodeComponent);
    setup_renderComponent(handle, renderComponentName, &renderComponent);
    // both components now in Idle state, no buffers, ports disabled

    // input port
    ilclient_enable_port_buffers(decodeComponent, 130,
                                 NULL, NULL, NULL);
    ilclient_enable_port(decodeComponent, 130);

    err = ilclient_change_component_state(decodeComponent,
                                          OMX_StateExecuting);
    if (err < 0) {
        fprintf(stderr, "Couldn't change state to Executing\n");
        exit(1);
    }
    printState(ilclient_get_handle(decodeComponent));

    // Read blocks and break out when we see a
    // PortSettingsChanged
    while (toread > 0) {
        buff_header =
            ilclient_get_input_buffer(decodeComponent,
                                      130,
                                      1 /* block */);
        if (buff_header != NULL) {
            read_into_buffer_and_empty(fp,
                                       decodeComponent,
                                       buff_header,
                                       &toread);

            // If all the file has been read in, then
            // we have to re-read this first block.
            // Broadcom bug?
            if (toread <= 0) {
                printf("Rewinding\n");
                // wind back to start and repeat
                fp = freopen(video_file, "r", fp);
                toread = get_file_size(video_file);
            }
        }

        if (toread > 0 && ilclient_remove_event(decodeComponent,
                                                OMX_EventPortSettingsChanged,
                                                131, 0, 0, 1) == 0) {
            printf("Removed port settings event\n");
            break;
        } else {
```

```
            printf("No portr settting seen yet\n");
        }
        // wait for first input block to set params for output port
        if (toread == 0) {
            // wait for first input block to set params for output port
            err = ilclient_wait_for_event(decodeComponent,
                                          OMX_EventPortSettingsChanged,
                                          131, 0, 0, 1,
                                          ILCLIENT_EVENT_ERROR | ILCLIENT_PARAMETER_CHANGED,
                                          2000);
            if (err < 0) {
                fprintf(stderr, "No port settings change\n");
                //exit(1);
            } else {
                printf("Port settings changed\n");
                break;
            }
        }
    }

    // set the decode component to idle and disable its ports
    err = ilclient_change_component_state(decodeComponent,
                                          OMX_StateIdle);
    if (err < 0) {
        fprintf(stderr, "Couldn't change state to Idle\n");
        exit(1);
    }
    ilclient_disable_port(decodeComponent, 131);
    ilclient_disable_port_buffers(decodeComponent, 131,
                                  NULL, NULL, NULL);

    // set up the tunnel between decode and render ports
    err = OMX_SetupTunnel(ilclient_get_handle(decodeComponent),
                          131,
                          ilclient_get_handle(renderComponent),
                          90);
    if (err != OMX_ErrorNone) {
        fprintf(stderr, "Error setting up tunnel %X\n", err);
        exit(1);
    } else {
        printf("Tunnel set up ok\n");
    }

    // Okay to go back to processing data
    // enable the decode output ports

    OMX_SendCommand(ilclient_get_handle(decodeComponent),
                    OMX_CommandPortEnable, 131, NULL);

    ilclient_enable_port(decodeComponent, 131);
```

```
    // enable the render output ports

    OMX_SendCommand(ilclient_get_handle(renderComponent),
                    OMX_CommandPortEnable, 90, NULL);

    ilclient_enable_port(renderComponent, 90);

    // set both components to executing state
    err = ilclient_change_component_state(decodeComponent,
                                          OMX_StateExecuting);
    if (err < 0) {
        fprintf(stderr, "Couldn't change state to Idle\n");
        exit(1);
    }
    err = ilclient_change_component_state(renderComponent,
                                          OMX_StateExecuting);
    if (err < 0) {
        fprintf(stderr, "Couldn't change state to Idle\n");
        exit(1);
    }

    // now work through the file
    while (toread > 0) {
        OMX_ERRORTYPE r;

        // do we have a decode input buffer we can fill and empty?
        buff_header =
            ilclient_get_input_buffer(decodeComponent,
                                      130,
                                      1 /* block */);
        if (buff_header != NULL) {
            read_into_buffer_and_empty(fp,
                                       decodeComponent,
                                       buff_header,
                                       &toread);
        }
    }

    ilclient_wait_for_event(renderComponent,
                            OMX_EventBufferFlag,
                            90, 0, OMX_BUFFERFLAG_EOS, 0,
                            ILCLIENT_BUFFER_FLAG_EOS, 10000);
    printf("EOS on render\n");

    sleep(10);

    exit(0);
}
```

The program takes the default H.264 but can be overridden by a command-line option. It plays the short segment and then terminates.

Going Full-Screen

The Broadcom version of OpenMAX has a nonstandard configuration mechanism that can be applied to video-rendering components. This can be done as follows:

```
OMX_CONFIG_DISPLAYREGIONTYPE display_region;
display_region.nSize = sizeof(OMX_CONFIG_DISPLAYREGIONTYPE);
display_region.nVersion.nVersion = OMX_VERSION;
display_region.nPortIndex = 90;

display_region.set = OMX_DISPLAY_SET_FULLSCREEN | OMX_DISPLAY_SET_NOASPECT;
display_region.fullscreen = OMX_TRUE;
display_region.noaspect = OMX_TRUE;
err = OMX_SetConfig(ilclient_get_handle(*renderComponent),
              OMX_IndexConfigDisplayRegion,
              &display_region);
if(err != OMX_ErrorNone) {
    fprintf(stderr, "Failed to set Full screen %x %s\n",
            err, err2str(err));
    exit(1);
}
```

Conclusion

This chapter has looked at decoding and rendering H.264 video streams. Playing MP4 files is covered in Chapter 16.

Resources

- *OpenMAX IL: The Standard for Media Library Portability*: https://www.khronos.org/openmax/il/
- *OpenMAX IL 1.1.2 Specification*: https://www.khronos.org/registry/omxil/specs/OpenMAX_IL_1_1_2_Specification.pdf
- *Raspberry Pi OpenMAX forum*: www.raspberrypi.org/forums/viewforum.php?f=70
- *VMCS-X OpenMAX IL Components*: www.jvcref.com/files/PI/documentation/ilcomponents/
- *Source code for ARM side libraries for interfacing to Raspberry Pi GPU*: https://github.com/raspberrypi/userland

CHAPTER 14

OpenMAX Audio on the Raspberry Pi

This chapter looks at audio processing on the Raspberry Pi. The support from OpenMAX is weaker: there is no satisfactory decode component. You have to resort to FFmpeg to decode and render audio.

Building Programs

Some programs in this chapter use libraries from the LibAV project, so development files from LibAV need to be installed.

```
sudo apt-get install libavcodec-dev
sudo apt-get install libavformat-dev
```

You can then build the programs in this chapter using the `Makefile`, which adds in the LibAV libraries.

```
DMX_INC =  -I/opt/vc/include/ -I /opt/vc/include/interface/vmcs_host/ -I/opt/vc/include/
interface/vcos/pthreads -I/opt/vc/include/interface/vmcs_host/linux
EGL_INC =
OMX_INC =  -I /opt/vc/include/IL
OMX_ILCLIENT_INC = -I/opt/vc/src/hello_pi/libs/ilclient
INCLUDES = $(DMX_INC) $(EGL_INC) $(OMX_INC) $(OMX_ILCLIENT_INC)

CFLAGS=-g -DRASPBERRY_PI -DOMX_SKIP64BIT $(INCLUDES)
CPPFLAGS =

DMX_LIBS =  -L/opt/vc/lib/ -lbcm_host -lvcos -lvchiq_arm -lpthread
EGL_LIBS = -L/opt/vc/lib/ -lEGL -lGLESv2
OMX_LIBS = -lopenmaxil
OMX_ILCLIENT_LIBS = -L/opt/vc/src/hello_pi/libs/ilclient -lilclient
AV_LIBS =  $(shell pkg-config --libs libavcodec libavformat libavutil)

LDLIBS =  $(DMX_LIBS) $(EGL_LIBS) $(OMX_LIBS) $(OMX_ILCLIENT_LIBS) $(AV_LIBS)

all:  api-example il_render_audio \
      il_ffmpeg_render_audio il_test_audio_encodings \
      il_ffmpeg_render_resample_audio
```


J. Newmarch, *Raspberry Pi GPU Audio Video Programming*, DOI 10.1007/978-1-4842-2472-4_14

Audio Components

The Raspberry Pi has a number of OpenMAX components specifically for audio processing.

- audio_capture
- audio_decode
- audio_encode
- audio_lowpower
- audio_mixer
- audio_processor
- audio_render
- audio_splitter

Audio Formats

OpenMAX has a number of data structures used to get and set information about components. You have seen some of these before.

- OMX_PORT_PARAM_TYPE: This is used with the index parameter OMX_IndexParamImageInit in a call to OMX_GetParameter. This gives the number of image ports and the port number of the first port.
- OMX_PARAM_PORTDEFINITIONTYPE: This is used with the index parameter OMX_PARAM_PORTDEFINITIONTYPE to give, for each port, the number of buffers, the size of each buffer, and the direction (input or output) of the port.
- OMX_AUDIO_PORTDEFINITIONTYPE: This is a field of the OMX_PARAM_PORTDEFINITIONTYPE for images.
- OMX_AUDIO_PARAM_PORTFORMATTYPE: This is used to get information about the different formats supported by each port.

Some of these were discussed in Chapter X.

You haven't looked at the field OMX_AUDIO_PORTDEFINITIONTYPE, which is part of the port definition information. It contains the following relevant fields:

```
typedef struct OMX_AUDIO_PORTDEFINITIONTYPE {
    OMX_STRING cMIMEType;
    OMX_NATIVE_DEVICETYPE pNativeRender;
    OMX_BOOL bFlagErrorConcealment;
    OMX_AUDIO_CODINGTYPE eEncoding;
} OMX_AUDIO_PORTDEFINITIONTYPE;
```

The last two fields are the current values set for the port. The *possible* values are obtained from the next structure, OMX_AUDIO_PARAM_PORTFORMATTYPE, so I will discuss it in the next paragraph. The major field you get here is the audio encoding.

OMX_AUDIO_PARAM_PORTFORMATTYPE is defined (in section 4.1.16 in the 1.1.2 specification) as follows:

```
typedef struct OMX_AUDIO_PARAM_PORTFORMATTYPE {
    OMX_U32 nSize;
    OMX_VERSIONTYPE nVersion;
    OMX_U32 nPortIndex;
    OMX_U32 nIndex;
    OMX_AUDIO_CODINGTYPE eEncoding;
} OMX_AUDIO_PARAM_PORTFORMATTYPE;
```

The first two fields are common to all OpenMAX structures. The nPortIndex field is the port you are looking at. The nIndex field is to distinguish between all the different format types supported by this port. The eEncoding field gives information about the format.

The values for OMX_AUDIO_CODINGTYPE are given in Table 4-66 of the 1.1.2 specification and on the RPi are given in the file /opt/vc/include/IL/OMX_Audio.h, as follows:

```
typedef enum OMX_AUDIO_CODINGTYPE {
    OMX_AUDIO_CodingUnused = 0,  /** Placeholder value when coding is N/A  */
    OMX_AUDIO_CodingAutoDetect,  /** auto detection of audio format */
    OMX_AUDIO_CodingPCM,         /** Any variant of PCM coding */
    OMX_AUDIO_CodingADPCM,       /** Any variant of ADPCM encoded data */
    OMX_AUDIO_CodingAMR,         /** Any variant of AMR encoded data */
    OMX_AUDIO_CodingGSMFR,       /** Any variant of GSM fullrate (i.e. GSM610) */
    OMX_AUDIO_CodingGSMEFR,      /** Any variant of GSM Enhanced Fullrate encoded data*/
    OMX_AUDIO_CodingGSMHR,       /** Any variant of GSM Halfrate encoded data */
    OMX_AUDIO_CodingPDCFR,       /** Any variant of PDC Fullrate encoded data */
    OMX_AUDIO_CodingPDCEFR,      /** Any variant of PDC Enhanced Fullrate encoded data */
    OMX_AUDIO_CodingPDCHR,       /** Any variant of PDC Halfrate encoded data */
    OMX_AUDIO_CodingTDMAFR,      /** Any variant of TDMA Fullrate encoded data (TIA/EIA-136-420) */
    OMX_AUDIO_CodingTDMAEFR,     /** Any variant of TDMA Enhanced Fullrate encoded data
(TIA/EIA-136-410) */
    OMX_AUDIO_CodingQCELP8,      /** Any variant of QCELP 8kbps encoded data */
    OMX_AUDIO_CodingQCELP13,     /** Any variant of QCELP 13kbps encoded data */
    OMX_AUDIO_CodingEVRC,        /** Any variant of EVRC encoded data */
    OMX_AUDIO_CodingSMV,         /** Any variant of SMV encoded data */
    OMX_AUDIO_CodingG711,        /** Any variant of G.711 encoded data */
    OMX_AUDIO_CodingG723,        /** Any variant of G.723 dot 1 encoded data */
    OMX_AUDIO_CodingG726,        /** Any variant of G.726 encoded data */
    OMX_AUDIO_CodingG729,        /** Any variant of G.729 encoded data */
    OMX_AUDIO_CodingAAC,         /** Any variant of AAC encoded data */
    OMX_AUDIO_CodingMP3,         /** Any variant of MP3 encoded data */
    OMX_AUDIO_CodingSBC,         /** Any variant of SBC encoded data */
    OMX_AUDIO_CodingVORBIS,      /** Any variant of VORBIS encoded data */
    OMX_AUDIO_CodingWMA,         /** Any variant of WMA encoded data */
    OMX_AUDIO_CodingRA,          /** Any variant of RA encoded data */
    OMX_AUDIO_CodingMIDI,        /** Any variant of MIDI encoded data */
    OMX_AUDIO_CodingKhronosExtensions = 0x6F000000, /** Reserved region for introducing
Khronos  Standard Extensions */
    OMX_AUDIO_CodingVendorStartUnused = 0x7F000000, /** Reserved region for introducing
Vendor Extensions */

    OMX_AUDIO_CodingFLAC,        /** Any variant of FLAC */
    OMX_AUDIO_CodingDDP,         /** Any variant of Dolby Digital Plus */
    OMX_AUDIO_CodingDTS,         /** Any variant of DTS */
```

```
    OMX_AUDIO_CodingWMAPRO,       /** Any variant of WMA Professional */
    OMX_AUDIO_CodingATRAC3,       /** Sony ATRAC-3 variants */
    OMX_AUDIO_CodingATRACX,       /** Sony ATRAC-X variants */
    OMX_AUDIO_CodingATRACAAL,     /** Sony ATRAC advanced-lossless variants  */

    OMX_AUDIO_CodingMax = 0x7FFFFFFF
} OMX_AUDIO_CODINGTYPE;
```

Running the program info from Chapter X shows the following for the audio_decode component:

```
Audio ports:
  Ports start on 120
  There are 2 open ports
  Port 120 has 128 buffers of size 16384
  Direction is input
    Port 120 requires 4 buffers
    Port 120 has min buffer size 16384 bytes
    Port 120 is an input port
    Port 120 is an audio port
    Port mimetype (null)
    Port encoding is MP3
      Supported audio formats are:
      Supported encoding is MP3
          MP3 default sampling rate 0
          MP3 default bits per sample 0
          MP3 default number of channels 0
      Supported encoding is PCM
          PCM default sampling rate 0
          PCM default bits per sample 0
          PCM default number of channels 0
      Supported encoding is AAC
      Supported encoding is WMA
      Supported encoding is Ogg Vorbis
      Supported encoding is RA
      Supported encoding is AMR
      Supported encoding is EVRC
      Supported encoding is G726
      Supported encoding is FLAC
      Supported encoding is DDP
      Supported encoding is DTS
      Supported encoding is WMAPRO
      Supported encoding is ATRAC3
      Supported encoding is ATRACX
      Supported encoding is ATRACAAL
      Supported encoding is MIDI
      No more formats supported
  Port 121 has 1 buffers of size 32768
  Direction is output
    Port 121 requires 1 buffers
    Port 121 has min buffer size 32768 bytes
    Port 121 is an output port
    Port 121 is an audio port
```

```
    Port mimetype (null)
    Port encoding is PCM
      Supported audio formats are:
      Supported encoding is PCM
          PCM default sampling rate 44100
          PCM default bits per sample 16
          PCM default number of channels 2
      Supported encoding is DDP
      Supported encoding is DTS
      No more formats supported
```

This looks really impressive, but regrettably, none of this is actually supported except for PCM. The following is according to jamesh in "OMX_AllocateBuffer fails for audio decoder component":

> *The way it works is that the component passes back success for all the codecs it can potentially support (i.e., all the codecs we've ever had going). That is then constrained by what codecs are actually installed. It would be better to run time detect which codecs are present, but that code has never been written since it's never been required. It's also unlikely ever to be done as Broadcom no longer supports audio codecs in this way—they have moved off the Videocore to the host CPU since they are now powerful enough to handle any audio decoding task.*

That's kind of sad, really.

So, how do you find out which encodings are really supported? You can get a partial answer by trying to allocate buffers for a particular encoding. So, for each port, you loop through the possible encodings, setting the encoding and trying to allocate the buffers. You can use `il_enable_port_buffers` for this as it will return -1 if the allocation fails.

The program to do this is `il_test_audio_encodings.c`.

```
#include <stdio.h>
#include <stdlib.h>
#include <sys/stat.h>

#include <OMX_Core.h>
#include <OMX_Component.h>

#include <bcm_host.h>
#include <ilclient.h>

FILE *outfp;

void printState(OMX_HANDLETYPE handle) {
    // elided
}

char *err2str(int err) {
    return "elided";
}

void eos_callback(void *userdata, COMPONENT_T *comp, OMX_U32 data) {
```

```
    fprintf(stderr, "Got eos event\n");
}

void error_callback(void *userdata, COMPONENT_T *comp, OMX_U32 data) {
    //fprintf(stderr, "OMX error %s\n", err2str(data));
}

int get_file_size(char *fname) {
    struct stat st;

    if (stat(fname, &st) == -1) {
        perror("Stat'ing img file");
        return -1;
    }
    return(st.st_size);
}

static void set_audio_decoder_input_format(COMPONENT_T *component,
                                           int port, int format) {
    // set input audio format
    //printf("Setting audio decoder format\n");
    OMX_AUDIO_PARAM_PORTFORMATTYPE audioPortFormat;
    //setHeader(&audioPortFormat,  sizeof(OMX_AUDIO_PARAM_PORTFORMATTYPE));
    memset(&audioPortFormat, 0, sizeof(OMX_AUDIO_PARAM_PORTFORMATTYPE));
    audioPortFormat.nSize = sizeof(OMX_AUDIO_PARAM_PORTFORMATTYPE);
    audioPortFormat.nVersion.nVersion = OMX_VERSION;

    audioPortFormat.nPortIndex = port;
    //audioPortFormat.eEncoding = OMX_AUDIO_CodingPCM;
    audioPortFormat.eEncoding = format;
    OMX_SetParameter(ilclient_get_handle(component),
                     OMX_IndexParamAudioPortFormat, &audioPortFormat);
    //printf("Format set ok to %d\n", format);
}

char *format2str(OMX_AUDIO_CODINGTYPE format) {
    switch(format) {
    case OMX_AUDIO_CodingUnused: return "OMX_AUDIO_CodingUnused";
    case OMX_AUDIO_CodingAutoDetect: return "OMX_AUDIO_CodingAutoDetect";
    case OMX_AUDIO_CodingPCM: return "OMX_AUDIO_CodingPCM";
    case OMX_AUDIO_CodingADPCM: return "OMX_AUDIO_CodingADPCM";
    case OMX_AUDIO_CodingAMR: return "OMX_AUDIO_CodingAMR";
    case OMX_AUDIO_CodingGSMFR: return "OMX_AUDIO_CodingGSMFR";
    case OMX_AUDIO_CodingGSMEFR: return "OMX_AUDIO_CodingGSMEFR" ;
    case OMX_AUDIO_CodingGSMHR: return "OMX_AUDIO_CodingGSMHR";
    case OMX_AUDIO_CodingPDCFR: return "OMX_AUDIO_CodingPDCFR";
    case OMX_AUDIO_CodingPDCEFR: return "OMX_AUDIO_CodingPDCEFR";
    case OMX_AUDIO_CodingPDCHR: return "OMX_AUDIO_CodingPDCHR";
    case OMX_AUDIO_CodingTDMAFR: return "OMX_AUDIO_CodingTDMAFR";
    case OMX_AUDIO_CodingTDMAEFR: return "OMX_AUDIO_CodingTDMAEFR";
    case OMX_AUDIO_CodingQCELP8: return "OMX_AUDIO_CodingQCELP8";
    case OMX_AUDIO_CodingQCELP13: return "OMX_AUDIO_CodingQCELP13";
```

```
    case OMX_AUDIO_CodingEVRC: return "OMX_AUDIO_CodingEVRC";
    case OMX_AUDIO_CodingSMV: return "OMX_AUDIO_CodingSMV";
    case OMX_AUDIO_CodingG711: return "OMX_AUDIO_CodingG711";
    case OMX_AUDIO_CodingG723: return "OMX_AUDIO_CodingG723";
    case OMX_AUDIO_CodingG726: return "OMX_AUDIO_CodingG726";
    case OMX_AUDIO_CodingG729: return "OMX_AUDIO_CodingG729";
    case OMX_AUDIO_CodingAAC: return "OMX_AUDIO_CodingAAC";
    case OMX_AUDIO_CodingMP3: return "OMX_AUDIO_CodingMP3";
    case OMX_AUDIO_CodingSBC: return "OMX_AUDIO_CodingSBC";
    case OMX_AUDIO_CodingVORBIS: return "OMX_AUDIO_CodingVORBIS";
    case OMX_AUDIO_CodingWMA: return "OMX_AUDIO_CodingWMA";
    case OMX_AUDIO_CodingRA: return "OMX_AUDIO_CodingRA";
    case OMX_AUDIO_CodingMIDI: return "OMX_AUDIO_CodingMIDI";
    case OMX_AUDIO_CodingFLAC: return "OMX_AUDIO_CodingFLAC";
    case OMX_AUDIO_CodingDDP: return "OMX_AUDIO_CodingDDP";
    case OMX_AUDIO_CodingDTS: return "OMX_AUDIO_CodingDTS";
    case OMX_AUDIO_CodingWMAPRO: return "OMX_AUDIO_CodingWMAPRO";
    case OMX_AUDIO_CodingATRAC3: return "OMX_AUDIO_CodingATRAC3";
    case OMX_AUDIO_CodingATRACX: return "OMX_AUDIO_CodingATRACX";
    case OMX_AUDIO_CodingATRACAAL: return "OMX_AUDIO_CodingATRACAAL" ;
    default: return "Unknown format";
    }
}

void test_audio_port_formats(COMPONENT_T *component, int port) {
    int n = 2;
    while (n <= OMX_AUDIO_CodingMIDI) {
        set_audio_decoder_input_format(component, port, n);

        // input port
        if (ilclient_enable_port_buffers(component, port,
                                         NULL, NULL, NULL) < 0) {
            printf("    Unsupported encoding is %s\n",
                   format2str(n));
        } else {
            printf("    Supported encoding is %s\n",
                  format2str(n));
            ilclient_disable_port_buffers(component, port,
                                          NULL, NULL, NULL);
        }
        n++;
    }
    n = OMX_AUDIO_CodingFLAC;
    while (n <= OMX_AUDIO_CodingATRACAAL) {
        set_audio_decoder_input_format(component, port, n);

        // input port
        if (ilclient_enable_port_buffers(component, port,
                                         NULL, NULL, NULL) < 0) {
            printf("    Unsupported encoding is %s\n",
                   format2str(n));
        } else {
```

```
            printf("    Supported encoding is %s\n",
                   format2str(n));
            ilclient_disable_port_buffers(component, port,
                                          NULL, NULL, NULL);
        }
        n++;
    }
}

void test_all_audio_ports(COMPONENT_T *component) {
    OMX_PORT_PARAM_TYPE param;
    OMX_PARAM_PORTDEFINITIONTYPE sPortDef;
    OMX_ERRORTYPE err;
    OMX_HANDLETYPE handle = ilclient_get_handle(component);

    int startPortNumber;
    int nPorts;
    int n;

    //setHeader(&param, sizeof(OMX_PORT_PARAM_TYPE));
    memset(&param, 0, sizeof(OMX_PORT_PARAM_TYPE));
    param.nSize = sizeof(OMX_PORT_PARAM_TYPE);
    param.nVersion.nVersion = OMX_VERSION;

    err = OMX_GetParameter(handle, OMX_IndexParamAudioInit, &param);
    if(err != OMX_ErrorNone){
        fprintf(stderr, "Error in getting audio OMX_PORT_PARAM_TYPE parameter\n");
        return;
    }
    printf("Audio ports:\n");

    startPortNumber = param.nStartPortNumber;
    nPorts = param.nPorts;
    if (nPorts == 0) {
        printf("No ports of this type\n");
        return;
    }

    printf("Ports start on %d\n", startPortNumber);
    printf("There are %d open ports\n", nPorts);

    for (n = 0; n < nPorts; n++) {
        memset(&sPortDef, 0, sizeof(OMX_PARAM_PORTDEFINITIONTYPE)) ;
        sPortDef.nSize = sizeof(OMX_PARAM_PORTDEFINITIONTYPE);
        sPortDef.nVersion.nVersion = OMX_VERSION;

        sPortDef.nPortIndex = startPortNumber + n;
        err = OMX_GetParameter(handle, OMX_IndexParamPortDefinition, &sPortDef);
        if(err != OMX_ErrorNone){
            fprintf(stderr, "Error in getting OMX_PORT_DEFINITION_TYPE parameter\n");
            exit(1);
        }
```

```
        printf("Port %d has %d buffers of size %d\n",
               sPortDef.nPortIndex,
               sPortDef.nBufferCountActual,
               sPortDef.nBufferSize);
        printf("Direction is %s\n",
               (sPortDef.eDir == OMX_DirInput ? "input" : "output"));
        test_audio_port_formats(component, sPortDef.nPortIndex);
    }
}

int main(int argc, char** argv) {
    char *componentName;
    int err;
    ILCLIENT_T  *handle;
    COMPONENT_T *component;

    componentName = "audio_decode";
    if (argc == 2) {
        componentName = argv[1];
    }

    bcm_host_init();

    handle = ilclient_init();
    if (handle == NULL) {
        fprintf(stderr, "IL client init failed\n") ;
        exit(1);
    }

    if (OMX_Init() != OMX_ErrorNone) {
        ilclient_destroy(handle);
        fprintf(stderr, "OMX init failed\n");
        exit(1);
    }

    ilclient_set_error_callback(handle,
                                error_callback,
                                NULL);
    ilclient_set_eos_callback(handle,
                              eos_callback,
                              NULL);

    err = ilclient_create_component(handle,
                                    &component,
                                    componentName,
                                    ILCLIENT_DISABLE_ALL_PORTS
                                    |
                                    ILCLIENT_ENABLE_INPUT_BUFFERS
                                    |
                                    ILCLIENT_ENABLE_OUTPUT_BUFFERS
                                    );
```

```
    if (err == -1) {
        fprintf(stderr, "Component create failed\n");
        exit(1);
    }
    printState(ilclient_get_handle(component));

    err = ilclient_change_component_state(component,
                                          OMX_StateIdle);
    if (err < 0) {
        fprintf(stderr, "Couldn't change state to Idle\n");
        exit(1);
    }
    printState(ilclient_get_handle(component));

    test_all_audio_ports(component);

    exit(0);
}
```

The program appears to be only partially successful. For the audio_decode component, it shows that only two possible formats can be decoded, PCM and ADPCM , but shows that they can be decoded to MP3, Vorbis, and so on, which seems most unlikely.

```
Audio ports:
Ports start on 120
There are 2 open ports
Port 120 has 128 buffers of size 16384
Direction is input
    Supported encoding is OMX_AUDIO_CodingPCM
    Supported encoding is OMX_AUDIO_CodingADPCM
    Unsupported encoding is OMX_AUDIO_CodingAMR
    Unsupported encoding is OMX_AUDIO_CodingGSMFR
    Unsupported encoding is OMX_AUDIO_CodingGSMEFR
    Unsupported encoding is OMX_AUDIO_CodingGSMHR
    Unsupported encoding is OMX_AUDIO_CodingPDCFR
    Unsupported encoding is OMX_AUDIO_CodingPDCEFR
    Unsupported encoding is OMX_AUDIO_CodingPDCHR
    Unsupported encoding is OMX_AUDIO_CodingTDMAFR
    Unsupported encoding is OMX_AUDIO_CodingTDMAEFR
    Unsupported encoding is OMX_AUDIO_CodingQCELP8
    Unsupported encoding is OMX_AUDIO_CodingQCELP13
    ...
Port 121 has 1 buffers of size 32768
Direction is output
    Supported encoding is OMX_AUDIO_CodingPCM
    Supported encoding is OMX_AUDIO_CodingADPCM
    Supported encoding is OMX_AUDIO_CodingAMR
    Supported encoding is OMX_AUDIO_CodingGSMFR
    Supported encoding is OMX_AUDIO_CodingGSMEFR
    Supported encoding is OMX_AUDIO_CodingGSMHR
    Supported encoding is OMX_AUDIO_CodingPDCFR
    Supported encoding is OMX_AUDIO_CodingPDCEFR
```

```
Supported encoding is OMX_AUDIO_CodingPDCHR
Supported encoding is OMX_AUDIO_CodingTDMAFR
Supported encoding is OMX_AUDIO_CodingTDMAEFR
Supported encoding is OMX_AUDIO_CodingQCELP8
Supported encoding is OMX_AUDIO_CodingQCELP13
Supported encoding is OMX_AUDIO_CodingEVRC
Supported encoding is OMX_AUDIO_CodingSMV
Supported encoding is OMX_AUDIO_CodingG711
Supported encoding is OMX_AUDIO_CodingG723
Supported encoding is OMX_AUDIO_CodingG726
Supported encoding is OMX_AUDIO_CodingG729
...
```

Decoding an Audio File Using audio_decode

The Broadcom `audio_decode` component will decode only Pulse Code Modulated (PCM) format data. It decodes it to...PCM format data. PCM is the binary format commonly used to represent unencoded audio data. In other words, unless Broadcom includes support for some of the audio codecs, then this component is pretty useless.

Rendering PCM Data

Now you'll learn how to render PCM data.

PCM Data

The following is according to Wikipedia:

> *[PCM] is a method used to digitally represent sampled analog signals. It is the standard form for digital audio in computers and various Blu-ray, DVD, and Compact Disc formats, as well as other uses such as digital telephone systems. A PCM stream is a digital representation of an analog signal, in which the magnitude of the analog signal is sampled regularly at uniform intervals, with each sample being quantized to the nearest value within a range of digital steps.*
>
> *PCM streams have two basic properties that determine their fidelity to the original analog signal: the sampling rate, which is the number of times per second that samples are taken; and the bit depth, which determines the number of possible digital values that each sample can take.*

PCM data can be stored in files as "raw" data. In this case, there is no header information to say what the sampling rate and bit depth are. Many tools such as `sox` use the file extension to determine these properties. The following is from `man soxformat`:

> *f32 and f64 indicate files encoded as 32- and 64-bit (IEEE single and double precision) floating-point PCM, respectively; s8, s16, s24, and s32 indicate 8-, 16-, 24-, and 32-bit signed integer PCM, respectively; u8, u16, u24, and u32 indicate 8-, 16-, 24-, and 32-bit unsigned integer PCM, respectively.*

But it should be noted that the file extension is only an aid to understanding some of the PCM codec parameters and how they are stored in the file.

Files can be converted into PCM by tools such as avconv. For example, to convert a WAV file to PCM, you use this:

```
avconv -i  enigma.wav -f s16le enigma.s16
```

The output will give information not saved in the file, which you will need to give to a processing program later.

```
Input #0, wav, from 'enigma.wav':
  Duration: 00:06:26.38, bitrate: 1411 kb/s
    Stream #0.0: Audio: pcm_s16le, 44100 Hz, stereo, s16, 1411 kb/s
Output #0, s16le, to 'enigma.s16':
  Metadata:
    encoder         : Lavf54.20.4
    Stream #0.0: Audio: pcm_s16le, 44100 Hz, stereo, s16, 1411 kb/s
```

From this you can see that the format is two channels, 44,100 Hz, and 16-bit little-endian. (The file I used was from a group called Enigma, who released an album as open content.)

To check that the encoding worked, you can use aplay as follows:

```
aplay -r 44100 -c 2 -f S16_LE enigma.s16
```

Choosing an Output Device

OpenMAX has a standard audio render component. But what device does it render to? The built-in sound card? A USB sound card? That is not part of OpenMAX IL—there isn't even a way to list the audio devices, only the audio components.

OpenMAX has an extension mechanism that can be used by an OpenMAX implementor to answer questions like this. The Broadcom core implementation has extension types OMX_CONFIG_BRCMAUDIODESTINATIONTYPE and OMX_CONFIG_BRCMAUDIOSOURCETYPE, which can be used to set the audio destination (source) device. Use the following code to do this:

```
void setOutputDevice(const char *name) {
   int32_t success = -1;
   OMX_CONFIG_BRCMAUDIODESTINATIONTYPE arDest;

   if (name && strlen(name) < sizeof(arDest.sName)) {
       setHeader(&arDest, sizeof(OMX_CONFIG_BRCMAUDIODESTINATIONTYPE));
       strcpy((char *)arDest.sName, name);

       err = OMX_SetParameter(handle, OMX_IndexConfigBrcmAudioDestination, &arDest);
       if (err != OMX_ErrorNone) {
           fprintf(stderr, "Error on setting audio destination\n");
           exit(1);
       }
   }
}
```

Here is where Broadcom becomes a bit obscure again. The header file `IL/OMX_Broadcom.h` states that the default value of `sName` is `local` but doesn't give any other values. The Raspberry Pi forums say that this refers to the 3.5 mm analog audio out and that HDMI is chosen by using the value `hdmi`. No other values are documented, and it seems that the Broadcom OpenMAX IL does not support any other audio devices. In particular, USB audio devices are not supported by the current Broadcom OpenMAX IL components for either input or output. So, you can't use OpenMAX IL for, say, audio capture on the Raspberry Pi since it has no Broadcom-supported audio input.

Setting PCM Format

You can use two functions to set the PCM format. The first contains nothing unusual.

```
void set_audio_render_input_format(COMPONENT_T *component) {
    // set input audio format
    printf("Setting audio render format\n");
    OMX_AUDIO_PARAM_PORTFORMATTYPE audioPortFormat;

    memset(&audioPortFormat, 0, sizeof(OMX_AUDIO_PARAM_PORTFORMATTYPE));
    audioPortFormat.nSize = sizeof(OMX_AUDIO_PARAM_PORTFORMATTYPE);
    audioPortFormat.nVersion.nVersion = OMX_VERSION;

    audioPortFormat.nPortIndex = 100;

    OMX_GetParameter(ilclient_get_handle(component),
                     OMX_IndexParamAudioPortFormat, &audioPortFormat);

    audioPortFormat.eEncoding = OMX_AUDIO_CodingPCM;
    OMX_SetParameter(ilclient_get_handle(component),
                     OMX_IndexParamAudioPortFormat, &audioPortFormat);

    setPCMMode(ilclient_get_handle(component), 100);

}
```

The second gets the current PCM parameters and then sets the required PCM parameters (which you know independently).

```
void setPCMMode(OMX_HANDLETYPE handle, int startPortNumber) {
    OMX_AUDIO_PARAM_PCMMODETYPE sPCMMode;
    OMX_ERRORTYPE err;

    memset(&sPCMMode, 0, sizeof(OMX_AUDIO_PARAM_PCMMODETYPE));
    sPCMMode.nSize = sizeof(OMX_AUDIO_PARAM_PCMMODETYPE);
    sPCMMode.nVersion.nVersion = OMX_VERSION;

    sPCMMode.nPortIndex = startPortNumber;

    err = OMX_GetParameter(handle, OMX_IndexParamAudioPcm, &sPCMMode);
    printf("Sampling rate %d, channels %d\n",
           sPCMMode.nSamplingRate,
           sPCMMode.nChannels);
```

```
    sPCMMode.nSamplingRate = 44100;
    sPCMMode.nChannels = 2;

    err = OMX_SetParameter(handle, OMX_IndexParamAudioPcm, &sPCMMode);
    if(err != OMX_ErrorNone){
        fprintf(stderr, "PCM mode unsupported\n");
        return;
    } else {
        fprintf(stderr, "PCM mode supported\n");
        fprintf(stderr, "PCM sampling rate %d\n", sPCMMode.nSamplingRate);
        fprintf(stderr, "PCM nChannels %d\n", sPCMMode.nChannels);
    }
}
```

The program is il_render_audio.c.

```
#include <stdio.h>
#include <stdlib.h>
#include <sys/stat.h>

#include <OMX_Core.h>
#include <OMX_Component.h>

#include <bcm_host.h>
#include <ilclient.h>

#define AUDIO  "enigma.s16"

/* For the RPi name can be "hdmi" or "local" */
void setOutputDevice(OMX_HANDLETYPE handle, const char *name) {
    OMX_ERRORTYPE err;
    OMX_CONFIG_BRCMAUDIODESTINATIONTYPE arDest;

    if (name && strlen(name) < sizeof(arDest.sName)) {
        memset(&arDest, 0, sizeof(OMX_CONFIG_BRCMAUDIODESTINATIONTYPE));
        arDest.nSize = sizeof(OMX_CONFIG_BRCMAUDIODESTINATIONTYPE);
        arDest.nVersion.nVersion = OMX_VERSION;

        strcpy((char *)arDest.sName, name);

        err = OMX_SetParameter(handle, OMX_IndexConfigBrcmAudioDestination, &arDest);
        if (err != OMX_ErrorNone) {
            fprintf(stderr, "Error on setting audio destination\n");
            exit(1);
        }
    }
}

void setPCMMode(OMX_HANDLETYPE handle, int startPortNumber) {
    OMX_AUDIO_PARAM_PCMMODETYPE sPCMMode;
    OMX_ERRORTYPE err;
```

```
    memset(&sPCMMode, 0, sizeof(OMX_AUDIO_PARAM_PCMMODETYPE));
    sPCMMode.nSize = sizeof(OMX_AUDIO_PARAM_PCMMODETYPE);
    sPCMMode.nVersion.nVersion = OMX_VERSION;

    sPCMMode.nPortIndex = startPortNumber;

    err = OMX_GetParameter(handle, OMX_IndexParamAudioPcm, &sPCMMode);
    printf("Sampling rate %d, channels %d\n",
           sPCMMode.nSamplingRate,
           sPCMMode.nChannels);

    sPCMMode.nSamplingRate = 44100;
    sPCMMode.nChannels = 2;

    err = OMX_SetParameter(handle, OMX_IndexParamAudioPcm, &sPCMMode);
    if(err != OMX_ErrorNone){
        fprintf(stderr, "PCM mode unsupported\n");
        return;
    } else {
        fprintf(stderr, "PCM mode supported\n");
        fprintf(stderr, "PCM sampling rate %d\n", sPCMMode.nSamplingRate);
        fprintf(stderr, "PCM nChannels %d\n", sPCMMode.nChannels);
    }
}

void printState(OMX_HANDLETYPE handle) {
    //elided
}

char *err2str(int err) {
    return "elided";
}

void eos_callback(void *userdata, COMPONENT_T *comp, OMX_U32 data) {
    fprintf(stderr, "Got eos event\n");
}

void error_callback(void *userdata, COMPONENT_T *comp, OMX_U32 data) {
    fprintf(stderr, "OMX error %s\n", err2str(data));
}

int get_file_size(char *fname) {
    struct stat st;

    if (stat(fname, &st) == -1) {
        perror("Stat'ing img file");
        return -1;
    }
    return(st.st_size);
}
```

```
static void set_audio_render_input_format(COMPONENT_T *component) {
    // set input audio format
    printf("Setting audio render format\n");
    OMX_AUDIO_PARAM_PORTFORMATTYPE audioPortFormat;
    //setHeader(&audioPortFormat,  sizeof(OMX_AUDIO_PARAM_PORTFORMATTYPE));
    memset(&audioPortFormat, 0, sizeof(OMX_AUDIO_PARAM_PORTFORMATTYPE));
    audioPortFormat.nSize = sizeof(OMX_AUDIO_PARAM_PORTFORMATTYPE);
    audioPortFormat.nVersion.nVersion = OMX_VERSION;

    audioPortFormat.nPortIndex = 100;

    OMX_GetParameter(ilclient_get_handle(component),
                     OMX_IndexParamAudioPortFormat, &audioPortFormat);

    audioPortFormat.eEncoding = OMX_AUDIO_CodingPCM;
    //audioPortFormat.eEncoding = OMX_AUDIO_CodingMP3;
    OMX_SetParameter(ilclient_get_handle(component),
                     OMX_IndexParamAudioPortFormat, &audioPortFormat);

    setPCMMode(ilclient_get_handle(component), 100);

}

OMX_ERRORTYPE read_into_buffer_and_empty(FILE *fp,
                                         COMPONENT_T *component,
                                         OMX_BUFFERHEADERTYPE *buff_header,
                                         int *toread) {
    OMX_ERRORTYPE r;

    int buff_size = buff_header->nAllocLen;
    int nread = fread(buff_header->pBuffer, 1, buff_size, fp);

    printf("Read %d\n", nread);

    buff_header->nFilledLen = nread;
    *toread -= nread;
    if (*toread <= 0) {
        printf("Setting EOS on input\n");
        buff_header->nFlags |= OMX_BUFFERFLAG_EOS;
    }
    r = OMX_EmptyThisBuffer(ilclient_get_handle(component),
                            buff_header);
    if (r != OMX_ErrorNone) {
        fprintf(stderr, "Empty buffer error %s\n",
                err2str(r));
    }
    return r;
}

int main(int argc, char** argv) {
```

```
int i;
char *componentName;
int err;
ILCLIENT_T  *handle;
COMPONENT_T *component;

char *audio_file = AUDIO;
if (argc == 2) {
    audio_file = argv[1];
}

FILE *fp = fopen(audio_file, "r");
int toread = get_file_size(audio_file);

OMX_BUFFERHEADERTYPE *buff_header;

componentName = "audio_render";

bcm_host_init();

handle = ilclient_init();
if (handle == NULL) {
    fprintf(stderr, "IL client init failed\n");
    exit(1);
}

if (OMX_Init() != OMX_ErrorNone) {
    ilclient_destroy(handle);
    fprintf(stderr, "OMX init failed\n");
    exit(1);
}

ilclient_set_error_callback(handle,
                            error_callback,
                            NULL);
ilclient_set_eos_callback(handle,
                          eos_callback,
                          NULL);

err = ilclient_create_component(handle,
                                &component,
                                componentName,
                                ILCLIENT_DISABLE_ALL_PORTS
                                |
                                ILCLIENT_ENABLE_INPUT_BUFFERS
                                );
if (err == -1) {
    fprintf(stderr, "Component create failed\n");
    exit(1);
}
printState(ilclient_get_handle(component));
```

```
    err = ilclient_change_component_state(component,
                                           OMX_StateIdle);
    if (err < 0) {
        fprintf(stderr, "Couldn't change state to Idle\n");
        exit(1);
    }
    printState(ilclient_get_handle(component));

    // must be before we enable buffers
    set_audio_render_input_format(component);

    setOutputDevice(ilclient_get_handle(component), "local");

    // input port
    ilclient_enable_port_buffers(component, 100,
                                 NULL, NULL, NULL);
    ilclient_enable_port(component, 100);

    err = ilclient_change_component_state(component,
                                           OMX_StateExecuting);
    if (err < 0) {
        fprintf(stderr, "Couldn't change state to Executing\n");
        exit(1);
    }
    printState(ilclient_get_handle(component));

    // now work through the file
    while (toread > 0) {
        OMX_ERRORTYPE r;

        // do we have an input buffer we can fill and empty?
        buff_header =
            ilclient_get_input_buffer(component,
                                      100,
                                      1 /* block */);
        if (buff_header != NULL) {
            read_into_buffer_and_empty(fp,
                                       component,
                                       buff_header,
                                       &toread);
        }
    }

    exit(0);
}
```

The program can be run with the command line of any file with two-channel 16-bit PCM with a sampling rate of 44,100 Hz. This is hard-coded into the program. It would be easy to add command-line parsing along the same lines as aplay.

Decoding an MP3 File Using FFmpeg or Avconv

If you want to play a compressed file such as MP3 or Ogg, it has to be decoded, and as noted earlier, the `audio_decode` component doesn't do this. So, you have to turn to another system. The prominent audio-decoding systems are FFmpeg (forked to LibAV) and GStreamer. I will use LibAV as that is the default install on the RPi.

FFmpeg was started in 2000. LibAV forked from it in 2011. Over time, both of the libraries and the file formats have evolved. Consequently, there are code examples on the Web that are no longer appropriate. Generally, FFmpeg and LibAV follow the same API and are generally interchangeable at the API code level—but not always.

The current FFmpeg source distro includes a program called `doc/examples/decoding_encoding.c`, while the LibAV distro has a similar example, called `avcodec.c`, which can decode MP3 files to PCM format. These *almost* work on the MP3 files I tried because the output format from the decoder has changed from *interleaved* to *planar*, and the examples have not been updated to reflect this.

The difference is easily illustrated with stereo: interleaved means LRLRLR. With planar, a set of consecutive Rs are given after a set of consecutive Ls, as in LLLLLL...RRRRR. Interleaved is a degenerate case of planar with a run length of 1.

A frame of video/audio that is decoded by FFmpeg/LibAV is built in a struct `AVFrame`. This includes the following fields:

```
typedef struct AVFrame {
    uint8_t *data[AV_NUM_DATA_POINTERS];
    int linesize[AV_NUM_DATA_POINTERS];
    int nb_samples;
}
```

In the interleaved case, all the samples are in `data[0]`. In the planar case, they are in `data[0]`, `data[1]`, ... There does not seem to be an explicit indicator of how many planar streams there are, but if a data element is non-null, it seems to contain a stream. So, by walking the `data` array until you find `NULL`, you can find the number of streams.

Many tools such as `aplay` will accept only interleaved samples. So, given multiple planar streams, you have to interleave them yourself. This isn't hard once you know the sample size in bytes, the length of each stream, and the number of streams (a more robust way is given in a later section).

```
int data_size = av_samples_get_buffer_size(NULL, c->channels,
                                           decoded_frame->nb_samples,
                                           c->sample_fmt, 1);
// first time: count the number of  planar streams
if (num_streams == 0) {
    while (num_streams < AV_NUM_DATA_POINTERS &&
           decoded_frame->data[num_streams] != NULL)
        num_streams++;
}

// first time: set sample_size from 0 to e.g 2 for 16-bit data
if (sample_size == 0) {
    sample_size =
        data_size / (num_streams * decoded_frame->nb_samples);
}

int m, n;
```

```
            for (n = 0; n < decoded_frame->nb_samples; n++) {
                // interleave the samples from the planar streams
                for (m = 0; m < num_streams; m++) {
                    fwrite(&decoded_frame->data[m][n*sample_size],
                           1, sample_size, outfile);
                }
            }
```

The revised program, which reads from an MP3 file and writes decoded data to `/tmp/test.sw`, is `api-example.c`.

```
/*
 * copyright (c) 2001 Fabrice Bellard
 *
 * This file is part of Libav.
 *
 * Libav is free software; you can redistribute it and/or
 * modify it under the terms of the GNU Lesser General Public
 * License as published by the Free Software Foundation; either
 * version 2.1 of the License, or (at your option) any later version.
 *
 * Libav is distributed in the hope that it will be useful,
 * but WITHOUT ANY WARRANTY; without even the implied warranty of
 * MERCHANTABILITY or FITNESS FOR A PARTICULAR PURPOSE.  See the GNU
 * Lesser General Public License for more details.
 *
 * You should have received a copy of the GNU Lesser General Public
 * License along with Libav; if not, write to the Free Software
 * Foundation, Inc., 51 Franklin Street, Fifth Floor, Boston, MA 02110-1301 USA
 */

// From http://code.haskell.org/~thielema/audiovideo-example/cbits/

/**
 * @file
 * libavcodec API use example.
 *
 * @example libavcodec/api-example.c
 * Note that this library only handles codecs (mpeg, mpeg4, etc...),
 * not file formats (avi, vob, etc...). See library 'libavformat' for the
 * format handling
 */

#include <stdlib.h>
#include <stdio.h>
#include <string.h>

#ifdef HAVE_AV_CONFIG_H
#undef HAVE_AV_CONFIG_H
#endif
```

```
#include "libavcodec/avcodec.h"
#include <libavformat/avformat.h>
#include "libavutil/mathematics.h"
#include "libavutil/samplefmt.h"

#define INBUF_SIZE 4096
#define AUDIO_INBUF_SIZE 20480
#define AUDIO_REFILL_THRESH 4096

/*
 * Audio decoding.
 */
static void audio_decode_example(const char *outfilename, const char *filename)
{
    AVCodec *codec;
    AVCodecContext *c = NULL;
    int len;
    FILE *f, *outfile;
    uint8_t inbuf[AUDIO_INBUF_SIZE + FF_INPUT_BUFFER_PADDING_SIZE];
    AVPacket avpkt;
    AVFrame *decoded_frame = NULL;
    int num_streams = 0;
    int sample_size = 0;

    av_init_packet(&avpkt);

    printf("Audio decoding\n");

    /* find the mpeg audio decoder */
    codec = avcodec_find_decoder(AV_CODEC_ID_MP3);
    if (!codec) {
        fprintf(stderr, "codec not found\n");
        exit(1);/home/httpd/html/RPi-hidden/OpenMAX/Audio/BST.mp3
    }

    c = avcodec_alloc_context3(codec);;

    /* open it */
    if (avcodec_open2(c, codec, NULL) < 0) {
        fprintf(stderr, "could not open codec\n");
        exit(1);
    }

    f = fopen(filename, "rb");
    if (!f) {
        fprintf(stderr, "could not open %s\n", filename);
        exit(1);
    }
    outfile = fopen(outfilename, "wb");
    if (!outfile) {
        av_free(c);
```

```
        exit(1);
    }

    /* decode until eof */
    avpkt.data = inbuf;
    avpkt.size = fread(inbuf, 1, AUDIO_INBUF_SIZE, f);

    while (avpkt.size > 0) {
        int got_frame = 0;

        if (!decoded_frame) {
            if (!(decoded_frame = avcodec_alloc_frame())) {
                fprintf(stderr, "out of memory\n");
                exit(1);
            }
        } else
            avcodec_get_frame_defaults(decoded_frame);

        len = avcodec_decode_audio4(c, decoded_frame, &got_frame, &avpkt);
        if (len < 0) {
            fprintf(stderr, "Error while decoding\n");
            exit(1);
        }
        if (got_frame) {
            printf("Decoded frame nb_samples %d, format %d\n",
                   decoded_frame->nb_samples,
                   decoded_frame->format);
            if (decoded_frame->data[1] != NULL)
                printf("Data[1] not null\n");
            elseAV_LIBS =  $(shell pkg-config --libs libavcodec libavformat libavutil)
                printf("Data[1] is null\n");
            /* if a frame has been decoded, output it */
            int data_size = av_samples_get_buffer_size(NULL, c->channels,
                                                        decoded_frame->nb_samples,
                                                        c->sample_fmt, 1);
            // first time: count the number of  planar streams
            if (num_streams == 0) {
                while (num_streams < AV_NUM_DATA_POINTERS &&
                       decoded_frame->data[num_streams] != NULL)
                    num_streams++;
            }

            // first time: set sample_size from 0 to e.g 2 for 16-bit data
            if (sample_size == 0) {
                sample_size =
                    data_size / (num_streams * decoded_frame->nb_samples);
            }

            int m, n;
            for (n = 0; n < decoded_frame->nb_samples; n++) {
                // interleave the samples from the planar streams
```

```
                for (m = 0; m < num_streams; m++) {
                    fwrite(&decoded_frame->data[m][n*sample_size],
                           1, sample_size, outfile);
                }
            }
        }
        avpkt.size -= len;
        avpkt.data += len;
        if (avpkt.size < AUDIO_REFILL_THRESH) {
            /* Refill the input buffer, to avoid trying to decode
             * incomplete frames. Instead of this, one could also use
             * a parser, or use a proper container format through
             * libavformat. */
            memmove(inbuf, avpkt.data, avpkt.size);
            avpkt.data = inbuf;
            len = fread(avpkt.data + avpkt.size, 1,
                        AUDIO_INBUF_SIZE - avpkt.size, f);
            if (len > 0)
                avpkt.size += len;
        }
    }

    fclose(outfile);
    fclose(f);

    avcodec_close(c);
    av_free(c);
    av_free(decoded_frame);
}

int main(int argc, char **argv)
{
    const char *filename = "BST.mp3";
    AVFormatContext *pFormatCtx = NULL;

    if (argc == 2) {
        filename = argv[1];
    }

    // Register all formats and codecs
    av_register_all();
    if(avformat_open_input(&pFormatCtx, filename, NULL, NULL)!=0) {
        fprintf(stderr, "Can't get format\n");
        return -1; // Couldn't open file
    }
    // Retrieve stream information
    if(avformat_find_stream_info(pFormatCtx, NULL)<0)
        return -1; // Couldn't find stream information
    av_dump_format(pFormatCtx, 0, filename, 0);
    printf("Num streams %d\n", pFormatCtx->nb_streams);
    printf("Bit rate %d\n", pFormatCtx->bit_rate);
    audio_decode_example("/tmp/test.sw", filename);
```

```
    return 0;
}
```

You can build the program using make. As it does not depend on OpenMAX, it can also be built independently using the compile command.

```
cc -c -g api-example.c
cc api-example.o -lavutil -lavcodec -lavformat -o api-example -lm
```

It can be run with an MP3 file being the command-line parameter.

You can test the result as follows (you may need to change parameters):

```
aplay -r 44100 -c 2 -f S16_LE /tmp/test.sw
```

Rendering MP3 Using FFmpeg or LibAV and OpenMAX

Since the Broadcom audio_decode component is apparently of little use, if you actually want to play MP3, Ogg, or other encoded formats, you have to use FFmpeg/LibAV (or GStreamer) to decode the audio to PCM and then pass it to the Broadcom audio_render component.

Essentially this means taking the last two programs and mashing them together. It isn't hard, just a bit messy. The only tricky point is that the buffers returned from FFmpeg/LibAV and the buffers used by audio_render are different sizes, and you don't really know which will be bigger. If the audio_render input buffers are bigger, then you just copy (and interleave) the FFmpeg data across; if smaller, then you have to keep fetching new buffers as each one is filled.

The resultant program is il_ffmpeg_render_audio.c.

```
#include <stdio.h>
#include <stdlib.h>
#include <sys/stat.h>

#include <OMX_Core.h>
#include <OMX_Component.h>

#include <bcm_host.h>
#include <ilclient.h>

#include "libavcodec/avcodec.h"
#include <libavformat/avformat.h>
#include "libavutil/mathematics.h"
#include "libavutil/samplefmt.h"

#define INBUF_SIZE 4096
#define AUDIO_INBUF_SIZE 20480
#define AUDIO_REFILL_THRESH 4096

#define AUDIO  "BST.mp3"

/* For the RPi name can be "hdmi" or "local" */
void setOutputDevice(OMX_HANDLETYPE handle, const char *name) {
    OMX_ERRORTYPE err;
    OMX_CONFIG_BRCMAUDIODESTINATIONTYPE arDest;
```

```
    if (name && strlen(name) < sizeof(arDest.sName)) {
        memset(&arDest, 0, sizeof(OMX_CONFIG_BRCMAUDIODESTINATIONTYPE));
        arDest.nSize = sizeof(OMX_CONFIG_BRCMAUDIODESTINATIONTYPE);
        arDest.nVersion.nVersion = OMX_VERSION;

        strcpy((char *)arDest.sName, name);

        err = OMX_SetParameter(handle, OMX_IndexConfigBrcmAudioDestination, &arDest);
        if (err != OMX_ErrorNone) {
            fprintf(stderr, "Error on setting audio destination\n");
            exit(1);
        }
    }
}

void setPCMMode(OMX_HANDLETYPE handle, int startPortNumber) {
    OMX_AUDIO_PARAM_PCMMODETYPE sPCMMode;
    OMX_ERRORTYPE err;

    memset(&sPCMMode, 0, sizeof(OMX_AUDIO_PARAM_PCMMODETYPE));
    sPCMMode.nSize = sizeof(OMX_AUDIO_PARAM_PCMMODETYPE);
    sPCMMode.nVersion.nVersion = OMX_VERSION;

    sPCMMode.nPortIndex = startPortNumber;

    err = OMX_GetParameter(handle, OMX_IndexParamAudioPcm, &sPCMMode);
    printf("Sampling rate %d, channels %d\n",
           sPCMMode.nSamplingRate,
           sPCMMode.nChannels);

    sPCMMode.nSamplingRate = 44100;
    sPCMMode.nChannels = 2; // assumed for now - should be checked

    err = OMX_SetParameter(handle, OMX_IndexParamAudioPcm, &sPCMMode);
    if(err != OMX_ErrorNone){
        fprintf(stderr, "PCM mode unsupported\n");
        return;
    } else {
        fprintf(stderr, "PCM mode supported\n");
        fprintf(stderr, "PCM sampling rate %d\n", sPCMMode.nSamplingRate);
        fprintf(stderr, "PCM nChannels %d\n", sPCMMode.nChannels);
    }
}

void printState(OMX_HANDLETYPE handle) {
    // elided
}

char *err2str(int err) {
    return "error elided";
}
```

```
void eos_callback(void *userdata, COMPONENT_T *comp, OMX_U32 data) {
    fprintf(stderr, "Got eos event\n");
}

void error_callback(void *userdata, COMPONENT_T *comp, OMX_U32 data) {
    fprintf(stderr, "OMX error %s\n", err2str(data));
}

int get_file_size(char *fname) {
    struct stat st;

    if (stat(fname, &st) == -1) {
        perror("Stat'ing img file");
        return -1;
    }
    return(st.st_size);
}

AVPacket avpkt;
AVCodecContext *c = NULL;

/*
 * Audio decoding.
 */
static void audio_decode_example(const char *filename)
{
    AVCodec *codec;

    av_init_packet(&avpkt);

    printf("Audio decoding\n");

    /* find the mpeg audio decoder */
    codec = avcodec_find_decoder(AV_CODEC_ID_MP3);
    if (!codec) {
        fprintf(stderr, "codec not found\n");
        exit(1);
    }

    c = avcodec_alloc_context3(codec);;

    /* open it */
    if (avcodec_open2(c, codec, NULL) < 0) {
        fprintf(stderr, "could not open codec\n");
        exit(1);
    }
}

static void set_audio_render_input_format(COMPONENT_T *component) {
    // set input audio format
    printf("Setting audio render format\n");
    OMX_AUDIO_PARAM_PORTFORMATTYPE audioPortFormat;
```

```
    //setHeader(&audioPortFormat,  sizeof(OMX_AUDIO_PARAM_PORTFORMATTYPE));
    memset(&audioPortFormat, 0, sizeof(OMX_AUDIO_PARAM_PORTFORMATTYPE));
    audioPortFormat.nSize = sizeof(OMX_AUDIO_PARAM_PORTFORMATTYPE);
    audioPortFormat.nVersion.nVersion = OMX_VERSION;

    audioPortFormat.nPortIndex = 100;

    OMX_GetParameter(ilclient_get_handle(component),
                     OMX_IndexParamAudioPortFormat, &audioPortFormat);

    audioPortFormat.eEncoding = OMX_AUDIO_CodingPCM;
    //audioPortFormat.eEncoding = OMX_AUDIO_CodingMP3;
    OMX_SetParameter(ilclient_get_handle(component),
                     OMX_IndexParamAudioPortFormat, &audioPortFormat);

    setPCMMode(ilclient_get_handle(component), 100);

}

int num_streams = 0;
int sample_size = 0;

OMX_ERRORTYPE read_into_buffer_and_empty(AVFrame *decoded_frame,
                                         COMPONENT_T *component,
                                         // OMX_BUFFERHEADERTYPE *buff_header,
                                         int total_len) {
    OMX_ERRORTYPE r;
    OMX_BUFFERHEADERTYPE *buff_header = NULL;
    int k, m, n;

    if (total_len <= 4096) { //buff_header->nAllocLen) {
        // all decoded frame fits into one OpenMAX buffer
        buff_header =
            ilclient_get_input_buffer(component,
                                      100,
                                      1 /* block */);
        for (k = 0, n = 0; n < decoded_frame->nb_samples; n++) {
            for (m = 0; m < num_streams; m++) {
                memcpy(&buff_header->pBuffer[k],
                       &decoded_frame->data[m][n*sample_size],
                       sample_size);
                k += sample_size;
            }
        }

        buff_header->nFilledLen = k;
        r = OMX_EmptyThisBuffer(ilclient_get_handle(component),
                                buff_header);
        if (r != OMX_ErrorNone) {
            fprintf(stderr, "Empty buffer error %s\n",
                    err2str(r));
        }
```

```
        return r;
    }

    // more than one OpenMAX buffer required
    for (k = 0, n = 0; n < decoded_frame->nb_samples; n++) {

        if (k == 0) {
             buff_header =
                ilclient_get_input_buffer(component,
                                          100,
                                          1 /* block */);
        }

        // interleave the samples from the planar streams
        for (m = 0; m < num_streams; m++) {
            memcpy(&buff_header->pBuffer[k],
                   &decoded_frame->data[m][n*sample_size],
                   sample_size);
            k += sample_size;
        }

        if (k >= buff_header->nAllocLen) {
            // this buffer is full
            buff_header->nFilledLen = k;
            r = OMX_EmptyThisBuffer(ilclient_get_handle(component),
                                    buff_header);
            if (r != OMX_ErrorNone) {
                fprintf(stderr, "Empty buffer error %s\n",
                        err2str(r));
            }
            k = 0;
            buff_header = NULL;
        }
    }
    if (buff_header != NULL) {
            buff_header->nFilledLen = k;
            r = OMX_EmptyThisBuffer(ilclient_get_handle(component),
                                    buff_header);
            if (r != OMX_ErrorNone) {
                fprintf(stderr, "Empty buffer error %s\n",
                        err2str(r));
            }
    }
    return r;
}

int main(int argc, char** argv) {

    int i;
    char *componentName;
    int err;
```

```
ILCLIENT_T  *handle;
COMPONENT_T *component;

AVFormatContext *pFormatCtx = NULL;

char *audio_file = AUDIO;
if (argc == 2) {
    audio_file = argv[1];
}

FILE *fp = fopen(audio_file, "r");
int toread = get_file_size(audio_file);

OMX_BUFFERHEADERTYPE *buff_header;

componentName = "audio_render";

bcm_host_init();

handle = ilclient_init();
if (handle == NULL) {
    fprintf(stderr, "IL client init failed\n");
    exit(1);
}

if (OMX_Init() != OMX_ErrorNone) {
    ilclient_destroy(handle);
    fprintf(stderr, "OMX init failed\n");
    exit(1);
}

ilclient_set_error_callback(handle,
                            error_callback,
                            NULL);
ilclient_set_eos_callback(handle,
                          eos_callback,
                          NULL);

err = ilclient_create_component(handle,
                                &component,
                                componentName,
                                ILCLIENT_DISABLE_ALL_PORTS
                                |
                                ILCLIENT_ENABLE_INPUT_BUFFERS
                                );
if (err == -1) {
    fprintf(stderr, "Component create failed\n");
    exit(1);
}
printState(ilclient_get_handle(component));
```

```
    err = ilclient_change_component_state(component,
                                          OMX_StateIdle);
    if (err < 0) {
        fprintf(stderr, "Couldn't change state to Idle\n");
        exit(1);
    }
    printState(ilclient_get_handle(component));

    // FFmpeg init
    av_register_all();
    if(avformat_open_input(&pFormatCtx, audio_file, NULL, NULL)!=0) {
        fprintf(stderr, "Can't get format\n");
        return -1; // Couldn't open file
    }
    // Retrieve stream information
    if(avformat_find_stream_info(pFormatCtx, NULL)<0)
        return -1; // Couldn't find stream information
    av_dump_format(pFormatCtx, 0, audio_file, 0);

    audio_decode_example(audio_file);

    // must be before we enable buffers
    set_audio_render_input_format(component);

    setOutputDevice(ilclient_get_handle(component), "local");

    // input port
    ilclient_enable_port_buffers(component, 100,
                                 NULL, NULL, NULL);
    ilclient_enable_port(component, 100);

    err = ilclient_change_component_state(component,
                                          OMX_StateExecuting);
    if (err < 0) {
        fprintf(stderr, "Couldn't change state to Executing\n");
        exit(1);
    }
    printState(ilclient_get_handle(component));

    // now work through the file

    int len;
    FILE *f;
    uint8_t inbuf[AUDIO_INBUF_SIZE + FF_INPUT_BUFFER_PADDING_SIZE];

    AVFrame *decoded_frame = NULL;

    f = fopen(audio_file, "rb");
    if (!f) {
        fprintf(stderr, "could not open %s\n", audio_file);
        exit(1);
    }
```

```
/* decode until eof */
avpkt.data = inbuf;
avpkt.size = fread(inbuf, 1, AUDIO_INBUF_SIZE, f);

while (avpkt.size > 0) {
    int got_frame = 0;

    if (!decoded_frame) {
        if (!(decoded_frame = avcodec_alloc_frame())) {
            fprintf(stderr, "out of memory\n");
            exit(1);
        }
    } else
        avcodec_get_frame_defaults(decoded_frame);

    len = avcodec_decode_audio4(c, decoded_frame, &got_frame, &avpkt);
    if (len < 0) {
        fprintf(stderr, "Error while decoding\n");
        exit(1);
    }
    if (got_frame) {
        /* if a frame has been decoded, we want to send it to OpenMAX */
        int data_size = av_samples_get_buffer_size(NULL, c->channels,
                                                   decoded_frame->nb_samples,
                                                   c->sample_fmt, 1);
        // first time: count the number of  planar streams
        if (num_streams == 0) {
            while (num_streams < AV_NUM_DATA_POINTERS &&
                   decoded_frame->data[num_streams] != NULL)
                num_streams++;
        }

        // first time: set sample_size from 0 to e.g 2 for 16-bit data
        if (sample_size == 0) {
            sample_size =
                data_size / (num_streams * decoded_frame->nb_samples);
        }

        // Empty into render_audio input buffers
        read_into_buffer_and_empty(decoded_frame,
                                   component,
                                   data_size
                                   );
    }

    avpkt.size -= len;
    avpkt.data += len;
    if (avpkt.size < AUDIO_REFILL_THRESH) {
        /* Refill the input buffer, to avoid trying to decode
         * incomplete frames. Instead of this, one could also use
         * a parser, or use a proper container format through
         * libavformat. */
```

```
            memmove(inbuf, avpkt.data, avpkt.size);
            avpkt.data = inbuf;
            len = fread(avpkt.data + avpkt.size, 1,
                        AUDIO_INBUF_SIZE - avpkt.size, f);
            if (len > 0)
                avpkt.size += len;
        }
    }

    printf("Finished decoding MP3\n");
    // clean up last empty buffer with EOS
    buff_header =
        ilclient_get_input_buffer(component,
                                  100,
                                  1 /* block */);
    buff_header->nFilledLen = 0;
    int r;
    buff_header->nFlags |= OMX_BUFFERFLAG_EOS;
    r = OMX_EmptyThisBuffer(ilclient_get_handle(component),
                            buff_header);
    if (r != OMX_ErrorNone) {
        fprintf(stderr, "Empty buffer error %s\n",
                err2str(r));
    } else {
        printf("EOS sent\n");
    }

    fclose(f);

    avcodec_close(c);
    av_free(c);
    av_free(decoded_frame);

    sleep(10);
    exit(0);
}
```

The compiled program will take a command-line argument that is the name of an MP3 file and play it to the analog audio port.

Rendering MP3 with ID3 Extensions Using FFmpeg or LibAV and OpenMAX

MP3 was originally designed as a stereo format without metadata information (artist, date of recording, and so on). Now you have 5.1 and 6.1 formats with probably more coming. The MP3 Surround extension looks after this. Metadata is typically added as an ID3 extension. (I discovered this only after the previous program broke badly on some newer MP3 files I have: no MP3 header.)

The command `file` will identify such files with the following:

```
$ file Beethoven.mp3
Beethoven.mp3: Audio file with ID3 version 2.3.0
```

An MP3 file will usually have two *channels* for stereo. An MP3+ID3 file containing, say, an image file, will have two *streams*, one for the audio and one for the image. The av_dump_format will show all the IDE metadata, plus the image stream.

```
Input #0, mp3, from 'Beethoven.mp3':
  Metadata:
    copyright       : 2013 Naxos Digital Services Ltd.
    album           : String Quartets - BEETHOVEN L van HAYDN FJ MOZART WA SCHUBERT F
JANACEK L (Petersen Quar~1
    TSRC            : US2TL0937001
    title           : String Quartet No 6 in B-Flat Major Op 18 No 6 I Allegro con brio
    TIT1            : String Quartets - BEETHOVEN L van HAYDN FJ MOZART WA SCHUBERT F
JANACEK L (Petersen Quar~1
    disc            : 1
    TLEN            : 522200
    track           : 1
    publisher       : Capriccio-(C51147)
    encoder         : LAME 32bits version 3.98.2 (http://www.mp3dev.org/)
    album_artist    : Petersen Quartet
    artist          : Petersen Quartet
    TEXT            : Beethoven, Ludwig van
    TOFN            : 729325.mp3
    genre           : Classical Music
    composer        : Beethoven, Ludwig van
    date            : 2009
  Duration: 00:08:42.24, start: 0.000000, bitrate: 323 kb/s
    Stream #0.0: Audio: mp3, 44100 Hz, 2 channels, s16p, 320 kb/s
    Stream #0.1: Video: mjpeg, yuvj444p, 500x509 [PAR 300:300 DAR 500:509], 90k tbn
    Metadata:
      title           :
      comment         : Cover (front)
```

Stream 0 is the audio stream, while an image is in stream 1.

You want to pass the MP3 stream to the FFmpeg/LibAV audio decoder but *not* the image stream. An AV frame contains the field stream_index that can be used to distinguish between them. If it is the audio stream, pass it to the OMX audio renderer; otherwise, skip it. You will see similar behavior when you look at rendering audio and video MPEG files. You find the audio stream index by asking av_find_best_stream for the AVMEDIA_TYPE_AUDIO stream.

In the previous sections, you read chunks in from the audio file and then relied on the decoder to break that into frames. That is now apparently on the way out. Instead, you should use the function av_read_frame, which reads only one frame at a time.

Unfortunately, on the RPi distro I was using, libavcodec-extra was at package 52, and that has a broken implementation of av_read_frame. You will need to do a package upgrade of LibAV to version 53 or later in order for the following code to work (the original version got the stream_index wrong).

So, by this stage, you read frames using av_read_frame, find the audio stream index of the frame, and use this to distinguish between audio and other. You still get frames with the audio samples in the wrong format such as AV_SAMPLE_FMT_S16P instead of AV_SAMPLE_FMT_S16. In the previous sections, you reformatted the stream by hand. But of course, the formats might change again, so your code would break.

The Audio Resample package gives a general-purpose way of managing this. It requires more setup, but once done will be more stable...

...except for a little glitch: this is one of the few areas in which FFmpeg and LibAV have different APIs. How to do it with FFmpeg is shown in "How to convert sample rate from AV_SAMPLE_FMT_FLTP to AV_SAMPLE_FMT_S16?" (http://stackoverflow.com/questions/14989397/how-to-convert-sample-rate-from-av-sample-fmt-fltp-to-av-sample-fmt-s16). You are using LibAV, which is basically the same but with differently named types. The following sets up the conversion parameters:

```
AVAudioResampleContext *swr = avresample_alloc_context();
av_opt_set_int(swr, "in_channel_layout",  audio_dec_ctx->channel_layout, 0);
av_opt_set_int(swr, "out_channel_layout", audio_dec_ctx->channel_layout,  0);
av_opt_set_int(swr, "in_sample_rate",     audio_dec_ctx->sample_rate, 0);
av_opt_set_int(swr, "out_sample_rate",    audio_dec_ctx->sample_rate, 0);
av_opt_set_int(swr, "in_sample_fmt",  audio_dec_ctx->sample_fmt, 0);
av_opt_set_int(swr, "out_sample_fmt", AV_SAMPLE_FMT_S16,  0);
avresample_open(swr);
```

The following performs the conversion:

```
uint8_t *buffer;
av_samples_alloc(&buffer, &out_linesize, 2, decoded_frame->nb_samples,
                 AV_SAMPLE_FMT_S16, 0);
avresample_convert(swr, &buffer,
                   decoded_frame->linesize[0],
                   decoded_frame->nb_samples,
                   decoded_frame->data,
                   decoded_frame->linesize[0],
                   decoded_frame->nb_samples);
```

The revised program is il_ffmpeg_render_resample_audio.c.

```
#include <stdio.h>
#include <stdlib.h>
#include <sys/stat.h>

#include <OMX_Core.h>
#include <OMX_Component.h>

#include <bcm_host.h>
#include <ilclient.h>

#include "libavcodec/avcodec.h"
#include <libavformat/avformat.h>
#include "libavutil/mathematics.h"
#include "libavutil/samplefmt.h"
#include "libavutil/opt.h"
#include "libavresample/avresample.h"

#define INBUF_SIZE 4096
#define AUDIO_INBUF_SIZE 20480
#define AUDIO_REFILL_THRESH 4096

#define AUDIO  "Bethoven.mp3"
```

```
AVCodecContext *audio_dec_ctx;
int audio_stream_idx;

/* For the RPi name can be "hdmi" or "local" */
void setOutputDevice(OMX_HANDLETYPE handle, const char *name) {
    OMX_ERRORTYPE err;
    OMX_CONFIG_BRCMAUDIODESTINATIONTYPE arDest;

    if (name && strlen(name) < sizeof(arDest.sName)) {
        memset(&arDest, 0, sizeof(OMX_CONFIG_BRCMAUDIODESTINATIONTYPE));
        arDest.nSize = sizeof(OMX_CONFIG_BRCMAUDIODESTINATIONTYPE);
        arDest.nVersion.nVersion = OMX_VERSION;

        strcpy((char *)arDest.sName, name);

        err = OMX_SetParameter(handle, OMX_IndexConfigBrcmAudioDestination, &arDest);
        if (err != OMX_ErrorNone) {
            fprintf(stderr, "Error on setting audio destination\n");
            exit(1);
        }
    }
}

void setPCMMode(OMX_HANDLETYPE handle, int startPortNumber) {
    OMX_AUDIO_PARAM_PCMMODETYPE sPCMMode;
    OMX_ERRORTYPE err;

    memset(&sPCMMode, 0, sizeof(OMX_AUDIO_PARAM_PCMMODETYPE));
    sPCMMode.nSize = sizeof(OMX_AUDIO_PARAM_PCMMODETYPE);
    sPCMMode.nVersion.nVersion = OMX_VERSION;

    sPCMMode.nPortIndex = startPortNumber;

    err = OMX_GetParameter(handle, OMX_IndexParamAudioPcm, &sPCMMode);
    printf("Sampling rate %d, channels %d\n",
           sPCMMode.nSamplingRate,
           sPCMMode.nChannels);

    sPCMMode.nSamplingRate = 44100;
    sPCMMode.nChannels = 2; // assumed for now - should be checked

    err = OMX_SetParameter(handle, OMX_IndexParamAudioPcm, &sPCMMode);
    if(err != OMX_ErrorNone){
        fprintf(stderr, "PCM mode unsupported\n");
        return;
    } else {
        fprintf(stderr, "PCM mode supported\n");
        fprintf(stderr, "PCM sampling rate %d\n", sPCMMode.nSamplingRate);
        fprintf(stderr, "PCM nChannels %d\n", sPCMMode.nChannels);
    }
}
```

```
void printState(OMX_HANDLETYPE handle) {
    // elided
}

char *err2str(int err) {
    return "error elided";
}

void eos_callback(void *userdata, COMPONENT_T *comp, OMX_U32 data) {
    fprintf(stderr, "Got eos event\n");
}

void error_callback(void *userdata, COMPONENT_T *comp, OMX_U32 data) {
    fprintf(stderr, "OMX error %s\n", err2str(data));
}

int get_file_size(char *fname) {
    struct stat st;

    if (stat(fname, &st) == -1) {
        perror("Stat'ing img file");
        return -1;
    }
    return(st.st_size);
}

AVPacket avpkt;

static void set_audio_render_input_format(COMPONENT_T *component) {
    // set input audio format
    printf("Setting audio render format\n");
    OMX_AUDIO_PARAM_PORTFORMATTYPE audioPortFormat;
    //setHeader(&audioPortFormat,  sizeof(OMX_AUDIO_PARAM_PORTFORMATTYPE));
    memset(&audioPortFormat, 0, sizeof(OMX_AUDIO_PARAM_PORTFORMATTYPE));
    audioPortFormat.nSize = sizeof(OMX_AUDIO_PARAM_PORTFORMATTYPE);
    audioPortFormat.nVersion.nVersion = OMX_VERSION;

    audioPortFormat.nPortIndex = 100;

    OMX_GetParameter(ilclient_get_handle(component),
                     OMX_IndexParamAudioPortFormat, &audioPortFormat);

    audioPortFormat.eEncoding = OMX_AUDIO_CodingPCM;
    //audioPortFormat.eEncoding = OMX_AUDIO_CodingMP3;
    OMX_SetParameter(ilclient_get_handle(component),
                     OMX_IndexParamAudioPortFormat, &audioPortFormat);

    setPCMMode(ilclient_get_handle(component), 100);

}
```

```
int num_streams = 0;
int sample_size = 0;

OMX_ERRORTYPE read_into_buffer_and_empty(AVFrame *decoded_frame,
                                         COMPONENT_T *component,
                                         // OMX_BUFFERHEADERTYPE *buff_header,
                                         int total_len) {
    OMX_ERRORTYPE r;
    OMX_BUFFERHEADERTYPE *buff_header = NULL;

    // do this once only
    AVAudioResampleContext *swr = avresample_alloc_context();
    av_opt_set_int(swr, "in_channel_layout",  audio_dec_ctx->channel_layout, 0);
    av_opt_set_int(swr, "out_channel_layout", audio_dec_ctx->channel_layout,  0);
    av_opt_set_int(swr, "in_sample_rate",     audio_dec_ctx->sample_rate, 0);
    av_opt_set_int(swr, "out_sample_rate",    audio_dec_ctx->sample_rate, 0);
    av_opt_set_int(swr, "in_sample_fmt",  audio_dec_ctx->sample_fmt, 0);
    av_opt_set_int(swr, "out_sample_fmt", AV_SAMPLE_FMT_S16,  0);
    avresample_open(swr);

    int required_decoded_size = 0;

    int out_linesize;
    required_decoded_size =
        av_samples_get_buffer_size(&out_linesize, 2,
                                   decoded_frame->nb_samples,
                                   AV_SAMPLE_FMT_S16, 0);
    uint8_t *buffer;
    av_samples_alloc(&buffer, &out_linesize, 2, decoded_frame->nb_samples,
                     AV_SAMPLE_FMT_S16, 0);
    avresample_convert(swr, &buffer,
                       decoded_frame->linesize[0],
                       decoded_frame->nb_samples,
                       // decoded_frame->extended_data,
                       decoded_frame->data,
                       decoded_frame->linesize[0],
                       decoded_frame->nb_samples);

    while (required_decoded_size >= 0) {
        buff_header =
            ilclient_get_input_buffer(component,
                                      100,
                                      1 /* block */);
        if (required_decoded_size > 4096) {
            memcpy(buff_header->pBuffer,
                   buffer, 4096);
            buff_header->nFilledLen = 4096;
            buffer += 4096;
        } else {
             memcpy(buff_header->pBuffer,
                   buffer, required_decoded_size);
            buff_header->nFilledLen = required_decoded_size;
```

```
        }
        required_decoded_size -= 4096;

        r = OMX_EmptyThisBuffer(ilclient_get_handle(component),
                                buff_header);
        if (r != OMX_ErrorNone) {
            fprintf(stderr, "Empty buffer error %s\n",
                    err2str(r));
            return r;
        }
    }
    return r;
}

FILE *favpkt = NULL;

int main(int argc, char** argv) {

    int i;
    char *componentName;
    int err;
    ILCLIENT_T  *handle;
    COMPONENT_T *component;

    AVFormatContext *pFormatCtx = NULL;

    char *audio_file = AUDIO;
    if (argc == 2) {
        audio_file = argv[1];
    }

    OMX_BUFFERHEADERTYPE *buff_header;

    componentName = "audio_render";

    bcm_host_init();

    handle = ilclient_init();
    if (handle == NULL) {
        fprintf(stderr, "IL client init failed\n");
        exit(1);
    }

    if (OMX_Init() != OMX_ErrorNone) {
        ilclient_destroy(handle);
        fprintf(stderr, "OMX init failed\n");
        exit(1);
    }

    ilclient_set_error_callback(handle,
                                error_callback,
                                NULL);
```

```
ilclient_set_eos_callback(handle,
                          eos_callback,
                          NULL);

err = ilclient_create_component(handle,
                                &component,
                                componentName,
                                ILCLIENT_DISABLE_ALL_PORTS
                                |
                                ILCLIENT_ENABLE_INPUT_BUFFERS
                                );
if (err == -1) {
    fprintf(stderr, "Component create failed\n");
    exit(1);
}
printState(ilclient_get_handle(component));

err = ilclient_change_component_state(component,
                                      OMX_StateIdle);
if (err < 0) {
    fprintf(stderr, "Couldn't change state to Idle\n");
    exit(1);
}
printState(ilclient_get_handle(component));

// FFmpeg init
av_register_all();
if(avformat_open_input(&pFormatCtx, audio_file, NULL, NULL)!=0) {
    fprintf(stderr, "Can't get format\n");
    return -1; // Couldn't open file
}
// Retrieve stream information
if(avformat_find_stream_info(pFormatCtx, NULL)<0)
    return -1; // Couldn't find stream information
av_dump_format(pFormatCtx, 0, audio_file, 0);

int ret;
if ((ret = av_find_best_stream(pFormatCtx, AVMEDIA_TYPE_AUDIO, -1, -1, NULL, 0)) >= 0) {
    //AVCodecContext* codec_context;
    AVStream *audio_stream;
    int sample_rate;

    audio_stream_idx = ret;
    fprintf(stderr, "Audio stream index is %d\n", ret);

    audio_stream = pFormatCtx->streams[audio_stream_idx];
    audio_dec_ctx = audio_stream->codec;

    sample_rate = audio_dec_ctx->sample_rate;
    printf("Sample rate is %d\n", sample_rate);
    printf("Sample format is %d\n", audio_dec_ctx->sample_fmt);
    printf("Num channels %d\n", audio_dec_ctx->channels);
```

```
        if (audio_dec_ctx->channel_layout == 0) {
            audio_dec_ctx->channel_layout =
                av_get_default_channel_layout(audio_dec_ctx->channels);
        }

        AVCodec *codec = avcodec_find_decoder(audio_stream->codec->codec_id);
        if (avcodec_open2(audio_dec_ctx, codec, NULL) < 0) {
            fprintf(stderr, "could not open codec\n");
            exit(1);
        }

        if (codec) {
            printf("Codec name %s\n", codec->name);
        }
    }

    av_init_packet(&avpkt);

    // must be before we enable buffers
    set_audio_render_input_format(component);

    setOutputDevice(ilclient_get_handle(component), "local");

    // input port
    ilclient_enable_port_buffers(component, 100,
                                 NULL, NULL, NULL);
    ilclient_enable_port(component, 100);

    err = ilclient_change_component_state(component,
                                          OMX_StateExecuting);
    if (err < 0) {
        fprintf(stderr, "Couldn't change state to Executing\n");
        exit(1);
    }
    printState(ilclient_get_handle(component));

    // now work through the file

    int len;
    uint8_t inbuf[AUDIO_INBUF_SIZE + FF_INPUT_BUFFER_PADDING_SIZE];

    AVFrame *decoded_frame = NULL;

    /* decode until eof */
    avpkt.data = inbuf;
    av_read_frame(pFormatCtx, &avpkt);

    while (avpkt.size > 0) {
        printf("Packet size %d\n", avpkt.size);
        printf("Stream idx is %d\n", avpkt.stream_index);
        printf("Codec type %d\n", pFormatCtx->streams[1]->codec->codec_type);
```

```
    if (avpkt.stream_index != audio_stream_idx) {
        // it's an image, subtitle, etc
        av_read_frame(pFormatCtx, &avpkt);
        continue;
    }

    int got_frame = 0;

    if (favpkt == NULL) {
        favpkt = fopen("tmp.mp3", "wb");
    }
    fwrite(avpkt.data, 1, avpkt.size, favpkt);

    if (!decoded_frame) {
        if (!(decoded_frame = avcodec_alloc_frame())) {
            fprintf(stderr, "out of memory\n");
            exit(1);
        }
    }

    len = avcodec_decode_audio4(audio_dec_ctx,
                                decoded_frame, &got_frame, &avpkt);
    if (len < 0) {
        fprintf(stderr, "Error while decoding\n");
        exit(1);
    }
    if (got_frame) {
        /* if a frame has been decoded, we want to send it to OpenMAX */
        int data_size =
            av_samples_get_buffer_size(NULL, audio_dec_ctx->channels,
                                       decoded_frame->nb_samples,
                                       audio_dec_ctx->sample_fmt, 1);

        // Empty into render_audio input buffers
        read_into_buffer_and_empty(decoded_frame,
                                   component,
                                   data_size
                                   );
    }
    av_read_frame(pFormatCtx, &avpkt);
    continue;
}

printf("Finished decoding MP3\n");
// clean up last empty buffer with EOS
buff_header =
    ilclient_get_input_buffer(component,
                              100,
                              1 /* block */);
buff_header->nFilledLen = 0;
int r;
```

```
    buff_header->nFlags |= OMX_BUFFERFLAG_EOS;
    r = OMX_EmptyThisBuffer(ilclient_get_handle(component),
                            buff_header);
    if (r != OMX_ErrorNone) {
        fprintf(stderr, "Empty buffer error %s\n",
                err2str(r));
    } else {
        printf("EOS sent\n");
    }

    avcodec_close(audio_dec_ctx);
    av_free(audio_dec_ctx);
    av_free(decoded_frame);

    sleep(10);
    exit(0);
}
```

The program takes an MP3 file as a command-line argument, defaulting to a file called Beethoven.mp3.

Conclusion

Audio support is not so good with the RPi. You have to use tools such as FFmpeg/LibAV or Gstreamer.

Resources

- *OpenMAX IL: The Standard for Media Library Portability*: https://www.khronos.org/openmax/il/
- *OpenMAX IL 1.1.2 Specification*: https://www.khronos.org/registry/omxil/specs/OpenMAX_IL_1_1_2_Specification.pdf
- *Raspberry Pi OpenMAX forum*: www.raspberrypi.org/forums/viewforum.php?f=70
- *VMCS-X OpenMAX IL Components*: www.jvcref.com/files/PI/documentation/ilcomponents/
- *Source code for ARM side libraries for interfacing to Raspberry Pi GPU*: https://github.com/raspberrypi/userland

CHAPTER 15

Rendering OpenMAX to OpenGL on the Raspberry Pi

OpenMAX generally forms a self-contained system that can display videos using the GPU. OpenGL ES is a distinct system that can use the GPU to draw graphics. There is a hook to allow OpenMAX videos to be displayed on OpenGL ES surfaces on the Raspberry Pi, which is discussed in this chapter.

EGLImage

EGL is the platform-independant way of drawing windows for OpenVG and OpenGL ES. OpenMAX has a buffer type corresponding to an EGL window, and this allows OpenMAX to render into windows also used by OpenGL ES.

OpenMAX

OpenMAX uses buffers to pass information into and out of components. Some components such as `video_render` communicate directly with the hardware to render their output buffers. Also, OpenMAX has an additional type of `EGLImage` that can be used as a buffer by some components. Images written to an `EGLImage` can then be rendered by OpenGL or OpenVG.

The OpenMAX specification says exactly *nothing* about how to create an `EGLImage`; this data type is not part of OpenMAX. All it talks about is how a component can be given an `EGLImage` to use for a buffer, and that is by using the call `UseEGLImageBuffer`.

```
OMX_UseEGLImage(hComponent,
                ppBufferHdr,
                nPortIndex,
                pAppPrivate,
                eglImage);
```

Even if you have an `EGLImage`, there is no guarantee that a component will be able to use it. It may not be set up to use this type of buffer, and if it can't, then the call to `OMX_UseEGLImageBuffer` will return `OMX_ErrorNotImplemented`.

OpenMAX is silent about which components will be able to handle this buffer type. It is not part of the OpenMAX specification.

J. Newmarch, *Raspberry Pi GPU Audio Video Programming*, DOI 10.1007/978-1-4842-2472-4_15

OpenGL

The type `EGLImage` is not part of the OpenGL ES specification either; it doesn't talk about how to display images from other sources.

eglCreateImageKHR

The function `eglCreateImageKHR` provides the missing link. The `KHR_image_base` specification defines a Khronos standard extension to OpenGL ES.

> *This extension defines a new EGL resource type that is suitable for sharing 2D arrays of image data between client APIs and the EGLImage. Although the intended purpose is sharing 2D image data, the underlying interface makes no assumptions about the format or purpose of the resource being shared, leaving those decisions to the application and associated client APIs.*

The specification defines these functions:

```
EGLImageKHR eglCreateImageKHR(
                        EGLDisplay dpy,
                        EGLContext ctx,
                        EGLenum target,
                        EGLClientBuffer buffer,
                        const EGLint *attrib_list)

EGLBoolean eglDestroyImageKHR(
                        EGLDisplay dpy,
                        EGLImageKHR image)
```

The display and context are standard OpenGL ES. The target is not tightly specified:

> *<target> specifies the type of resource being used as the EGLImage source (examples include two-dimensional textures in OpenGL ES contexts and VGImage objects in OpenVG contexts).*

The buffer is the resource used, cast to type `EGLClientBuffer`.

For OpenGL ES, a value of `EGL_GL_TEXTURE_2D_KHR` is given in `eglext.h` and can be used to specify that the resource is an OpenGL ES texture. The buffer itself is an OpenGL ES texture ID.

Broadcom GPU

There are two marvelous examples in the `hello_pi` source tree: `hello_videocube` and `hello_teapot`. The examples given later are based on these. However, they use OpenGL ES version 1, whereas new applications should use OpenGL ES version 2. An examination of these examples also reveals a number of issues.

Components

The Broadcom video_render component does not support EGLImage. Instead, there is a new component called egl_render. This takes video or image inputs but also has an output buffer, which must be set to the EGLImage.

Threads

OpenGL ES has a processing loop, which typically draws frames as quickly as possible. OpenMAX also has a processing loop as it feeds data through a component pipeline. One of these can be run in the main thread, but the other requires its own thread, easily given by pthreads.

The OpenMAX thread will be filling the EGLImage buffer; the OpenGL ES thread will be using this to draw a texture. Should there be synchronization? The RPi examples have none, and it doesn't seem to be a problem.

Rendering a Video into an OpenGL ES Texture

Rendering using an EGLImage falls naturally into two sections: setting up the OpenGL ES environment and setting up the OpenMAX environment. These two are essentially disjoint, with the connection point being the creation of the EGLImage in the OpenGL ES part and setting that as a buffer in the OpenMAX part. The two parts then each run in their own thread.

The OpenGL ES you use is borrowed from the image-drawing program of Chapter 12. The only substantive change is to the function CreateSimpleTexture2D where you create an EGLImage and hand that to a new POSIX thread that runs the OpenMAX code.

```
GLuint CreateSimpleTexture2D(ESContext *esContext)
{
   // Texture object handle
   GLuint textureId;
   UserData *userData = esContext->userData;

   //userData->width = esContext->width;
   //userData->height = esContext->height;

   // Generate a texture object
   glGenTextures ( 1, &textureId );

   // Bind the texture object
   glBindTexture ( GL_TEXTURE_2D, textureId );

   // Load the texture
   glTexImage2D ( GL_TEXTURE_2D, 0, GL_RGBA,
                  IMAGE_SIZE_WIDTH, IMAGE_SIZE_HEIGHT,
                  0, GL_RGBA, GL_UNSIGNED_BYTE, NULL );

   // Set the filtering mode
   glTexParameteri ( GL_TEXTURE_2D, GL_TEXTURE_MIN_FILTER, GL_NEAREST );
   glTexParameteri ( GL_TEXTURE_2D, GL_TEXTURE_MAG_FILTER, GL_NEAREST );
   //glTexParameteri(GL_TEXTURE_2D, GL_TEXTURE_WRAP_S, GL_CLAMP_TO_EDGE);
   //glTexParameteri(GL_TEXTURE_2D, GL_TEXTURE_WRAP_T, GL_CLAMP_TO_EDGE);
```

```
   /* Create EGL Image */
   eglImage = eglCreateImageKHR(
                esContext->eglDisplay,
                esContext->eglContext,
                EGL_GL_TEXTURE_2D_KHR,
                textureId, // (EGLClientBuffer)esContext->texture, 0);

   if (eglImage == EGL_NO_IMAGE_KHR)
   {
      printf("eglCreateImageKHR failed.\n");
      exit(1);
   }

   // Start rendering
   pthread_create(&thread1, NULL, video_decode_test, eglImage);

   return textureId;
}
```

The full file for the OpenGL ES code is square.c.

```
//
// Book:      OpenGL(R) ES 2.0 Programming Guide
// Authors:   Aaftab Munshi, Dan Ginsburg, Dave Shreiner
// ISBN-10:   0321502795
// ISBN-13:   9780321502797
// Publisher: Addison-Wesley Professional
// URLs:      http://safari.informit.com/9780321563835
//            http://www.opengles-book.com
//

// Simple_Texture2D.c
//
//    This is a simple example that draws a quad with a 2D
//    texture image. The purpose of this example is to demonstrate
//    the basics of 2D texturing
//
#include <stdlib.h>
#include <stdio.h>
#include "esUtil.h"

#include "EGL/eglext.h"

#include "triangle.h"

typedef struct
{
   // Handle to a program object
   GLuint programObject;
```

```
   // Attribute locations
   GLint  positionLoc;
   GLint  texCoordLoc;

   // Sampler location
   GLint samplerLoc;

   // Texture handle
   GLuint textureId;

   GLubyte *image;
    int width, height;
} UserData;

static void* eglImage = 0;
static pthread_t thread1;

#define IMAGE_SIZE_WIDTH 1920
#define IMAGE_SIZE_HEIGHT 1080

GLuint CreateSimpleTexture2D(ESContext *esContext)
{
   // Texture object handle
   GLuint textureId;
   UserData *userData = esContext->userData;

   //userData->width = esContext->width;
   //userData->height = esContext->height;

   // Generate a texture object
   glGenTextures ( 1, &textureId );

   // Bind the texture object
   glBindTexture ( GL_TEXTURE_2D, textureId );

   // Load the texture
   glTexImage2D ( GL_TEXTURE_2D, 0, GL_RGBA,
                  IMAGE_SIZE_WIDTH, IMAGE_SIZE_HEIGHT,
                  0, GL_RGBA, GL_UNSIGNED_BYTE, NULL );

   // Set the filtering mode
   glTexParameteri ( GL_TEXTURE_2D, GL_TEXTURE_MIN_FILTER, GL_NEAREST );
   glTexParameteri ( GL_TEXTURE_2D, GL_TEXTURE_MAG_FILTER, GL_NEAREST );
   //glTexParameteri(GL_TEXTURE_2D, GL_TEXTURE_WRAP_S, GL_CLAMP_TO_EDGE);
   //glTexParameteri(GL_TEXTURE_2D, GL_TEXTURE_WRAP_T, GL_CLAMP_TO_EDGE);

   /* Create EGL Image */
   eglImage = eglCreateImageKHR(
                esContext->eglDisplay,
                esContext->eglContext,
```

```
                EGL_GL_TEXTURE_2D_KHR,
                textureId, // (EGLClientBuffer)esContext->texture, 0);

   if (eglImage == EGL_NO_IMAGE_KHR)
   {
      printf("eglCreateImageKHR failed.\n");
      exit(1);
   }

   // Start rendering
   pthread_create(&thread1, NULL, video_decode_test, eglImage);

   return textureId;
}

///
// Initialize the shader and program object
//
int Init ( ESContext *esContext )
{
    UserData *userData = esContext->userData;
    GLbyte vShaderStr[] =
      "attribute vec4 a_position;   \n"
      "attribute vec2 a_texCoord;   \n"
      "varying vec2 v_texCoord;     \n"
      "void main()                  \n"
      "{                            \n"
      "   gl_Position = a_position; \n"
      "   v_texCoord = a_texCoord;  \n"
      "}                            \n";

    GLbyte fShaderStr[] =
      "precision mediump float;                            \n"
      "varying vec2 v_texCoord;                            \n"
      "uniform sampler2D s_texture;                        \n"
      "void main()                                         \n"
      "{                                                   \n"
      "   gl_FragColor = texture2D( s_texture, v_texCoord );\n"
      "}                                                   \n";

   // Load the shaders and get a linked program object
   userData->programObject = esLoadProgram ( vShaderStr, fShaderStr );

   // Get the attribute locations
   userData->positionLoc = glGetAttribLocation ( userData->programObject, "a_position" );
   userData->texCoordLoc = glGetAttribLocation ( userData->programObject, "a_texCoord" );

   // Get the sampler location
   userData->samplerLoc = glGetUniformLocation ( userData->programObject, "s_texture" );
```

```
   // Load the texture
   userData->textureId = CreateSimpleTexture2D (esContext);

   glClearColor ( 0.5f, 0.5f, 0.5f, 1.0f );

   return GL_TRUE;
}

///
// Draw a triangle using the shader pair created in Init()
//
void Draw ( ESContext *esContext )
{
   UserData *userData = esContext->userData;
   GLfloat vVertices[] = { -1.0f,  1.0f, 0.0f,  // Position 0
                            0.0f,  0.0f,        // TexCoord 0
                           -1.0f, -1.0f, 0.0f,  // Position 1
                            0.0f,  1.0f,        // TexCoord 1
                            1.0f, -1.0f, 0.0f,  // Position 2
                            1.0f,  1.0f,        // TexCoord 2
                            1.0f,  1.0f, 0.0f,  // Position 3
                            1.0f,  0.0f         // TexCoord 3
                         };
   GLushort indices[] = { 0, 1, 2, 0, 2, 3 };

   // Set the viewport
   glViewport ( 0, 0, 1920, 1080); //esContext->width, esContext->height );

   // Clear the color buffer
   glClear ( GL_COLOR_BUFFER_BIT );

   // Use the program object
   glUseProgram ( userData->programObject );

   // Load the vertex position
   glVertexAttribPointer ( userData->positionLoc, 3, GL_FLOAT,
                           GL_FALSE, 5 * sizeof(GLfloat), vVertices );
   // Load the texture coordinate
   glVertexAttribPointer ( userData->texCoordLoc, 2, GL_FLOAT,
                           GL_FALSE, 5 * sizeof(GLfloat), &vVertices[3] );

   glEnableVertexAttribArray ( userData->positionLoc );
   glEnableVertexAttribArray ( userData->texCoordLoc );

   // Bind the texture
   glActiveTexture ( GL_TEXTURE0 );
   glBindTexture ( GL_TEXTURE_2D, userData->textureId );

   // Set the sampler texture unit to 0
   glUniform1i ( userData->samplerLoc, 0 );
```

```
   glDrawElements ( GL_TRIANGLES, 6, GL_UNSIGNED_SHORT, indices );

}

///
// Cleanup
//
void ShutDown ( ESContext *esContext )
{
   UserData *userData = esContext->userData;

   // Delete texture object
   glDeleteTextures ( 1, &userData->textureId );

   // Delete program object
   glDeleteProgram ( userData->programObject );

   free(esContext->userData);
}

int main ( int argc, char *argv[] )
{
   ESContext esContext;
   UserData  userData;

   int width = 1920, height = 1080;
   GLubyte *image;

   esInitContext ( &esContext );
   esContext.userData = &userData;

   esCreateWindow ( &esContext, "Simple Texture 2D", width, height, ES_WINDOW_RGB );

   if ( !Init ( &esContext ) )
      return 0;

   esRegisterDrawFunc ( &esContext, Draw );

   esMainLoop ( &esContext );

   ShutDown ( &esContext );
}
```

The code on the OpenMAX side is a bit more complicated. The rendering component changes from `video_render` to `egl_render`, and this component has an output port, 221. This needs the `EGLImage` attached with the following:

```
OMX_UseEGLImage(ILC_GET_HANDLE(egl_render), &eglBuffer, 221, NULL, eglImage)
```

This should be attached after the `PortSettingsChanged` event has been received by the `video_decode` component and the tunnel has been set up between the two components.

You need to fill the EGLImage buffer. This is done by the usual OMX_FillThisBuffer call. But what happens when it is full? There is no specification for this. What the Broadcom component appears to do is to render its contents onto the EGL surface. The examples do not show any synchronization technique, so does it just happen?

After it has been filled (and rendered?), the buffer should be filled again. When a buffer is filled, it generates a FilledBuffer event that is caught by the IL Client library. The library includes a hook called ilclient_set_fill_buffer_done_callback whereby you can add a callback function that just refills the buffer.

```
void my_fill_buffer_done(void* data, COMPONENT_T* comp)
{
    if (OMX_FillThisBuffer(ilclient_get_handle(egl_render), eglBuffer) != OMX_ErrorNone)
        {
            printf("OMX_FillThisBuffer failed in callback\n");
            exit(1);
        }
}
```

With these additions, the code to render the video to the EGLImage is in the file video.c.

```
/*
  Copyright (c) 2012, Broadcom Europe Ltd
  Copyright (c) 2012, OtherCrashOverride
  All rights reserved.

  Redistribution and use in source and binary forms, with or without
  modification, are permitted provided that the following conditions are met:
  * Redistributions of source code must retain the above copyright
  notice, this list of conditions and the following disclaimer.
  * Redistributions in binary form must reproduce the above copyright
  notice, this list of conditions and the following disclaimer in the
  documentation and/or other materials provided with the distribution.
  * Neither the name of the copyright holder nor the
  names of its contributors may be used to endorse or promote products
  derived from this software without specific prior written permission.

  THIS SOFTWARE IS PROVIDED BY THE COPYRIGHT HOLDERS AND CONTRIBUTORS "AS IS" AND
  ANY EXPRESS OR IMPLIED WARRANTIES, INCLUDING, BUT NOT LIMITED TO, THE IMPLIED
  WARRANTIES OF MERCHANTABILITY AND FITNESS FOR A PARTICULAR PURPOSE ARE
  DISCLAIMED. IN NO EVENT SHALL THE COPYRIGHT HOLDER OR CONTRIBUTORS BE LIABLE FOR ANY
  DIRECT, INDIRECT, INCIDENTAL, SPECIAL, EXEMPLARY, OR CONSEQUENTIAL DAMAGES
  (INCLUDING, BUT NOT LIMITED TO, PROCUREMENT OF SUBSTITUTE GOODS OR SERVICES;
  LOSS OF USE, DATA, OR PROFITS; OR BUSINESS INTERRUPTION) HOWEVER CAUSED AND
  ON ANY THEORY OF LIABILITY, WHETHER IN CONTRACT, STRICT LIABILITY, OR TORT
  (INCLUDING NEGLIGENCE OR OTHERWISE) ARISING IN ANY WAY OUT OF THE USE OF THIS
  SOFTWARE, EVEN IF ADVISED OF THE POSSIBILITY OF SUCH DAMAGE.
*/

// Video decode demo using OpenMAX IL though the ilcient helper library

#include <stdio.h>
#include <stdlib.h>
```

```
#include <string.h>
#include <sys/stat.h>

#include "bcm_host.h"
#include "ilclient.h"

static OMX_BUFFERHEADERTYPE* eglBuffer = NULL;
static COMPONENT_T* egl_render = NULL;

static void* eglImage = 0;

void my_fill_buffer_done(void* data, COMPONENT_T* comp)
{
    if (OMX_FillThisBuffer(ilclient_get_handle(egl_render), eglBuffer) != OMX_ErrorNone)
        {
            printf("OMX_FillThisBuffer failed in callback\n");
            exit(1);
        }
}

int get_file_size(char *fname) {
    struct stat st;

    if (stat(fname, &st) == -1) {
        perror("Stat'ing img file");
        return -1;
    }
    return(st.st_size);
}

#define err2str(x) ""

OMX_ERRORTYPE read_into_buffer_and_empty(FILE *fp,
                                         COMPONENT_T *component,
                                         OMX_BUFFERHEADERTYPE *buff_header,
                                         int *toread) {
    OMX_ERRORTYPE r;

    int buff_size = buff_header->nAllocLen;
    int nread = fread(buff_header->pBuffer, 1, buff_size, fp);

    buff_header->nFilledLen = nread;
    *toread -= nread;

    if (*toread <= 0) {
        printf("Setting EOS on input\n");
        buff_header->nFlags |= OMX_BUFFERFLAG_EOS;
    }
    r = OMX_EmptyThisBuffer(ilclient_get_handle(component),
                            buff_header);
    if (r != OMX_ErrorNone) {
```

```
        fprintf(stderr, "Empty buffer error %s\n",
                err2str(r));
    }
    return r;
}

// Modified function prototype to work with pthreads
void *video_decode_test(void* arg)
{
    const char* filename = "/opt/vc/src/hello_pi/hello_video/test.h264";
    eglImage = arg;

    if (eglImage == 0)
        {
            printf("eglImage is null.\n");
            exit(1);
        }

    OMX_VIDEO_PARAM_PORTFORMATTYPE format;
    OMX_TIME_CONFIG_CLOCKSTATETYPE cstate;

    COMPONENT_T *video_decode = NULL;
    COMPONENT_T *list[3];  // last entry should be null
    TUNNEL_T tunnel[2]; // last entry should be null

    ILCLIENT_T *client;
    FILE *in;
    int status = 0;
    unsigned int data_len = 0;

    memset(list, 0, sizeof(list));
    memset(tunnel, 0, sizeof(tunnel));

    if((in = fopen(filename, "rb")) == NULL)
        return (void *)-2;

    if((client = ilclient_init()) == NULL)
        {
            fclose(in);
            return (void *)-3;
        }

    if(OMX_Init() != OMX_ErrorNone)
        {
            ilclient_destroy(client);
            fclose(in);
            return (void *)-4;
        }

    // callback
    ilclient_set_fill_buffer_done_callback(client, my_fill_buffer_done, 0);
```

```
    // create video_decode
    if(ilclient_create_component(client, &video_decode, "video_decode", ILCLIENT_DISABLE_
ALL_PORTS | ILCLIENT_ENABLE_INPUT_BUFFERS) != 0)
        status = -14;
    list[0] = video_decode;

    // create egl_render
    if(status == 0 && ilclient_create_component(client, &egl_render, "egl_render", ILCLIENT_
DISABLE_ALL_PORTS | ILCLIENT_ENABLE_OUTPUT_BUFFERS) != 0)
        status = -14;
    list[1] = egl_render;

    set_tunnel(tunnel, video_decode, 131, egl_render, 220);
    ilclient_change_component_state(video_decode, OMX_StateIdle);

    memset(&format, 0, sizeof(OMX_VIDEO_PARAM_PORTFORMATTYPE));
    format.nSize = sizeof(OMX_VIDEO_PARAM_PORTFORMATTYPE);
    format.nVersion.nVersion = OMX_VERSION;
    format.nPortIndex = 130;
    format.eCompressionFormat = OMX_VIDEO_CodingAVC;

    if (status != 0) {
        fprintf(stderr, "Error has occurred %d\n", status);
        exit(1);
    }

    if(OMX_SetParameter(ILC_GET_HANDLE(video_decode),
                        OMX_IndexParamVideoPortFormat, &format) != OMX_ErrorNone) {
        fprintf(stderr, "Error setting port format\n");
        exit(1);
    }

    if(ilclient_enable_port_buffers(video_decode, 130, NULL, NULL, NULL) != 0) {
        fprintf(stderr, "Error enablng port buffers\n");
        exit(1);
    }

    OMX_BUFFERHEADERTYPE *buf;
    int port_settings_changed = 0;
    int first_packet = 1;

    ilclient_change_component_state(video_decode, OMX_StateExecuting);

    int toread = get_file_size(filename);
    // Read the first block so that the video_decode can get
    // the dimensions of the video and call port settings
    // changed on the output port to configure it
    while (toread > 0) {
        buf =
            ilclient_get_input_buffer(video_decode,
                                      130,
                                      1 /* block */);
```

```
        if (buf != NULL) {
            read_into_buffer_and_empty(in,
                                       video_decode,
                                       buf,
                                       &toread);

            // If all the file has been read in, then
            // we have to re-read this first block.
            // Broadcom bug?
            if (toread <= 0) {
                printf("Rewinding\n");
                // wind back to start and repeat
                //fp = freopen(IMG, "r", fp);
                rewind(in);
                toread = get_file_size(filename);
            }
        }

        if (toread > 0 && ilclient_remove_event(video_decode,
                                                OMX_EventPortSettingsChanged,
                                                131, 0, 0, 1) == 0) {
            printf("Removed port settings event\n");
            break;
        } else {
            // printf("No portr settting seen yet\n");
        }
        // wait for first input block to set params for output port
        if (toread == 0) {
            int err;
            // wait for first input block to set params for output port
            err = ilclient_wait_for_event(video_decode,
                                          OMX_EventPortSettingsChanged,
                                          131, 0, 0, 1,
                                          ILCLIENT_EVENT_ERROR | ILCLIENT_PARAMETER_CHANGED,
                                          2000);
            if (err < 0) {
                fprintf(stderr, "No port settings change\n");
                //exit(1);
            } else {
                printf("Port settings changed\n");
                break;
            }
        }
    }

    if(ilclient_setup_tunnel(tunnel, 0, 0) != 0)
        {
            status = -7;
            exit(1);
        }
```

```
    // Set egl_render to idle
    ilclient_change_component_state(egl_render, OMX_StateIdle);

    // Enable the output port and tell egl_render to use the texture as a buffer
    //ilclient_enable_port(egl_render, 221); THIS BLOCKS SO CANT BE USED
    if (OMX_SendCommand(ILC_GET_HANDLE(egl_render), OMX_CommandPortEnable, 221, NULL) !=
OMX_ErrorNone)
        {
            printf("OMX_CommandPortEnable failed.\n");
            exit(1);
        }

    if (OMX_UseEGLImage(ILC_GET_HANDLE(egl_render), &eglBuffer, 221, NULL, eglImage) !=
OMX_ErrorNone)
        {
            printf("OMX_UseEGLImage failed.\n");
            exit(1);
        }

    // Set egl_render to executing
    ilclient_change_component_state(egl_render, OMX_StateExecuting);

    // Request egl_render to write data to the texture buffer
    if(OMX_FillThisBuffer(ILC_GET_HANDLE(egl_render), eglBuffer) != OMX_ErrorNone)
        {
            printf("OMX_FillThisBuffer failed.\n");
            exit(1);
        }

   // now work through the file
    while (toread > 0) {
        OMX_ERRORTYPE r;

        // do we have a decode input buffer we can fill and empty?
        buf =
            ilclient_get_input_buffer(video_decode,
                                      130,
                                      1 /* block */);
        if (buf != NULL) {
            read_into_buffer_and_empty(in,
                                       video_decode,
                                       buf,
                                       &toread);
        }
    }

    sleep(2);

    // need to flush the renderer to allow video_decode to disable its input port
    ilclient_flush_tunnels(tunnel, 1);

    ilclient_disable_port_buffers(video_decode, 130, NULL, NULL, NULL);
```

```
    fclose(in);

    ilclient_disable_tunnel(tunnel);
    ilclient_teardown_tunnels(tunnel);

    ilclient_state_transition(list, OMX_StateIdle);
    ilclient_state_transition(list, OMX_StateLoaded);

    ilclient_cleanup_components(list);

    OMX_Deinit();

    ilclient_destroy(client);
    return (void *)status;
}
```

Conclusion

This chapter showed you how to render a video from OpenMAX onto an EGL surface using OpenGL ES. By changing the OpenGL ES program, complex effects such as spinning the image or rendering to a "teapot" texture can be achieved.

Resources

- *OpenMAX IL: The Standard for Media Library Portability*: `https://www.khronos.org/openmax/il/`
- *OpenMAX IL 1.1.2 Specification*: `https://www.khronos.org/registry/omxil/specs/OpenMAX_IL_1_1_2_Specification.pdf`
- *Raspberry Pi OpenMAX forum*: `www.raspberrypi.org/forums/viewforum.php?f=70`
- *VMCS-X OpenMAX IL Components* : `www.jvcref.com/files/PI/documentation/ilcomponents/`
- *OpenGL ES Shading Language*: `https://www.khronos.org/files/opengles_shading_language.pdf`
- *OpenGL ES Specification*: `https://www.khronos.org/registry/gles/specs/2.0/es_full_spec_2.0.25.pdf`
- *KHR_image_base defines KHR extensions for OpenGL ES*: `https://www.khronos.org/registry/egl/extensions/KHR/EGL_KHR_image_base.txt`
- *Changing the eglImage used by egl_render*: `www.raspberrypi.org/forums/viewtopic.php?t=41958&p=414954`
- *Demo H264 video clips*: `www.h264info.com/clips.html`

CHAPTER 16

Playing Multimedia Files on the Raspberry Pi

This chapter covers how to play multiplexed files on the Raspberry Pi such as MP4 files, which consist of both audio and video streams. OpenMAX does not support demultiplexing such files, so you must use a software demuxer such as FFmpeg or LibAV. The libraries available on the RPi do not support the decoding of encoded audio streams, so you need to use a software decoder. Finally, you need to manage the synchronization of video and audio streams.

This chapter would not have been possible without seeing what was done in `omxplayer`. *Many thanks to its authors!*

Desperate Debugging

Sometimes nothing works. Debugging using `gdb` or similar gets you nowhere. Broadcom has a command called `vcdbg` that can report on what the GPU is doing. It is run with the following:

```
vcgencmd cache_flush
vcgencmd set_logging level=0x40
sudo vcdbg log msg
```

In code, you can call the following:

```
vcos_log_set_level(VCOS_LOG_CATEGORY, VCOS_LOG_TRACE);
```

You can find more information at `http://elinux.org/RPI_vcgencmd_usage` and `https://github.com/nezticle/RaspberryPi-BuildRoot/wiki/VideoCore-Tools`.

Multimedia Files

Multimedia files containing both audio and video are *container* files. Typically they will have header information and then streams of audio and video data, usually interleaved in some way, where each stream has its own appropriate format.

There are many such formats (see, for example, Wikipedia's "Digital container format" article, at `http://en.wikipedia.org/wiki/Digital_container_format#Multimedia_container_formats`). You will just look at the MP4 format here. An MP4 file will typically contain both audio and video streams but may also contain subtitles and still images.

J. Newmarch, *Raspberry Pi GPU Audio Video Programming*, DOI 10.1007/978-1-4842-2472-4_16

Although an MP4 file can contain audio and video streams of many different formats, the most common are Advanced Audio Coding (AAC) for audio and H.264 for video.

The program `mediainfo` can show a huge amount of information about a file. For example, for the sample file Big Buck Bunny, it shows the following:

```
General
Complete name                            : big_buck_bunny.mp4
Format                                   : MPEG-4
Format profile                           : Base Media / Version 2
Codec ID                                 : mp42
File size                                : 5.26 MiB
Duration                                 : 1mn 0s
Overall bit rate                         : 734 Kbps
Encoded date                             : UTC 2010-02-09 01:55:39
Tagged date                              : UTC 2010-02-09 01:55:40

Video
ID                                       : 2
Format                                   : AVC
Format/Info                              : Advanced Video Codec
Format profile                           : Baseline@L3.0
Format settings, CABAC                   : No
Format settings, ReFrames                : 2 frames
Format settings, GOP                     : M=1, N=64
Codec ID                                 : avc1
Codec ID/Info                            : Advanced Video Coding
Duration                                 : 1mn 0s
Duration_LastFrame                       : 95ms
Bit rate                                 : 613 Kbps
Width                                    : 640 pixels
Height                                   : 360 pixels
Display aspect ratio                     : 16:9
Frame rate mode                          : Constant
Frame rate                               : 24.000 fps
Color space                              : YUV
Chroma subsampling                       : 4:2:0
Bit depth                                : 8 bits
Scan type                                : Progressive
Bits/(Pixel*Frame)                       : 0.111
Stream size                              : 4.39 MiB (83%)
Language                                 : English
Encoded date                             : UTC 2010-02-09 01:55:39
Tagged date                              : UTC 2010-02-09 01:55:40
Color primaries                          : BT.601 NTSC
Transfer characteristics                 : BT.709
Matrix coefficients                      : BT.601

Audio
ID                                       : 1
Format                                   : AAC
Format/Info                              : Advanced Audio Codec
```

```
Format profile                          : LC
Codec ID                                : 40
Duration                                : 1mn 0s
Source duration                         : 1mn 0s
Bit rate mode                           : Constant
Bit rate                                : 64.0 Kbps
Channel(s)                              : 2 channels
Channel positions                       : Front: L R
Sampling rate                           : 22.05 KHz
Compression mode                        : Lossy
Stream size                             : 478 KiB (9%)
Source stream size                      : 478 KiB (9%)
Language                                : English
Encoded date                            : UTC 2010-02-09 01:55:39
Tagged date                             : UTC 2010-02-09 01:55:40
```

The main points for now are the video format Advanced Video Codec (H.264) and the audio format AAC.

Demuxing Multimedia Files

To play a multimedia file, the file must be split into two streams, or *demultiplexed*. Typically the streams will be interleaved as a sequence of packets of each type. Because of the large variety of possible multimedia file formats, *none* of them are supported by the Broadcom GPU, although there have been repeated requests for at least some to be supported (for example, see `www.raspberrypi.org/forums/viewtopic.php?t=32601&p=280735`).

To demux a file, a software package such as FFmpeg or LibAV must be used. FFmpeg and LibAV forked some years ago, through a vitriolic and ongoing brawl. The Debian package for the RPi uses LibAV, while the superb multimedia `omxplayer` uses FFmpeg. In addition, there is a continual evolution of code, formats, codecs, and so on, so it can be traumatic figuring out what code to build. Added to the difficulties of OpenMAX itself, this is not an easy area to navigate.

Decoding the Audio Stream Using LibAV

In this section, you'll just look at managing an audio stream using LibAV/FFmpeg. This is a continually changing area, and the working version as of March 2015 uses LibAV codec version 54.35. Adding in OpenMAX will be done in the following section.

LibAV forked from FFmpeg acrimoniously (use Google to dig up the details). Debian and most Debian-derived systems use LibAV. `omxplayer` doesn't: it uses FFmpeg. The two forks align much of the time but not always. The headers are different, sometimes the APIs differ, and of course there may be semantic differences. I'm going to use LibAV since that is what is installed in the Raspbian package. I take no sides, other than to comment that I know the packages must keep evolving with developments in AV codecs, containers, and so on, but to add the complexity of two parallel packages is an added burden on the developer.

Whatever you do, using either package involves setting up a *format context*. This is initialized with a call to `avformat_open`, which takes a file name as a parameter. A (container) file will contain one or more streams, and you can find information about all of this with `avformat_find_stream_info`. This can be printed out with `av_dump_format`.

You can find the audio stream through the call `av_find_best_stream`, which returns an index into the array of streams maintained by the context.

If you want to demultiplex only the streams, this is sufficient for setup. This is the case for the video AVC streams as the Broadcom GPU can decode and render them. For audio, the situation is not so good: the Broadcom GPU will handle only WAV or raw PCM data. So, you have to decode the AAC stream in software.

You can find the decoder and its context from additional calls to avcodec_find_decoder and avcodec_alloc_context3.

Some codecs pull out extra information from the audio stream that the codec context didn't see. However, it turns out that the context may need those extra pieces of information. This is a bit yucky: the extra data has to be installed from the codec into its context (shouldn't this be done with avcodec_alloc_context?). After this has been done, the codec can be opened by avcodec_open2.

This initialization code is as follows:

```
int setup_demuxer(const char *filename) {
    // Register all formats and codecs
    av_register_all();
    if(avformat_open_input(&pFormatCtx, filename, NULL, NULL)!=0) {
        fprintf(stderr, "Can't get format\n");
        return -1; // Couldn't open file
    }
    // Retrieve stream information
    if (avformat_find_stream_info(pFormatCtx, NULL) < 0) {
        return -1; // Couldn't find stream information
    }
    printf("Format:\n");
    av_dump_format(pFormatCtx, 0, filename, 0);

    int ret;
    ret = av_find_best_stream(pFormatCtx, AVMEDIA_TYPE_AUDIO, -1, -1, NULL, 0);
    if (ret >= 0) {
        audio_stream_idx = ret;

        audio_stream = pFormatCtx->streams[audio_stream_idx];
        audio_dec_ctx = audio_stream->codec;

        AVCodec *codec = avcodec_find_decoder(audio_stream->codec->codec_id);
        codec_context = avcodec_alloc_context3(codec);

        // copy across info from codec about extradata
        codec_context->extradata =  audio_stream->codec->extradata;
        codec_context->extradata_size =  audio_stream->codec->extradata_size;

        if (codec) {
            printf("Codec name %s\n", codec->name);
        }

        if (!avcodec_open2(codec_context, codec, NULL) < 0) {
            fprintf(stderr, "Could not find open the needed codec");
            exit(1);
        }
    }
    return 0;
}
```

The main AV processing loop reads packets using av_read_frame. If they are audio packets, then they need to be decoded into frames in a slightly convoluted way by using avcodec_decode_audio4. The returned value is *not* the size of the decoded frame but is the size of the packet *consumed* by the decoding. Strictly, you should loop until the amount consumed equals the size of the read packet, but you can ignore that here, assuming that the total packet is consumed.

At one time it appeared that LibAV/FFmpeg decoded AAC frames into raw PCM data, with each channel interleaved by each sample, AV_SAMPLE_FMT_S16. Now they decode into a format called AV_SAMPLE_FMT_FLTP, which is a fixed-point planar format (multiple samples in each channel are together). These need to be converted, using the package AVResample, which uses a different API in LibAV and FFmpeg. Using LibAV, you create a buffer for a stereo big enough to hold enough 16-bit samples with av_samples_alloc and then convert the samples with avresample_convert.

For this exercise, you then save the 16-bit stereo samples to a file. The code is as follows:

```
FILE *out = fopen("tmp.s16", "w");

// for converting from AV_SAMPLE_FMT_FLTP to AV_SAMPLE_FMT_S16
// Set up SWR context once you've got codec information
AVAudioResampleContext *swr = avresample_alloc_context();
av_opt_set_int(swr, "in_channel_layout",  audio_dec_ctx->channel_layout, 0);
av_opt_set_int(swr, "out_channel_layout", audio_dec_ctx->channel_layout,  0);
av_opt_set_int(swr, "in_sample_rate",     audio_dec_ctx->sample_rate, 0);
av_opt_set_int(swr, "out_sample_rate",    audio_dec_ctx->sample_rate, 0);
av_opt_set_int(swr, "in_sample_fmt",  AV_SAMPLE_FMT_FLTP, 0);
av_opt_set_int(swr, "out_sample_fmt", AV_SAMPLE_FMT_S16,  0);
avresample_open(swr);

int buffer_size;
int num_ret;

/* read frames from the file */
AVFrame *frame = avcodec_alloc_frame(); // av_frame_alloc
while (av_read_frame(pFormatCtx, &pkt) >= 0) {
    // printf("Read pkt %d\n", pkt.size);

    AVPacket orig_pkt = pkt;
    if (pkt.stream_index == audio_stream_idx) {
        // printf("  read audio pkt %d\n", pkt.size);
        // fwrite(pkt.data, 1, pkt.size, out);

        AVPacket avpkt;
        int got_frame;
        av_init_packet(&avpkt);
        avpkt.data = pkt.data;
        avpkt.size = pkt.size;

        uint8_t *buffer;
        if (((err = avcodec_decode_audio4(codec_context,
                                          frame,
                                          &got_frame,
                                          &avpkt)) < 0)  || !got_frame) {
            fprintf(stderr, "Error decoding %d\n", err);
```

```
            continue;
        }

        int out_linesize;
        /*
        av_samples_get_buffer_size(&out_linesize, 2, frame->nb_samples,
                                   AV_SAMPLE_FMT_S16, 0);
        */
        av_samples_alloc(&buffer, &out_linesize, 2, frame->nb_samples,
                         AV_SAMPLE_FMT_S16, 0);

        avresample_convert(swr, &buffer, frame->linesize[0],
                           frame->nb_samples, frame->extended_data,
                           frame->linesize[0], frame->nb_samples);
        /*
        printf("Pkt size %d, decoded to %d line size %d\n",
               pkt.size, err, out_linesize);
        printf("Samples: pkt size %d nb_samples %d\n",
              pkt.size, frame->nb_samples);
        printf("Buffer is (decoded size %d)  %d %d %d %d %d\n",
               required_decoded_size,
               buffer[0], buffer[1], buffer[2], buffer[3], err);
        */
        fwrite(buffer, 1, frame->nb_samples*4, out);
    }
    av_free_packet(Vig_pkt);
}
```

The complete program is ffmpeg_demux_decode_audio.c.

The resultant raw PCM file can be played with the following:

```
aplay -c 2 -r 22050 -f  S16_LE tmp.s16
```

Or it can be converted to WAV format, for example, and then played with any player such as mplayer.

```
sox -c 2 -r 22050 tmp.s16 tmp.wav
```

Note that the audio file contains raw PCM data only, and this does not contain any information such as the sampling rate of 22050 Hz. You have to supply this yourself, using information from av_dump_format or mediainfo. Playing using aplay requires this explicitly, while conversion to WAV format by sox adds in WAV header information to the PCM data.

Rendering an Audio Stream

Now that you have successfully demuxed and decoded the audio stream into a stream of PCM data, sending it to the Broadcom audio render device is straightforward using the techniques covered in Chapter 14.

Essentially, you set up the OMX audio renderer using information from the LibAV decoder and then feed the decoded packets into the audio renderer. It's just a looong program by this stage!

The full program is il_ffmpeg_demux_render_audio.c.

```
#include <stdio.h>
#include <stdlib.h>
#include <sys/stat.h>

#include <OMX_Core.h>
#include <OMX_Component.h>

#include <bcm_host.h>
#include <vcos_logging.h>

#define VCOS_LOG_CATEGORY (&il_ffmpeg_log_category)
static VCOS_LOG_CAT_T il_ffmpeg_log_category;

#include <ilclient.h>

#include "libavcodec/avcodec.h"
#include <libavformat/avformat.h>
#include "libavutil/mathematics.h"
#include "libavutil/samplefmt.h"
#include "libavresample/avresample.h"

#define INBUF_SIZE 4096
#define AUDIO_INBUF_SIZE 20480
#define AUDIO_REFILL_THRESH 4096

char *IMG = "taichi.mp4";

static AVCodecContext *audio_dec_ctx = NULL;
static AVStream *audio_stream = NULL;
static AVPacket pkt;
AVFormatContext *pFormatCtx = NULL;
AVAudioResampleContext *swr;

int sample_rate;
int channels;

static int audio_stream_idx = -1;

AVCodec *codec;

void setPCMMode(OMX_HANDLETYPE handle, int startPortNumber) {
    OMX_AUDIO_PARAM_PCMMODETYPE sPCMMode;
    OMX_ERRORTYPE err;

    memset(&sPCMMode, 0, sizeof(OMX_AUDIO_PARAM_PCMMODETYPE));
    sPCMMode.nSize = sizeof(OMX_AUDIO_PARAM_PCMMODETYPE);
    sPCMMode.nVersion.nVersion = OMX_VERSION;

    sPCMMode.nPortIndex = startPortNumber;

    err = OMX_GetParameter(handle, OMX_IndexParamAudioPcm, &sPCMMode);
```

```
    printf("Sampling rate %d, channels %d\n",
           sPCMMode.nSamplingRate,
           sPCMMode.nChannels);

    //sPCMMode.nSamplingRate = 44100; // for taichi
    //sPCMMode.nSamplingRate = 22050; // for big buck bunny
    sPCMMode.nSamplingRate = sample_rate; // for anything
    sPCMMode.nChannels = channels;

    err = OMX_SetParameter(handle, OMX_IndexParamAudioPcm, &sPCMMode);
    if(err != OMX_ErrorNone){
        fprintf(stderr, "PCM mode unsupported\n");
        return;
    } else {
        fprintf(stderr, "PCM mode supported\n");
        fprintf(stderr, "PCM sampling rate %d\n", sPCMMode.nSamplingRate);
        fprintf(stderr, "PCM nChannels %d\n", sPCMMode.nChannels);
    }
}

static void set_audio_render_input_format(COMPONENT_T *component) {
    // set input audio format
    printf("Setting audio render format\n");
    OMX_AUDIO_PARAM_PORTFORMATTYPE audioPortFormat;
    //setHeader(&audioPortFormat,  sizeof(OMX_AUDIO_PARAM_PORTFORMATTYPE));
    memset(&audioPortFormat, 0, sizeof(OMX_AUDIO_PARAM_PORTFORMATTYPE));
    audioPortFormat.nSize = sizeof(OMX_AUDIO_PARAM_PORTFORMATTYPE);
    audioPortFormat.nVersion.nVersion = OMX_VERSION;

    audioPortFormat.nPortIndex = 100;

    OMX_GetParameter(ilclient_get_handle(component),
                     OMX_IndexParamAudioPortFormat, &audioPortFormat);

    audioPortFormat.eEncoding = OMX_AUDIO_CodingPCM;
    //audioPortFormat.eEncoding = OMX_AUDIO_CodingMP3;
    OMX_SetParameter(ilclient_get_handle(component),
                     OMX_IndexParamAudioPortFormat, &audioPortFormat);

    setPCMMode(ilclient_get_handle(component), 100);

}

void printState(OMX_HANDLETYPE handle) {
    // elided
}

char *err2str(int err) {
    return "error elided";
}
```

```
void eos_callback(void *userdata, COMPONENT_T *comp, OMX_U32 data) {
    printf("Got eos event\n");
}

void error_callback(void *userdata, COMPONENT_T *comp, OMX_U32 data) {
    printf("OMX error %s\n", err2str(data));
}

void port_settings_callback(void *userdata, COMPONENT_T *comp, OMX_U32 data) {
    printf("Got port Settings event\n");
    // exit(0);
}

void empty_buffer_done_callback(void *userdata, COMPONENT_T *comp) {
    // printf("Got empty buffer done\n");
}

int get_file_size(char *fname) {
    struct stat st;

    if (stat(fname, &st) == -1) {
        perror("Stat'ing img file");
        return -1;
    }
    return(st.st_size);
}

OMX_ERRORTYPE read_audio_into_buffer_and_empty(AVFrame *decoded_frame,
                                               COMPONENT_T *component,
                                               // OMX_BUFFERHEADERTYPE *buff_header,
                                               int total_len) {

    OMX_ERRORTYPE r;
    OMX_BUFFERHEADERTYPE *buff_header = NULL;

#if AUDIO_DECODE
    int port_index = 120;
#else
    int port_index = 100;
#endif

    int required_decoded_size = 0;

    int out_linesize;
    required_decoded_size =
        av_samples_get_buffer_size(&out_linesize, 2,
                                   decoded_frame->nb_samples,
                                   AV_SAMPLE_FMT_S16, 0);
    uint8_t *buffer, *start_buffer;
    av_samples_alloc(&buffer, &out_linesize, 2, decoded_frame->nb_samples,
                     AV_SAMPLE_FMT_S16, 0);
    start_buffer = buffer;
```

```
avresample_convert(swr, &buffer,
                   decoded_frame->linesize[0],
                   decoded_frame->nb_samples,
                   // decoded_frame->extended_data,
                   decoded_frame->data,
                   decoded_frame->linesize[0],
                   decoded_frame->nb_samples);
// printf("Decoded audio size %d\n", required_decoded_size);

/* printf("Audio timestamp %lld\n", (decoded_frame->pkt_pts * USECS_IN_SEC *
   atime_base_num /
   atime_base_den));
*/
//print_clock_time("OMX Audio empty timestamp");

while (required_decoded_size > 0) {
    buff_header =
        ilclient_get_input_buffer(component,
                                  port_index,
                                  1 /* block */);

    /*
    buff_header->nTimeStamp = ToOMXTime((uint64_t)
                                        (decoded_frame->pkt_pts * USECS_IN_SEC *
                                         atime_base_num /
                                         atime_base_den));
    */
    int len = buff_header->nAllocLen;

    if (required_decoded_size > len) {
        // fprintf(stderr, "Buffer not big enough %d, looping\n", len);
        memcpy(buff_header->pBuffer,
               buffer, len);
        buff_header->nFilledLen = len;
        buffer += len;
    } else {
        memcpy(buff_header->pBuffer,
               buffer, required_decoded_size);
        buff_header->nFilledLen = required_decoded_size;
    }
    // gettimeofday(&tv, NULL);
    // printf("Time audio empty start %ld\n",
    // tv.tv_sec * USECS_IN_SEC + tv.tv_usec - starttime);
    r = OMX_EmptyThisBuffer(ilclient_get_handle(component),
                            buff_header);

    // gettimeofday(&tv, NULL);
    // printf("Time audio empty stop  %ld\n",
    //  tv.tv_sec * USECS_IN_SEC + tv.tv_usec - starttime);

    //exit_on_omx_error(r, "Empty buffer error %s\n");
```

```
        required_decoded_size -= len;
    }
    //av_freep(&start_buffer[0]);
    av_free(&start_buffer[0]);
    return r;
}

AVCodecContext* codec_context;

int setup_demuxer(const char *filename) {
    // Register all formats and codecs
    av_register_all();
    if(avformat_open_input(&pFormatCtx, filename, NULL, NULL)!=0) {
        fprintf(stderr, "Can't get format\n");
        return -1; // Couldn't open file
    }
    // Retrieve stream information
    if (avformat_find_stream_info(pFormatCtx, NULL) < 0) {
        return -1; // Couldn't find stream information
    }
    printf("Format:\n");
    av_dump_format(pFormatCtx, 0, filename, 0);

    int ret;
    ret = av_find_best_stream(pFormatCtx, AVMEDIA_TYPE_AUDIO, -1, -1, NULL, 0);
    if (ret >= 0) {
        audio_stream_idx = ret;
        // audio_stream_idx = 2; // JN

        audio_stream = pFormatCtx->streams[audio_stream_idx];
        audio_dec_ctx = audio_stream->codec;

        sample_rate = audio_dec_ctx->sample_rate;
        channels =  audio_dec_ctx->channels;
        printf("Sample rate is %d channels %d\n", sample_rate, channels);

        AVCodec *codec = avcodec_find_decoder(audio_stream->codec->codec_id);
        codec_context = avcodec_alloc_context3(codec);

        // copy across info from codec about extradata
        codec_context->extradata =  audio_stream->codec->extradata;
        codec_context->extradata_size =  audio_stream->codec->extradata_size;

        if (codec) {
            printf("Codec name %s\n", codec->name);
        }

        if (!avcodec_open2(codec_context, codec, NULL) < 0) {
            fprintf(stderr, "Could not find open the needed codec");
            exit(1);
        }
```

```
    }
    return 0;
}

/* For the RPi name can be "hdmi" or "local" */
void setOutputDevice(OMX_HANDLETYPE handle, const char *name) {
    OMX_ERRORTYPE err;
    OMX_CONFIG_BRCMAUDIODESTINATIONTYPE arDest;

    if (name && strlen(name) < sizeof(arDest.sName)) {
        memset(&arDest, 0, sizeof(OMX_CONFIG_BRCMAUDIODESTINATIONTYPE));
        arDest.nSize = sizeof(OMX_CONFIG_BRCMAUDIODESTINATIONTYPE);
        arDest.nVersion.nVersion = OMX_VERSION;

        strcpy((char *)arDest.sName, name);

        err = OMX_SetParameter(handle, OMX_IndexConfigBrcmAudioDestination, &arDest);
        if (err != OMX_ErrorNone) {
            fprintf(stderr, "Error on setting audio destination\n");
            exit(1);
        }
    }
}

void setup_audio_renderComponent(ILCLIENT_T  *handle,
                                 char *componentName,
                                 COMPONENT_T **component) {
    int err;
    err = ilclient_create_component(handle,
                                    component,
                                    componentName,
                                    ILCLIENT_DISABLE_ALL_PORTS
                                    |
                                    ILCLIENT_ENABLE_INPUT_BUFFERS
                                    );
    if (err == -1) {
        fprintf(stderr, "Component create failed\n");
        exit(1);
    }
    printState(ilclient_get_handle(*component));

    err = ilclient_change_component_state(*component,
                                          OMX_StateIdle);
    if (err < 0) {
        fprintf(stderr, "Couldn't change state to Idle\n");
        exit(1);
    }
    printState(ilclient_get_handle(*component));

    // must be before we enable buffers
    set_audio_render_input_format(*component);
```

```
    setOutputDevice(ilclient_get_handle(*component), "local");
    //setOutputDevice(ilclient_get_handle(*component), "hdmi");

    // input port
    ilclient_enable_port_buffers(*component, 100,
                                 NULL, NULL, NULL);
    ilclient_enable_port(*component, 100);

    err = ilclient_change_component_state(*component,
                                          OMX_StateExecuting);
    if (err < 0) {
        fprintf(stderr, "Couldn't change state to Executing\n");
        exit(1);
    }
    printState(ilclient_get_handle(*component));
}

int main(int argc, char** argv) {

    char *renderComponentName;

    int err;
    ILCLIENT_T  *handle;
    COMPONENT_T *renderComponent;

    if (argc > 1) {
        IMG = argv[1];
    }

    OMX_BUFFERHEADERTYPE *buff_header;

    setup_demuxer(IMG);

    renderComponentName = "audio_render";

    bcm_host_init();

    handle = ilclient_init();
    // vcos_log_set_level(VCOS_LOG_CATEGORY, VCOS_LOG_TRACE);
    if (handle == NULL) {
        fprintf(stderr, "IL client init failed\n");
        exit(1);
    }

    if (OMX_Init() != OMX_ErrorNone) {
        ilclient_destroy(handle);
        fprintf(stderr, "OMX init failed\n");
        exit(1);
    }

    ilclient_set_error_callback(handle,
```

```
                                error_callback,
                                NULL);
    ilclient_set_eos_callback(handle,
                              eos_callback,
                              NULL);
    ilclient_set_port_settings_callback(handle,
                                        port_settings_callback,
                                        NULL);
    ilclient_set_empty_buffer_done_callback(handle,
                                            empty_buffer_done_callback,
                                            NULL);

    setup_audio_renderComponent(handle, renderComponentName, &renderComponent);

    FILE *out = fopen("tmp.se16", "wb");

    // for converting from AV_SAMPLE_FMT_FLTP to AV_SAMPLE_FMT_S16
    // Set up SWR context once you've got codec information
    //AVAudioResampleContext *swr = avresample_alloc_context();
    swr = avresample_alloc_context();

    av_opt_set_int(swr, "in_channel_layout",
                   av_get_default_channel_layout(audio_dec_ctx->channels) , 0);
    av_opt_set_int(swr, "out_channel_layout", AV_CH_LAYOUT_STEREO,  0);
    av_opt_set_int(swr, "in_sample_rate",     audio_dec_ctx->sample_rate, 0);
    av_opt_set_int(swr, "out_sample_rate",    audio_dec_ctx->sample_rate, 0);
    av_opt_set_int(swr, "in_sample_fmt", audio_dec_ctx->sample_fmt, 0);
    av_opt_set_int(swr, "out_sample_fmt", AV_SAMPLE_FMT_S16,  0);
    avresample_open(swr);

    fprintf(stderr, "Num channels for resmapling %d\n",
           av_get_channel_layout_nb_channels(AV_CH_LAYOUT_STEREO));

    int buffer_size;
    int num_ret;

    /* read frames from the file */
    AVFrame *frame = avcodec_alloc_frame(); // av_frame_alloc
    while (av_read_frame(pFormatCtx, &pkt) >= 0) {
        // printf("Read pkt %d\n", pkt.size);

        AVPacket orig_pkt = pkt;
        if (pkt.stream_index == audio_stream_idx) {
            // printf("  read audio pkt %d\n", pkt.size);
            //fwrite(pkt.data, 1, pkt.size, out);

            AVPacket avpkt;
            int got_frame;
            av_init_packet(&avpkt);
            avpkt.data = pkt.data;
            avpkt.size = pkt.size;
```

```
        uint8_t *buffer;
        if (((err = avcodec_decode_audio4(codec_context,
                                          frame,
                                          &got_frame,
                                          &avpkt)) < 0)  || !got_frame) {
            fprintf(stderr, "Error decoding %d\n", err);
            continue;
        }
        int required_decoded_size = 0;

        if (audio_dec_ctx->sample_fmt == AV_SAMPLE_FMT_S16) {
            buffer = frame->data;
        } else {
            read_audio_into_buffer_and_empty(frame,
                                             renderComponent,
                                             required_decoded_size
                                             );
        }
        fwrite(buffer, 1, frame->nb_samples*4, out);
    }

    av_free_packet(&orig_pkt);
  }

  exit(0);
}
```

Rendering the Video Stream (at Full Speed)

An MP4 file will usually contain an H.264 video stream, so I will restrict discussion to that type. This can be decoded by the GPU using the Broadcom component `OMX.broadcom.video_decode`. There is no need to decode this in software. So, a first cut at rendering the video stream is to first demux it using FFmpeg/LibAV, pass it into the Broadcom video decoder, and then pass it into the Broadcom video renderer, `OMX.broadcom.video_render`.

It works, but with no controls, it will play the video as fast as it can render it, just as though it was playing every frame on fast-forward. Nevertheless, it is worth looking at as there are some wrinkles arising from the decoding to be addressed.

Demuxing an MP4 file takes place, as described earlier. Instead of selecting on audio frames, you now select on video frames. But on feeding it to an OMX component, you are now back in the situation of needing to wait for a `PortSettingsChanged` event to occur before setting up the tunnel between the decode and render components. The code follows the structure of earlier examples, looping through the file until you get the event, setting up the tunnel, and then looping through the rest of the file. It's nothing new.

Setting up the demuxer, however, requires code specific to video rather than audio. Once you have found the "best video stream," you can extract video data from it such as image width and height. But also, you may get *extradata*, which is more information about the video stream. (This may include things such as Huffman tables, but you don't need to look into the details, fortunately.)

This is similar to passing extra data from the audio codec to the audio context, but it isn't so simple. First you have to capture it and then send it into the video decoder through an OMX input buffer.

The code to set up the demuxer is as follows:

```
int setup_demuxer(const char *filename) {
    // Register all formats and codecs
    av_register_all();
    if(avformat_open_input(&pFormatCtx, filename, NULL, NULL)!=0) {
        fprintf(stderr, "Can't get format\n");
        return -1; // Couldn't open file
    }
    // Retrieve stream information
    if (avformat_find_stream_info(pFormatCtx, NULL) < 0) {
        return -1; // Couldn't find stream information
    }
    printf("Format:\n");
    av_dump_format(pFormatCtx, 0, filename, 0);

    int ret;
    ret = av_find_best_stream(pFormatCtx, AVMEDIA_TYPE_VIDEO, -1, -1, NULL, 0);
    if (ret >= 0) {
        video_stream_idx = ret;

        video_stream = pFormatCtx->streams[video_stream_idx];
        video_dec_ctx = video_stream->codec;

        img_width         = video_stream->codec->width;
        img_height        = video_stream->codec->height;
        extradata         = video_stream->codec->extradata;
        extradatasize     = video_stream->codec->extradata_size;

        AVCodec *codec = avcodec_find_decoder(video_stream->codec->codec_id);

        if (codec) {
            printf("Codec name %s\n", codec->name);
        }
    }
    return 0;
}
```

The extra data needs to be fed into the video decoder, just like the extra data for audio needed to be fed into the audio decoder. This time, the video decoder is the OpenMAX component, and sending the data to it means getting a buffer, copying the data into it, and emptying the buffer, before sending any data frames.

```
int SendDecoderConfig(COMPONENT_T *component)
{
  OMX_ERRORTYPE omx_err   = OMX_ErrorNone;

  /* send decoder config */
  if(extradatasize > 0 && extradata != NULL)
  {
      OMX_BUFFERHEADERTYPE *omx_buffer = ilclient_get_input_buffer(component,
                                              130,
                                              1 /* block */);
```

```
    if(omx_buffer == NULL)
    {
        fprintf(stderr, "%s - buffer error 0x%08x", __func__, omx_err);
      return 0;
    }

    omx_buffer->nOffset = 0;
    omx_buffer->nFilledLen = extradatasize;
    if(omx_buffer->nFilledLen > omx_buffer->nAllocLen)
    {
        fprintf(stderr, "%s - omx_buffer->nFilledLen > omx_buffer-nAllocLen",  __func__);
      return 0;
    }

    memset((unsigned char *)omx_buffer->pBuffer, 0x0, omx_buffer->nAllocLen);
    memcpy((unsigned char *)omx_buffer->pBuffer, extradata, omx_buffer->nFilledLen);
    omx_buffer->nFlags = OMX_BUFFERFLAG_CODECCONFIG | OMX_BUFFERFLAG_ENDOFFRAME;

    omx_err =  OMX_EmptyThisBuffer(ilclient_get_handle(component),
                                   omx_buffer);
    if (omx_err != OMX_ErrorNone)
    {
        fprintf(stderr, "%s - OMX_EmptyThisBuffer() failed with result(0x%x)\n", __func__,
        omx_err);
      return 0;
    } else {
        printf("Config sent, emptying buffer %d\n", extradatasize);
    }
  }
  return 1;
}
```

That is the essential extra point in decoding the video and sending it to the renderer. The code is at `il_ffmpeg_demux_render_video_full_speed.c` and, as mentioned before, plays the video as fast as it can.

Rendering the Video Stream (with Scheduling)

Playing the video stream at a reasonable speed adds substantial layers of complexity. You need to add a clock and a scheduler to begin with. The scheduler is interposed between the decoder and the renderer, and the scheduler takes times from the clock.

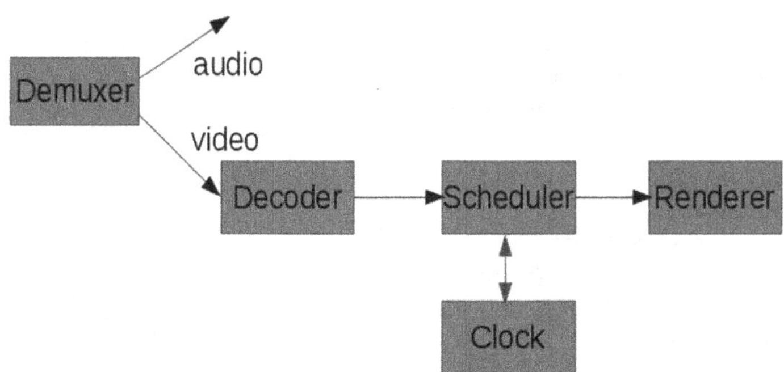

There are now four OpenMAX components to set up.

- video_decode
- video_render
- video_scheduler
- clock

The code for the first three is straightforward.

```
void setup_decodeComponent(ILCLIENT_T  *handle,
                           char *decodeComponentName,
                           COMPONENT_T **decodeComponent) {
    int err;

    err = ilclient_create_component(handle,
                                    decodeComponent,
                                    decodeComponentName,
                                    ILCLIENT_DISABLE_ALL_PORTS
                                    |
                                    ILCLIENT_ENABLE_INPUT_BUFFERS
                                    |
                                    ILCLIENT_ENABLE_OUTPUT_BUFFERS
                                    );
    if (err == -1) {
        fprintf(stderr, "DecodeComponent create failed\n");
        exit(1);
    }
    printState(ilclient_get_handle(*decodeComponent));

    err = ilclient_change_component_state(*decodeComponent,
                                          OMX_StateIdle);
    if (err < 0) {
        fprintf(stderr, "Couldn't change state to Idle\n");
        exit(1);
    }
    printState(ilclient_get_handle(*decodeComponent));

    // must be before we enable buffers
    set_video_decoder_input_format(*decodeComponent);
}

void setup_renderComponent(ILCLIENT_T  *handle,
                           char *renderComponentName,
                           COMPONENT_T **renderComponent) {
    int err;

    err = ilclient_create_component(handle,
                                    renderComponent,
                                    renderComponentName,
                                    ILCLIENT_DISABLE_ALL_PORTS
                                    |
```

```
                                        ILCLIENT_ENABLE_INPUT_BUFFERS
                                        );
    if (err == -1) {
        fprintf(stderr, "RenderComponent create failed\n");
        exit(1);
    }
    printState(ilclient_get_handle(*renderComponent));

    err = ilclient_change_component_state(*renderComponent,
                                          OMX_StateIdle);
    if (err < 0) {
        fprintf(stderr, "Couldn't change state to Idle\n");
        exit(1);
    }
    printState(ilclient_get_handle(*renderComponent));
}

void setup_schedulerComponent(ILCLIENT_T  *handle,
                           char *schedulerComponentName,
                           COMPONENT_T **schedulerComponent) {
    int err;

    err = ilclient_create_component(handle,
                                    schedulerComponent,
                                    schedulerComponentName,
                                    ILCLIENT_DISABLE_ALL_PORTS
                                    |
                                    ILCLIENT_ENABLE_INPUT_BUFFERS
                                    );
    if (err == -1) {
        fprintf(stderr, "SchedulerComponent create failed\n");
        exit(1);
    }
    printState(ilclient_get_handle(*schedulerComponent));

    err = ilclient_change_component_state(*schedulerComponent,
                                          OMX_StateIdle);
    if (err < 0) {
        fprintf(stderr, "Couldn't change state to Idle\n");
        exit(1);
    }
    printState(ilclient_get_handle(*schedulerComponent));
}
```

The clock component is new. The clock can use a variety of sources for its timing information. For this case, you are dealing with video (and later audio). The clock can use a video clock or an audio clock as its time reference source. Here you want to deal only with video, so you use a video clock.

```
OMX_TIME_CONFIG_ACTIVEREFCLOCKTYPE refClock;
refClock.nSize = sizeof(OMX_TIME_CONFIG_ACTIVEREFCLOCKTYPE);
refClock.nVersion.nVersion = OMX_VERSION;
refClock.eClock = OMX_TIME_RefClockVideo;
```

```
OMX_SetConfig(ilclient_get_handle(*clockComponent),
              OMX_IndexConfigTimeActiveRefClock, &refClock);
```

How the clock locates its reference clock doesn't seem to be specified anywhere. It seems to be a bit of "behind-the-scenes" magic done implicitly by OMX implementors. There are hints in the 1.1.2 specification in section 6.2.6, "Audio-Video File Playback Example Use Case," which shows that it is probably the scheduler that plays this role.

The clock needs to have a rate set. This is set in the field `xScale` of an `OMX_TIME_CONFIG_SCALETYPE` struct. The values do not seem to be clearly specified by OpenMAX. The wonderful `omxplayer` seems to have tried various values for this in different versions. The only clue is that section 3.2.2.11.2, "Sample Code Showing Calling Sequence," of the 1.1.2 specification shows the following:

```
oScale.xScale = 0x00020000; /*2x*/
```

From this you can guess that the value of normal speed could be 0x00010000, but by juggling other parameters, this need not be the case (which is what `omxplayer` currently does). Assuming this value, the code is as follows:

```
OMX_TIME_CONFIG_SCALETYPE scaleType;
scaleType.nSize = sizeof(OMX_TIME_CONFIG_SCALETYPE);
scaleType.nVersion.nVersion = OMX_VERSION;
scaleType.xScale = 0x00010000;

OMX_SetConfig(ilclient_get_handle(*clockComponent),
                          OMX_IndexConfigTimeScale, &scaleType);
```

The complete clock setup is as follows:

```
void setup_clockComponent(ILCLIENT_T  *handle,
                          char *clockComponentName,
                          COMPONENT_T **clockComponent) {
    int err;

    err = ilclient_create_component(handle,
                                    clockComponent,
                                    clockComponentName,
                                    ILCLIENT_DISABLE_ALL_PORTS
                                    );
    if (err == -1) {
        fprintf(stderr, "ClockComponent create failed\n");
        exit(1);
    }
    printState(ilclient_get_handle(*clockComponent));

    err = ilclient_change_component_state(*clockComponent,
                                          OMX_StateIdle);
    if (err < 0) {
        fprintf(stderr, "Couldn't change state to Idle\n");
        exit(1);
    }
    printState(ilclient_get_handle(*clockComponent));
    printClockState(*clockComponent);
```

```
    OMX_TIME_CONFIG_ACTIVEREFCLOCKTYPE refClock;
    refClock.nSize = sizeof(OMX_TIME_CONFIG_ACTIVEREFCLOCKTYPE);
    refClock.nVersion.nVersion = OMX_VERSION;
    refClock.eClock = OMX_TIME_RefClockVideo;

    err = OMX_SetConfig(ilclient_get_handle(*clockComponent),
                        OMX_IndexConfigTimeActiveRefClock, &refClock);
    if(err != OMX_ErrorNone) {
        fprintf(stderr, "COMXCoreComponent::SetConfig - %s failed with omx_err(0x%x)\n",
              "clock", err);
    }

    OMX_TIME_CONFIG_SCALETYPE scaleType;
    scaleType.nSize = sizeof(OMX_TIME_CONFIG_SCALETYPE);
    scaleType.nVersion.nVersion = OMX_VERSION;
    scaleType.xScale = (1 << 16);

    err = OMX_SetConfig(ilclient_get_handle(*clockComponent),
                        OMX_IndexConfigTimeScale, &scaleType);
    if(err != OMX_ErrorNone) {
        fprintf(stderr, "COMXCoreComponent::SetConfig - %s failed with omx_err(0x%x)\n",
              "clock", err);
    }

}
```

Execution reads from the file, demuxing frames using FFmpeg/LibAV and sending frames to the decoder until a PortSettingsChanged event occurs. Then it sets up tunnels. There are three of these:

- From the decoder to the scheduler
- From the clock to the scheduler
- From the scheduler to the renderer

Here's an example:

```
TUNNEL_T decodeTunnel;
set_tunnel(&decodeTunnel, decodeComponent, 131, schedulerComponent, 10);
if ((err = ilclient_setup_tunnel(&decodeTunnel, 0, 0)) < 0) {
    fprintf(stderr, "Error setting up decode tunnel %X\n", err);
    exit(1);
} else {
    printf("Decode tunnel set up ok\n");
}

TUNNEL_T schedulerTunnel;
set_tunnel(&schedulerTunnel, schedulerComponent, 11, renderComponent, 90);
if ((err = ilclient_setup_tunnel(&schedulerTunnel, 0, 0)) < 0) {
    fprintf(stderr, "Error setting up scheduler tunnel %X\n", err);
    exit(1);
} else {
    printf("Scheduler tunnel set up ok\n");
}
```

```
    TUNNEL_T clockTunnel;
    set_tunnel(&clockTunnel, clockComponent, 80, schedulerComponent, 12);
    if ((err = ilclient_setup_tunnel(&clockTunnel, 0, 0)) < 0) {
        fprintf(stderr, "Error setting up clock tunnel %X\n", err);
        exit(1);
    } else {
        printf("Clock tunnel set up ok\n");
    }
    startClock(clockComponent);
```

The clock is started with the following:

```
void startClock(COMPONENT_T *clockComponent) {
    OMX_ERRORTYPE err = OMX_ErrorNone;
    OMX_TIME_CONFIG_CLOCKSTATETYPE clockState;

    memset(&clockState, 0, sizeof( OMX_TIME_CONFIG_CLOCKSTATETYPE));
    clockState.nSize = sizeof( OMX_TIME_CONFIG_CLOCKSTATETYPE);
    clockState.nVersion.nVersion = OMX_VERSION;

    err = OMX_GetConfig(ilclient_get_handle(clockComponent),
                        OMX_IndexConfigTimeClockState, &clockState);
    if (err != OMX_ErrorNone) {
        fprintf(stderr, "Error getting clock state %s\n", err2str(err));
        return;
    }
    clockState.eState = OMX_TIME_ClockStateRunning;
    err = OMX_SetConfig(ilclient_get_handle(clockComponent),
                        OMX_IndexConfigTimeClockState, &clockState);
    if (err != OMX_ErrorNone) {
        fprintf(stderr, "Error starting clock %s\n", err2str(err));
        return;
    }

}
```

Now you turn to the tricky part of the program: matching the timing of the LibAV/FFmpeg packets to the OMX decoder. The relevant information from LibAV/FFmpeg is contained in the following structures:

```
typedef struct AVPacket {
    int64_t pts;
    int64_t dts;
    ...
} AVPacket;

typedef struct AVStream {
    AVRational r_frame_rate;
    AVRational time_base;
    ...
} AVStream;
```

The field you have to match it to is as follows:

```
typedef struct OMX_BUFFERHEADERTYPE
{
    OMX_TICKS nTimeStamp;
    ...
} OMX_BUFFERHEADERTYPE;
```

The fields dts and pts refer to "decode timestamp" and "presentation timestamp," respectively. A good explanation is given by the tutorial at http://dranger.com/ffmpeg/tutorial05.html by dranger. The conclusion is that you aren't actually interested in the decode time, only the presentation time, pts.

The values of these from some typical files are useful in explaining these fields. Note that the type AVRational represents a number with numerator (num) and denominator (den).

	time_base		r_frame_rate			
File	**num**	**den**	**num (fps rate)**	**den (fps scale)**	**pts Increment**	**Frame Rate**
Big Buck Bunny	1	600	24	1	25	23.96
Tai Chi	1	29969	29969	1000	1000	29.97
ClipCanvas	1	600	25	1	24	25
small	1	90000	30	1	3000	30
Wildlife	1	2997	2997	100	100	29.97
Sintel	1	24	24	1	1	24

The times for LibAV/FFmpeg are in seconds. So, for Big Buck Bunny, the timebase is 1/600 second, the frame rate is 24 frames per second, and the pts increment is 25 timebase units. The timebase and the frame rate are given from the demuxer.

```
video_stream = pFormatCtx->streams[video_stream_idx];

fpsscale        = video_stream->r_frame_rate.den;
fpsrate         = video_stream->r_frame_rate.num;
time_base_num   = video_stream->time_base.num;
time_base_den   = video_stream->time_base.den;
```

The pts values are given in the decoded packets. They may not be in order, so you may have to look at several values to obtain the highest common factor.

The relation between these fields is as follows:

```
pts increment × timebase = 1 / frame rate
```

so that the following values are true:

- For Big Buck Bunny, 25 × 1 / 600 = 1 /24.
- For Tai Chi, 1000 × 1 / 299969 = 1000 / 2997 = 1 / 29.969.
- For ClipCanvas, 24 × 1 / 600 = 1 / 25.

- For small, 3000 × 1 / 90000 = 1 / 30.
- For Wildlife, 100 × 1 / 2997 = 100 / 2997 = 1 / 29.97.

Each frame occurs at the time `pts * timebase` seconds.

Let's turn to OMX. Section 6.2.1, "Timestamps," of the 1.1.2 specification states that the timestamp structure will be "interpreted as a signed 64-bit value representing microseconds." This means any seconds value from LibAV/FFmpeg needs to be multiplied by 1,000,000 to convert it to OMX timestamp microseconds.

This relates LibAV/FFmpeg to OMX timestamps with the following:

```
buff_header->nTimeStamp = pts * 1000000 * time_base_num / time_base_den;
```

On 32-bit CPUs like the RPi, the 64-bit timestamp needs to be unpacked into a struct with two 32-bit fields. This is done with the following:

```
OMX_TICKS ToOMXTime(int64_t pts)
{
    OMX_TICKS ticks;
    ticks.nLowPart = pts;
    ticks.nHighPart = pts >> 32;
    return ticks;
}
```

Once the tunnels are set up and the clock is running, the program reads and demuxes frames and then sends them to the decoder. The complete program is `il_ffmpeg_demux_render_video.c`.

```
#include <stdio.h>
#include <stdlib.h>
#include <sys/stat.h>

#include <OMX_Core.h>
#include <OMX_Component.h>

#include <bcm_host.h>
#include <vcos_logging.h>

#define VCOS_LOG_CATEGORY (&il_ffmpeg_log_category)
static VCOS_LOG_CAT_T il_ffmpeg_log_category;

#include <ilclient.h>

#include "libavcodec/avcodec.h"
#include <libavformat/avformat.h>
#include "libavutil/mathematics.h"
#include "libavutil/samplefmt.h"

#define INBUF_SIZE 4096
#define AUDIO_INBUF_SIZE 20480
#define AUDIO_REFILL_THRESH 4096

char *IMG = "taichi.mp4";
```

```
static AVCodecContext *video_dec_ctx = NULL;
static AVStream *video_stream = NULL;
static AVPacket pkt;
AVFormatContext *pFormatCtx = NULL;

static int video_stream_idx = -1;

uint8_t extradatasize;
void *extradata;

AVCodec *codec;

void printState(OMX_HANDLETYPE handle) {
    // elided
}

char *err2str(int err) {
    return "error elided";
}

void printClockState(COMPONENT_T *clockComponent) {
    OMX_ERRORTYPE err = OMX_ErrorNone;
    OMX_TIME_CONFIG_CLOCKSTATETYPE clockState;

    memset(&clockState, 0, sizeof( OMX_TIME_CONFIG_CLOCKSTATETYPE));
    clockState.nSize = sizeof( OMX_TIME_CONFIG_CLOCKSTATETYPE);
    clockState.nVersion.nVersion = OMX_VERSION;

    err = OMX_GetConfig(ilclient_get_handle(clockComponent),
                        OMX_IndexConfigTimeClockState, &clockState);
    if (err != OMX_ErrorNone) {
        fprintf(stderr, "Error getting clock state %s\n", err2str(err));
        return;
    }
    switch (clockState.eState) {
    case OMX_TIME_ClockStateRunning:
        printf("Clock running\n");
        break;
    case OMX_TIME_ClockStateWaitingForStartTime:
        printf("Clock waiting for start time\n");
        break;
    case OMX_TIME_ClockStateStopped:
        printf("Clock stopped\n");
        break;
    default:
        printf("Clock in other state\n");
    }
}
```

```
void startClock(COMPONENT_T *clockComponent) {
    OMX_ERRORTYPE err = OMX_ErrorNone;
    OMX_TIME_CONFIG_CLOCKSTATETYPE clockState;

    memset(&clockState, 0, sizeof( OMX_TIME_CONFIG_CLOCKSTATETYPE));
    clockState.nSize = sizeof( OMX_TIME_CONFIG_CLOCKSTATETYPE);
    clockState.nVersion.nVersion = OMX_VERSION;

    err = OMX_GetConfig(ilclient_get_handle(clockComponent),
                        OMX_IndexConfigTimeClockState, &clockState);
    if (err != OMX_ErrorNone) {
        fprintf(stderr, "Error getting clock state %s\n", err2str(err));
        return;
    }
    clockState.eState = OMX_TIME_ClockStateRunning;
    err = OMX_SetConfig(ilclient_get_handle(clockComponent),
                        OMX_IndexConfigTimeClockState, &clockState);
    if (err != OMX_ErrorNone) {
        fprintf(stderr, "Error starting clock %s\n", err2str(err));
        return;
    }

}

void eos_callback(void *userdata, COMPONENT_T *comp, OMX_U32 data) {
    printf("Got eos event\n");
}

void error_callback(void *userdata, COMPONENT_T *comp, OMX_U32 data) {
    printf("OMX error %s\n", err2str(data));
}

void port_settings_callback(void *userdata, COMPONENT_T *comp, OMX_U32 data) {
    printf("Got port Settings event\n");
    // exit(0);
}

void empty_buffer_done_callback(void *userdata, COMPONENT_T *comp) {
    printf("Got empty buffer done\n");
}

int get_file_size(char *fname) {
    struct stat st;

    if (stat(fname, &st) == -1) {
        perror("Stat'ing img file");
        return -1;
    }
    return(st.st_size);
}
```

```
unsigned int uWidth;
unsigned int uHeight;

unsigned int fpsscale;
unsigned int fpsrate;
unsigned int time_base_num;
unsigned int time_base_den;

#ifdef OMX_SKIP64BIT
OMX_TICKS ToOMXTime(int64_t pts)
{
    OMX_TICKS ticks;
    ticks.nLowPart = pts;
    ticks.nHighPart = pts >> 32;
    return ticks;
}
#else
#define FromOMXTime(x) (x)
#endif

OMX_ERRORTYPE copy_into_buffer_and_empty(AVPacket *pkt,
                                         COMPONENT_T *component,
                                         OMX_BUFFERHEADERTYPE *buff_header) {
    OMX_ERRORTYPE r;

    int buff_size = buff_header->nAllocLen;
    int size = pkt->size;
    uint8_t *content = pkt->data;

    while (size > 0) {
        buff_header->nFilledLen = (size > buff_header->nAllocLen-1) ?
            buff_header->nAllocLen-1 : size;
        memset(buff_header->pBuffer, 0x0, buff_header->nAllocLen);
        memcpy(buff_header->pBuffer, content, buff_header->nFilledLen);
        size -= buff_header->nFilledLen;
        content += buff_header->nFilledLen;

        /*
        if (size < buff_size) {
            memcpy((unsigned char *)buff_header->pBuffer,
                   pkt->data, size);
        } else {
            printf("Buffer not big enough %d %d\n", buff_size, size);
            return -1;
        }

        buff_header->nFilledLen = size;
        */

        buff_header->nFlags = 0;
        if (size <= 0)
            buff_header->nFlags |= OMX_BUFFERFLAG_ENDOFFRAME;
```

```
        printf("  DTS is %s %ld\n", "str", pkt->dts);
        printf("  PTS is %s %ld\n", "str", pkt->pts);

        if (pkt->dts == 0) {
            buff_header->nFlags |= OMX_BUFFERFLAG_STARTTIME;
        } else {
            buff_header->nTimeStamp = ToOMXTime((uint64_t)
                                          (pkt->pts * 1000000/
                                           time_base_den));

            printf("Time stamp %d\n",   buff_header->nTimeStamp);
        }

        r = OMX_EmptyThisBuffer(ilclient_get_handle(component),
                                buff_header);
        if (r != OMX_ErrorNone) {
            fprintf(stderr, "Empty buffer error %s\n",
                    err2str(r));
        } else {
            printf("Emptying buffer %p\n", buff_header);
        }
        if (size > 0) {
             buff_header =
                ilclient_get_input_buffer(component,
                                          130,
                                          1 /* block */);
        }
    }
    return r;
}

int img_width, img_height;

int SendDecoderConfig(COMPONENT_T *component, FILE *out)
{
    OMX_ERRORTYPE omx_err   = OMX_ErrorNone;

    /* send decoder config */
    if(extradatasize > 0 && extradata != NULL)
        {
            fwrite(extradata, 1, extradatasize, out);

            OMX_BUFFERHEADERTYPE *omx_buffer = ilclient_get_input_buffer(component,
                                                                         130,
                                                                         1 /* block */);

            if(omx_buffer == NULL)
                {
                    fprintf(stderr, "%s - buffer error 0x%08x", __func__, omx_err);
                    return 0;
                }
```

```
        omx_buffer->nOffset = 0;
        omx_buffer->nFilledLen = extradatasize;
        if(omx_buffer->nFilledLen > omx_buffer->nAllocLen)
            {
                fprintf(stderr, "%s - omx_buffer->nFilledLen > omx_buffer->nAllocLen",
                __func__);
                return 0;
            }

        memset((unsigned char *)omx_buffer->pBuffer, 0x0, omx_buffer->nAllocLen);
        memcpy((unsigned char *)omx_buffer->pBuffer, extradata, omx_buffer->nFilledLen);
        omx_buffer->nFlags = OMX_BUFFERFLAG_CODECCONFIG | OMX_BUFFERFLAG_ENDOFFRAME;

        omx_err =  OMX_EmptyThisBuffer(ilclient_get_handle(component),
                                       omx_buffer);
        if (omx_err != OMX_ErrorNone)
            {
                fprintf(stderr, "%s - OMX_EmptyThisBuffer() failed with result(0x%x)\n",
                __func__, omx_err);
                return 0;
            } else {
            printf("Config sent, emptying buffer %d\n", extradatasize);
        }
    }
    return 1;
}

OMX_ERRORTYPE set_video_decoder_input_format(COMPONENT_T *component) {
    int err;

    // set input video format
    printf("Setting video decoder format\n");
    OMX_VIDEO_PARAM_PORTFORMATTYPE videoPortFormat;

    memset(&videoPortFormat, 0, sizeof(OMX_VIDEO_PARAM_PORTFORMATTYPE));
    videoPortFormat.nSize = sizeof(OMX_VIDEO_PARAM_PORTFORMATTYPE);
    videoPortFormat.nVersion.nVersion = OMX_VERSION;
    videoPortFormat.nPortIndex = 130;

    err = OMX_GetParameter(ilclient_get_handle(component),
                           OMX_IndexParamVideoPortFormat, &videoPortFormat);
    if (err != OMX_ErrorNone) {
        fprintf(stderr, "Error getting video decoder format %s\n", err2str(err));
        return err;
    }

    videoPortFormat.nPortIndex = 130;
    videoPortFormat.nIndex = 0;
    videoPortFormat.eCompressionFormat = OMX_VIDEO_CodingAVC;
    videoPortFormat.eColorFormat = OMX_COLOR_FormatUnused;
    videoPortFormat.xFramerate = 0;
```

```
#if 1 // doesn't seem to make any difference!!!
    if (fpsscale > 0 && fpsrate > 0) {
        videoPortFormat.xFramerate =
            (long long)(1<<16)*fpsrate / fpsscale;
    } else {
        videoPortFormat.xFramerate = 25 * (1<<16);
    }
    printf("FPS num %d den %d\n", fpsrate, fpsscale);
    printf("Set frame rate to %d\n", videoPortFormat.xFramerate);
#endif
    err = OMX_SetParameter(ilclient_get_handle(component),
                           OMX_IndexParamVideoPortFormat, &videoPortFormat);
    if (err != OMX_ErrorNone) {
        fprintf(stderr, "Error setting video decoder format %s\n", err2str(err));
        return err;
    } else {
        printf("Video decoder format set up ok\n");
    }

    OMX_PARAM_PORTDEFINITIONTYPE portParam;
    memset(&portParam, 0, sizeof( OMX_PARAM_PORTDEFINITIONTYPE));
    portParam.nSize = sizeof( OMX_PARAM_PORTDEFINITIONTYPE);
    portParam.nVersion.nVersion = OMX_VERSION;

    portParam.nPortIndex = 130;

    err =  OMX_GetParameter(ilclient_get_handle(component),
                            OMX_IndexParamPortDefinition, &portParam);
    if(err != OMX_ErrorNone)
        {
            fprintf(stderr, "COMXVideo::Open error OMX_IndexParamPortDefinition omx_
            err(0x%08x)\n", err);
            return err;
        }

    printf("Default framerate %d\n", portParam.format.video.xFramerate);

    portParam.nPortIndex = 130;

    portParam.format.video.nFrameWidth  = img_width;
    portParam.format.video.nFrameHeight = img_height;

    err =  OMX_SetParameter(ilclient_get_handle(component),
                            OMX_IndexParamPortDefinition, &portParam);
    if(err != OMX_ErrorNone)
        {
            fprintf(stderr, "COMXVideo::Open error OMX_IndexParamPortDefinition omx_
            err(0x%08x)\n", err);
            return err;
        }

    return OMX_ErrorNone;
}
```

```
int setup_demuxer(const char *filename) {
    // Register all formats and codecs
    av_register_all();
    if(avformat_open_input(&pFormatCtx, filename, NULL, NULL)!=0) {
        fprintf(stderr, "Can't get format\n");
        return -1; // Couldn't open file
    }
    // Retrieve stream information
    if (avformat_find_stream_info(pFormatCtx, NULL) < 0) {
        return -1; // Couldn't find stream information
    }
    printf("Format:\n");
    av_dump_format(pFormatCtx, 0, filename, 0);

    int ret;
    ret = av_find_best_stream(pFormatCtx, AVMEDIA_TYPE_VIDEO, -1, -1, NULL, 0);
    if (ret >= 0) {
        video_stream_idx = ret;

        video_stream = pFormatCtx->streams[video_stream_idx];
        video_dec_ctx = video_stream->codec;

        img_width           = video_stream->codec->width;
        img_height          = video_stream->codec->height;
        extradata           = video_stream->codec->extradata;
        extradatasize       = video_stream->codec->extradata_size;
        fpsscale            = video_stream->r_frame_rate.den;
        fpsrate             = video_stream->r_frame_rate.num;
        time_base_num           = video_stream->time_base.num;
        time_base_den           = video_stream->time_base.den;

        printf("Rate %d scale %d time base %d %d\n",
               video_stream->r_frame_rate.num,
               video_stream->r_frame_rate.den,
               video_stream->time_base.num,
               video_stream->time_base.den);

        AVCodec *codec = avcodec_find_decoder(video_stream->codec->codec_id);

        if (codec) {
            printf("Codec name %s\n", codec->name);
        }
    }
    return 0;
}

void setup_decodeComponent(ILCLIENT_T  *handle,
                           char *decodeComponentName,
                           COMPONENT_T **decodeComponent) {
    int err;
```

```
    err = ilclient_create_component(handle,
                                    decodeComponent,
                                    decodeComponentName,
                                    ILCLIENT_DISABLE_ALL_PORTS
                                    |
                                    ILCLIENT_ENABLE_INPUT_BUFFERS
                                    |
                                    ILCLIENT_ENABLE_OUTPUT_BUFFERS
                                    );
    if (err == -1) {
        fprintf(stderr, "DecodeComponent create failed\n");
        exit(1);
    }
    printState(ilclient_get_handle(*decodeComponent));

    err = ilclient_change_component_state(*decodeComponent,
                                          OMX_StateIdle);
    if (err < 0) {
        fprintf(stderr, "Couldn't change state to Idle\n");
        exit(1);
    }
    printState(ilclient_get_handle(*decodeComponent));

    // must be before we enable buffers
    set_video_decoder_input_format(*decodeComponent);
}

void setup_renderComponent(ILCLIENT_T  *handle,
                           char *renderComponentName,
                           COMPONENT_T **renderComponent) {
    int err;

    err = ilclient_create_component(handle,
                                    renderComponent,
                                    renderComponentName,
                                    ILCLIENT_DISABLE_ALL_PORTS
                                    |
                                    ILCLIENT_ENABLE_INPUT_BUFFERS
                                    );
    if (err == -1) {
        fprintf(stderr, "RenderComponent create failed\n");
        exit(1);
    }
    printState(ilclient_get_handle(*renderComponent));

    err = ilclient_change_component_state(*renderComponent,
                                          OMX_StateIdle);
    if (err < 0) {
        fprintf(stderr, "Couldn't change state to Idle\n");
        exit(1);
    }
    printState(ilclient_get_handle(*renderComponent));
}
```

```
void setup_schedulerComponent(ILCLIENT_T  *handle,
                              char *schedulerComponentName,
                              COMPONENT_T **schedulerComponent) {
    int err;

    err = ilclient_create_component(handle,
                                    schedulerComponent,
                                    schedulerComponentName,
                                    ILCLIENT_DISABLE_ALL_PORTS
                                    |
                                    ILCLIENT_ENABLE_INPUT_BUFFERS
                                    );
    if (err == -1) {
        fprintf(stderr, "SchedulerComponent create failed\n");
        exit(1);
    }
    printState(ilclient_get_handle(*schedulerComponent));

    err = ilclient_change_component_state(*schedulerComponent,
                                          OMX_StateIdle);
    if (err < 0) {
        fprintf(stderr, "Couldn't change state to Idle\n");
        exit(1);
    }
    printState(ilclient_get_handle(*schedulerComponent));
}

void setup_clockComponent(ILCLIENT_T  *handle,
                          char *clockComponentName,
                          COMPONENT_T **clockComponent) {
    int err;

    err = ilclient_create_component(handle,
                                    clockComponent,
                                    clockComponentName,
                                    ILCLIENT_DISABLE_ALL_PORTS
                                    );
    if (err == -1) {
        fprintf(stderr, "ClockComponent create failed\n");
        exit(1);
    }
    printState(ilclient_get_handle(*clockComponent));

    err = ilclient_change_component_state(*clockComponent,
                                          OMX_StateIdle);
    if (err < 0) {
        fprintf(stderr, "Couldn't change state to Idle\n");
        exit(1);
    }
```

```
    printState(ilclient_get_handle(*clockComponent));
    printClockState(*clockComponent);

    OMX_COMPONENTTYPE*clock = ilclient_get_handle(*clockComponent);

    OMX_TIME_CONFIG_ACTIVEREFCLOCKTYPE refClock;
    refClock.nSize = sizeof(OMX_TIME_CONFIG_ACTIVEREFCLOCKTYPE);
    refClock.nVersion.nVersion = OMX_VERSION;
    refClock.eClock = OMX_TIME_RefClockVideo; // OMX_CLOCKPORT0;

    err = OMX_SetConfig(ilclient_get_handle(*clockComponent),
                        OMX_IndexConfigTimeActiveRefClock, &refClock);
    if(err != OMX_ErrorNone) {
        fprintf(stderr, "COMXCoreComponent::SetConfig - %s failed with omx_err(0x%x)\n",
                "clock", err);
    }

    OMX_TIME_CONFIG_SCALETYPE scaleType;
    scaleType.nSize = sizeof(OMX_TIME_CONFIG_SCALETYPE);
    scaleType.nVersion.nVersion = OMX_VERSION;
    scaleType.xScale = 0x00010000;

    err = OMX_SetConfig(ilclient_get_handle(*clockComponent),
                        OMX_IndexConfigTimeScale, &scaleType);
    if(err != OMX_ErrorNone) {
        fprintf(stderr, "COMXCoreComponent::SetConfig - %s failed with omx_err(0x%x)\n",
                "clock", err);
    }
}

int main(int argc, char** argv) {

    char *decodeComponentName;
    char *renderComponentName;
    char *schedulerComponentName;
    char *clockComponentName;
    int err;
    ILCLIENT_T  *handle;
    COMPONENT_T *decodeComponent;
    COMPONENT_T *renderComponent;
    COMPONENT_T *schedulerComponent;
    COMPONENT_T *clockComponent;

    if (argc > 1) {
        IMG = argv[1];
    }

    OMX_BUFFERHEADERTYPE *buff_header;

    setup_demuxer(IMG);
```

```
decodeComponentName = "video_decode";
renderComponentName = "video_render";
schedulerComponentName = "video_scheduler";
clockComponentName = "clock";

bcm_host_init();

handle = ilclient_init();
vcos_log_set_level(VCOS_LOG_CATEGORY, VCOS_LOG_TRACE);
if (handle == NULL) {
    fprintf(stderr, "IL client init failed\n");
    exit(1);
}

if (OMX_Init() != OMX_ErrorNone) {
    ilclient_destroy(handle);
    fprintf(stderr, "OMX init failed\n");
    exit(1);
}

ilclient_set_error_callback(handle,
                            error_callback,
                            NULL);
ilclient_set_eos_callback(handle,
                          eos_callback,
                          NULL);
ilclient_set_port_settings_callback(handle,
                                    port_settings_callback,
                                    NULL);
ilclient_set_empty_buffer_done_callback(handle,
                                        empty_buffer_done_callback,
                                        NULL);

setup_decodeComponent(handle, decodeComponentName, &decodeComponent);
setup_renderComponent(handle, renderComponentName, &renderComponent);
setup_schedulerComponent(handle, schedulerComponentName, &schedulerComponent);
setup_clockComponent(handle, clockComponentName, &clockComponent);
// both components now in Idle state, no buffers, ports disabled

// input port
err = ilclient_enable_port_buffers(decodeComponent, 130,
                                   NULL, NULL, NULL);
if (err < 0) {
    fprintf(stderr, "Couldn't enable buffers\n");
    exit(1);
}
ilclient_enable_port(decodeComponent, 130);

err = ilclient_change_component_state(decodeComponent,
                                      OMX_StateExecuting);
```

```
    if (err < 0) {
        fprintf(stderr, "Couldn't change state to Executing\n");
        exit(1);
    }
    printState(ilclient_get_handle(decodeComponent));

    FILE *out = fopen("tmp.h264", "wb");
    SendDecoderConfig(decodeComponent, out);

    /* read frames from the file */
    while (av_read_frame(pFormatCtx, &pkt) >= 0) {
        printf("Read pkt %d\n", pkt.size);

        AVPacket orig_pkt = pkt;
        if (pkt.stream_index == video_stream_idx) {
            printf("  read video pkt %d\n", pkt.size);
            fwrite(pkt.data, 1, pkt.size, out);
            buff_header =
                ilclient_get_input_buffer(decodeComponent,
                                          130,
                                          1 /* block */);
            if (buff_header != NULL) {
                copy_into_buffer_and_empty(&pkt,
                                           decodeComponent,
                                           buff_header);
            } else {
                fprintf(stderr, "Couldn't get a buffer\n");
            }

            err = ilclient_wait_for_event(decodeComponent,
                                          OMX_EventPortSettingsChanged,
                                          131, 0, 0, 1,
                                          ILCLIENT_EVENT_ERROR | ILCLIENT_PARAMETER_CHANGED,
                                          0);
            if (err < 0) {
                printf("No port settings change\n");
                //exit(1);
            } else {
                printf("Port settings changed\n");
                // exit(0);
                break;
            }

            if (ilclient_remove_event(decodeComponent,
                                      OMX_EventPortSettingsChanged,
                                      131, 0, 0, 1) == 0) {
                printf("Removed port settings event\n");
                //exit(0);
                break;
            } else {
```

```
                printf("No portr settting seen yet\n");
            }
        }
        av_free_packet(&orig_pkt);
    }

    TUNNEL_T decodeTunnel;
    set_tunnel(&decodeTunnel, decodeComponent, 131, schedulerComponent, 10);
    if ((err = ilclient_setup_tunnel(&decodeTunnel, 0, 0)) < 0) {
        fprintf(stderr, "Error setting up decode tunnel %X\n", err);
        exit(1);
    } else {
        printf("Decode tunnel set up ok\n");
    }

    TUNNEL_T schedulerTunnel;
    set_tunnel(&schedulerTunnel, schedulerComponent, 11, renderComponent, 90);
    if ((err = ilclient_setup_tunnel(&schedulerTunnel, 0, 0)) < 0) {
        fprintf(stderr, "Error setting up scheduler tunnel %X\n", err);
        exit(1);
    } else {
        printf("Scheduler tunnel set up ok\n");
    }

    TUNNEL_T clockTunnel;
    set_tunnel(&clockTunnel, clockComponent, 80, schedulerComponent, 12);
    if ((err = ilclient_setup_tunnel(&clockTunnel, 0, 0)) < 0) {
        fprintf(stderr, "Error setting up clock tunnel %X\n", err);
        exit(1);
    } else {
        printf("Clock tunnel set up ok\n");
    }
    startClock(clockComponent);
    printClockState(clockComponent);

    // Okay to go back to processing data
    // enable the decode output ports

    OMX_SendCommand(ilclient_get_handle(decodeComponent),
                    OMX_CommandPortEnable, 131, NULL);

    ilclient_enable_port(decodeComponent, 131);

    // enable the clock output ports
    OMX_SendCommand(ilclient_get_handle(clockComponent),
                    OMX_CommandPortEnable, 80, NULL);

    ilclient_enable_port(clockComponent, 80);

    // enable the scheduler ports
    OMX_SendCommand(ilclient_get_handle(schedulerComponent),
                    OMX_CommandPortEnable, 10, NULL);
```

```
    ilclient_enable_port(schedulerComponent, 10);

    OMX_SendCommand(ilclient_get_handle(schedulerComponent),
                    OMX_CommandPortEnable, 11, NULL);

    ilclient_enable_port(schedulerComponent, 11);

    OMX_SendCommand(ilclient_get_handle(schedulerComponent),
                    OMX_CommandPortEnable, 12, NULL);

    ilclient_enable_port(schedulerComponent, 12);

    // enable the render input ports

    OMX_SendCommand(ilclient_get_handle(renderComponent),
                    OMX_CommandPortEnable, 90, NULL);

    ilclient_enable_port(renderComponent, 90);

    // set both components to executing state
    err = ilclient_change_component_state(decodeComponent,
                                          OMX_StateExecuting);
    if (err < 0) {
        fprintf(stderr, "Couldn't change state to Idle\n");
        exit(1);
    }
    err = ilclient_change_component_state(renderComponent,
                                          OMX_StateExecuting);
    if (err < 0) {
        fprintf(stderr, "Couldn't change state to Idle\n");
        exit(1);
    }

    err = ilclient_change_component_state(schedulerComponent,
                                          OMX_StateExecuting);
    if (err < 0) {
        fprintf(stderr, "Couldn't change state to Idle\n");
        exit(1);
    }

    err = ilclient_change_component_state(clockComponent,
                                          OMX_StateExecuting);
    if (err < 0) {
        fprintf(stderr, "Couldn't change state to Idle\n");
        exit(1);
    }

    // now work through the file
    while (av_read_frame(pFormatCtx, &pkt) >= 0) {
        printf("Read pkt after port settings %d\n", pkt.size);
        fwrite(pkt.data, 1, pkt.size, out);
```

```
        if (pkt.stream_index != video_stream_idx) {
            continue;
        }
        printf("  is video pkt\n");
        //printf("  Best timestamp is %d\n", );

        // do we have a decode input buffer we can fill and empty?
        buff_header =
            ilclient_get_input_buffer(decodeComponent,
                                      130,
                                      1 /* block */);
        if (buff_header != NULL) {
            copy_into_buffer_and_empty(&pkt,
                                       decodeComponent,
                                       buff_header);
        }

        err = ilclient_wait_for_event(decodeComponent,
                                      OMX_EventPortSettingsChanged,
                                      131, 0, 0, 1,
                                      ILCLIENT_EVENT_ERROR | ILCLIENT_PARAMETER_CHANGED,
                                      0);
        if (err >= 0) {
            printf("Another port settings change\n");
        }

    }

    ilclient_wait_for_event(renderComponent,
                            OMX_EventBufferFlag,
                            90, 0, OMX_BUFFERFLAG_EOS, 0,
                            ILCLIENT_BUFFER_FLAG_EOS, 10000);
    printf("EOS on render\n");

    exit(0);
}
```

Rendering Both Audio and Video

The code is getting long and messy. I've done some cleanup to bring in some macros and other functions to simplify creating and manipulating OMX components. I've also collapsed the two `while` loops into one, checking—usually redundantly—the port setting's changed event. This simplifies the code and doesn't cost much.

You now have both the audio and video decode/render streams. You can picture this as follows:

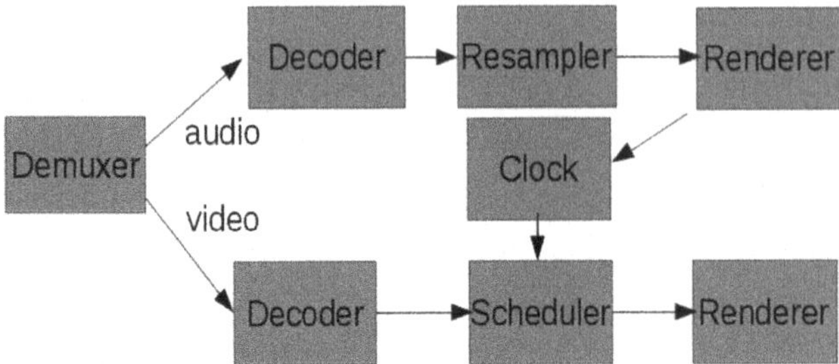

The audio stream does the following:

- Decodes in software using `avcodec_decode_audio4`
- Resamples this to get it into the right format with `avresample_convert`
- Sends it to the OMX audio render component

The video stream does the following:

- Decodes packets using the OMX video decoder component
- Schedules them using the OMX scheduler
- Passes packets to the OMX renderer

You also have an OMX clock. People are apparently much more sensitive to aberrations in audio than in video, so all recommendations are to set the clock to the audio stream rather than to the video stream.

- You should set the clock to the audio clock (probably the audio renderer).
- You should set the scheduler to the clock.

Use the Audio Decoder?

The RPi can apparently decode some audio formats, such as AC3 and DTS. It doesn't need to use software decoding for these. So, it can use the OMX audio decoder for some formats. The `omxplayer` includes the audio decoder in the audio pipeline, even for PCM data.

The audio decoder can't actually deal with PCM data, which is what will come out of the LibAV decoder. It can manage WAV data, which has the additional header information of sample size, and so on. So, if the OMX audio decoder is inserted into the pipeline, additional code must first send the appropriate WAV header information to the OMX decoder before sending the PCM data.

```
// from omxplayer linux/PlatformDefs.h
// header for PCM audio data
typedef unsigned short  WORD;
typedef unsigned int  DWORD;
typedef struct tWAVEFORMATEX {
    WORD    format_tag;
    WORD    channels;
    DWORD   samples_per_sec;
    DWORD   avg_bytes_per_sec;
    WORD    block_align;
```

```
    WORD    bits_per_sample;
    WORD    cb_size;
} __attribute__((__packed__)) WAVEFORMATEX, *PWAVEFORMATEX, *LPWAVEFORMATEX;

int send_audio_decoder_config(COMPONENT_T *component, FILE *out)
{
    WAVEFORMATEX wave_header =
        {.format_tag =  0x0001, //WAVE_FORMAT_PCM
         .channels = channels,
         .samples_per_sec = sample_rate,
         .avg_bytes_per_sec = 0,
         .block_align = block_align,
         .bits_per_sample = bits_per_sample,
         .cb_size = 0};

    OMX_ERRORTYPE omx_err   = OMX_ErrorNone;

    OMX_BUFFERHEADERTYPE *omx_buffer = ilclient_get_input_buffer(component,
                                                                 120,
                                                                 1 /* block */);

    if(omx_buffer == NULL) {
            fprintf(stderr, "%s - buffer error 0x%08x", __func__, omx_err);
            return 0;
    }

    omx_buffer->nOffset = 0;
    omx_buffer->nFilledLen = sizeof(wave_header);
    if(omx_buffer->nFilledLen > omx_buffer->nAllocLen) {
            fprintf(stderr, "%s - omx_buffer->nFilledLen > omx_buffer->nAllocLen",
            __func__);
            return 0;
    }

    memset((unsigned char *)omx_buffer->pBuffer, 0x0,
           omx_buffer->nAllocLen);
    memcpy((unsigned char *)omx_buffer->pBuffer, &wave_header,
           sizeof(wave_header));
    omx_buffer->nFlags = OMX_BUFFERFLAG_CODECCONFIG | OMX_BUFFERFLAG_ENDOFFRAME;

    omx_err =  OMX_EmptyThisBuffer(ilclient_get_handle(component),
                                   omx_buffer);
    exit_on_omx_error(omx_err, "Error setting up audio config %s\n");

    return 1;
}
```

In the program that follows, a switch called AUDIO_DECODE is used to toggle whether the audio decode component is inserted into the audio pipeline. I've done that in case you want to use the code specifically in one or the other modes. The program is il_ffmpeg_demux_render_audio_video.c.

```
#include <stdio.h>
#include <stdlib.h>
#include <sys/stat.h>

#include <OMX_Core.h>
#include <OMX_Component.h>

#include <bcm_host.h>
#include <vcos_logging.h>

//#define VCOS_LOG_CATEGORY (&il_ffmpeg_log_category)
// VCOS_LOG_CAT_T il_ffmpeg_log_category;

#include <ilclient.h>

// #include <sys/time.h>

#include "libavcodec/avcodec.h"
#include <libavformat/avformat.h>
#include "libavutil/mathematics.h"
#include "libavutil/samplefmt.h"
#include "libavutil/opt.h"
#include "libavresample/avresample.h"

#define USECS_IN_SEC 1000000

// from omxplayer OMXCore.h
// initialise any OMX struct
#define OMX_INIT_STRUCTURE(a) \
  memset(&(a), 0, sizeof(a)); \
  (a).nSize = sizeof(a); \
  (a).nVersion.s.nVersionMajor = OMX_VERSION_MAJOR; \
  (a).nVersion.s.nVersionMinor = OMX_VERSION_MINOR; \
  (a).nVersion.s.nRevision = OMX_VERSION_REVISION; \
  (a).nVersion.s.nStep = OMX_VERSION_STEP

// from omxplayer linux/PlatformDefs.h
// header for PCM audio data
typedef unsigned short  WORD;
typedef unsigned int  DWORD;
typedef struct tWAVEFORMATEX {
    WORD    format_tag;
    WORD    channels;
    DWORD   samples_per_sec;
    DWORD   avg_bytes_per_sec;
    WORD    block_align;
    WORD    bits_per_sample;
    WORD    cb_size;
} __attribute__((__packed__)) WAVEFORMATEX, *PWAVEFORMATEX, *LPWAVEFORMATEX;
```

```
//#define VCOS_LOG_CATEGORY (&il_ffmpeg_log_category)
// VCOS_LOG_CAT_T il_ffmpeg_log_category;

// struct timeval tv;
// time_t starttime = 0;

// default file to play
char *fname = "taichi.mp4";

// libav/ffmpeg vbls
AVCodecContext *video_dec_ctx = NULL;
AVStream *video_stream = NULL;
AVPacket pkt;
AVFormatContext *format_ctx = NULL;
AVCodecContext *audio_dec_ctx;
AVAudioResampleContext *swr;
AVCodec *codec;

 int video_stream_idx = -1;
int audio_stream_idx;

// data passed from video demux to video decoder
uint8_t extradatasize;
void *extradata;

// use the OMX audio decoder?
#define AUDIO_DECODE 1

// OMX/IL vbls
ILCLIENT_T  *handle;
COMPONENT_T *video_decode_component;
COMPONENT_T *video_render_component;
COMPONENT_T *audio_render_component;
COMPONENT_T *scheduler_component;
COMPONENT_T *clock_component;
#if AUDIO_DECODE
COMPONENT_T *audio_decode_component;
#endif

char *video_decode_component_name = "video_decode";
char *video_render_component_name = "video_render";
char *audio_render_component_name = "audio_render";
char *scheduler_component_name = "video_scheduler";
char *clock_component_name = "clock";
#if AUDIO_DECODE

// OMX/IL vbls
ILCLIENT_T  *handle;
COMPONENT_T *video_decode_component;
COMPONENT_T *video_render_component;
COMPONENT_T *audio_render_component;
```

```
COMPONENT_T *scheduler_component;
COMPONENT_T *clock_component;
#if AUDIO_DECODE
COMPONENT_T *audio_decode_component;
#endif

char *video_decode_component_name = "video_decode";
char *video_render_component_name = "video_render";
char *audio_render_component_name = "audio_render";
char *scheduler_component_name = "video_scheduler";
char *clock_component_name = "clock";
#if AUDIO_DECODE
char *audio_decode_component_name = "audio_decode";
#endif

// audio parameters
int sample_rate;
int channels;
int block_align;
int bits_per_sample;

// video parameters
int img_width, img_height;
unsigned int fpsscale;
unsigned int fpsrate;
unsigned int vtime_base_num;
unsigned int vtime_base_den;
unsigned int atime_base_num;
unsigned int atime_base_den;

#ifdef OMX_SKIP64BIT
OMX_TICKS ToOMXTime(int64_t pts)
{
    OMX_TICKS ticks;
    ticks.nLowPart = pts;
    ticks.nHighPart = pts >> 32;
    return ticks;
}
#else
#define FromOMXTime(x) (x)
#endif

int64_t convert_timestamp(int64_t pts, int den, int num) {
    return pts;
}

void exit_on_il_error(int err, char *err_str) {
    if (err < 0) {
        fprintf(stderr, err_str);
        exit(1);
    }
}
```

```
void print_state(OMX_HANDLETYPE handle) {
    // elided
}

char *err2str(int err) {
    return "error elided";
}

void exit_on_omx_error(int err, char *err_str) {
    if (err != OMX_ErrorNone) {
        fprintf(stderr, err_str, err2str(err));
        exit(1);
    }
}

void print_clock_state(COMPONENT_T *clock_component) {
    OMX_ERRORTYPE err = OMX_ErrorNone;
    OMX_TIME_CONFIG_CLOCKSTATETYPE clock_state;

    OMX_INIT_STRUCTURE(clock_state);

    err = OMX_GetConfig(ilclient_get_handle(clock_component),
                        OMX_IndexConfigTimeClockState, &clock_state);
    exit_on_omx_error(err, "Error getting clock state %s\n");

    switch (clock_state.eState) {
    case OMX_TIME_ClockStateRunning:
        printf("Clock running\n");
        break;
    case OMX_TIME_ClockStateWaitingForStartTime:
        printf("Clock waiting for start time\n");
        break;
    case OMX_TIME_ClockStateStopped:
        printf("Clock stopped\n");
        break;
    default:
        printf("Clock in other state\n");
    }
}

void start_clock(COMPONENT_T *clock_component) {
    OMX_ERRORTYPE err = OMX_ErrorNone;
    OMX_TIME_CONFIG_CLOCKSTATETYPE clock_state;

    OMX_INIT_STRUCTURE(clock_state);

    err = OMX_GetConfig(ilclient_get_handle(clock_component),
                        OMX_IndexConfigTimeClockState, &clock_state);
    exit_on_omx_error(err,  "Error getting clock state %s\n");
```

```
    clock_state.eState = OMX_TIME_ClockStateRunning;
    err = OMX_SetConfig(ilclient_get_handle(clock_component),
                        OMX_IndexConfigTimeClockState, &clock_state);
    exit_on_omx_error(err, "Error starting clock %s\n");
}

void eos_callback(void *userdata, COMPONENT_T *comp, OMX_U32 data) {
    printf("Got eos event\n");
}

void error_callback(void *userdata, COMPONENT_T *comp, OMX_U32 data) {
    printf("OMX error %s\n", err2str(data));
}

void port_settings_callback(void *userdata, COMPONENT_T *comp, OMX_U32 data) {
    fprintf(stderr, "Got port settings event\n");
}

void empty_buffer_done_callback(void *userdata, COMPONENT_T *comp) {
    // printf("Got empty buffer done\n");
}

void setup_callbacks() {
    ilclient_set_error_callback(handle,
                                error_callback,
                                NULL);
    ilclient_set_eos_callback(handle,
                              eos_callback,
                              NULL);
    ilclient_set_port_settings_callback(handle,
                                        port_settings_callback,
                                        NULL);
    ilclient_set_empty_buffer_done_callback(handle,
                                            empty_buffer_done_callback,
                                            NULL);
}

void print_clock_time(char *s) {
    OMX_TIME_CONFIG_TIMESTAMPTYPE stamp;

    OMX_INIT_STRUCTURE(stamp);

    stamp.nPortIndex = 81;

    OMX_GetParameter(ilclient_get_handle(clock_component),
                     OMX_IndexConfigTimeCurrentMediaTime,
                     //OMX_IndexConfigTimeCurrentAudioReference,
                     &stamp);
    //printf("%s %u%u\n", s, stamp.nTimestamp.nHighPart,
    //       stamp.nTimestamp.nLowPart);
}
```

```
OMX_ERRORTYPE copy_video_into_buffer_and_empty(AVPacket *pkt,
                                               COMPONENT_T *component) {
    OMX_ERRORTYPE r;
    OMX_BUFFERHEADERTYPE *buff_header = NULL;

    //int buff_size = buff_header->nAllocLen;
    int size = pkt->size;
    uint8_t *content = pkt->data;

    while (size > 0) {
        buff_header =
            ilclient_get_input_buffer(component,
                                      130,
                                      1 /* block */);
        buff_header->nFilledLen = (size > buff_header->nAllocLen-1) ?
            buff_header->nAllocLen-1 : size;
        memset(buff_header->pBuffer, 0x0, buff_header->nAllocLen);
        memcpy(buff_header->pBuffer, content, buff_header->nFilledLen);
        size -= buff_header->nFilledLen;
        content += buff_header->nFilledLen;

        buff_header->nFlags = 0;
        if (size <= 0)
            buff_header->nFlags |= OMX_BUFFERFLAG_ENDOFFRAME;

        if (pkt->dts == 0) {
            buff_header->nFlags |= OMX_BUFFERFLAG_STARTTIME;
        } else {
            buff_header->nTimeStamp = ToOMXTime((uint64_t)
                                                (pkt->pts * USECS_IN_SEC *
                                                 vtime_base_num /
                                                 vtime_base_den));

            // printf("Video timestamp %u\n",  buff_header->nTimeStamp.nLowPart);
        }

        r = OMX_EmptyThisBuffer(ilclient_get_handle(component),
                                buff_header);
        exit_on_omx_error(r,  "Empty buffer error %s\n");
    }
    return r;
}

OMX_ERRORTYPE read_audio_into_buffer_and_empty(AVFrame *decoded_frame,
                                               COMPONENT_T *component,
                                               // OMX_BUFFERHEADERTYPE *buff_header,
                                               int total_len) {
    OMX_ERRORTYPE r;
    OMX_BUFFERHEADERTYPE *buff_header = NULL;
```

```
#if AUDIO_DECODE
    int port_index = 120;
#else
    int port_index = 100;
#endif

    int required_decoded_size = 0;

    int out_linesize;
    required_decoded_size =
        av_samples_get_buffer_size(&out_linesize, 2,
                                   decoded_frame->nb_samples,
                                   AV_SAMPLE_FMT_S16, 0);
    uint8_t *buffer, *start_buffer;
    av_samples_alloc(&buffer, &out_linesize, 2, decoded_frame->nb_samples,
                     AV_SAMPLE_FMT_S16, 0);
    start_buffer = buffer;
    avresample_convert(swr, &buffer,
                       decoded_frame->linesize[0],
                       decoded_frame->nb_samples,
                       decoded_frame->data,
                       decoded_frame->linesize[0],
                       decoded_frame->nb_samples);

    while (required_decoded_size > 0) {
        buff_header =
            ilclient_get_input_buffer(component,
                                      port_index,
                                      1 /* block */);

        buff_header->nTimeStamp = ToOMXTime((uint64_t)
                                            (decoded_frame->pkt_pts * USECS_IN_SEC *
                                             atime_base_num /
                                             atime_base_den));

        int len = buff_header->nAllocLen;

        if (required_decoded_size > len) {
            memcpy(buff_header->pBuffer,
                   buffer, len);
            buff_header->nFilledLen = len;
            buffer += len;
        } else {
            memcpy(buff_header->pBuffer,
                   buffer, required_decoded_size);
            buff_header->nFilledLen = required_decoded_size;
        }

        r = OMX_EmptyThisBuffer(ilclient_get_handle(component),
                                buff_header);
        exit_on_omx_error(r, "Empty buffer error %s\n");
```

```
        required_decoded_size -= len;
    }
    av_free(&start_buffer[0]);
    return r;
}

int send_video_decoder_config(COMPONENT_T *component)
{
    OMX_ERRORTYPE omx_err   = OMX_ErrorNone;

    /* send decoder config */
    if(extradatasize > 0 && extradata != NULL) {
        OMX_BUFFERHEADERTYPE *omx_buffer =
            ilclient_get_input_buffer(component,
                                      130,
                                      1 /* block */);

        if(omx_buffer == NULL) {
            fprintf(stderr, "%s - buffer error 0x%08x", __func__, omx_err);
            return 0;
        }

        omx_buffer->nOffset = 0;
        omx_buffer->nFilledLen = extradatasize;
        if(omx_buffer->nFilledLen > omx_buffer->nAllocLen) {
            fprintf(stderr, "%s - omx_buffer->nFilledLen > omx_buffer->nAllocLen",
            __func__);
            return 0;
        }

        memset((unsigned char *)omx_buffer->pBuffer, 0x0,
               omx_buffer->nAllocLen);
        memcpy((unsigned char *)omx_buffer->pBuffer, extradata,
               omx_buffer->nFilledLen);
        omx_buffer->nFlags = OMX_BUFFERFLAG_CODECCONFIG | OMX_BUFFERFLAG_ENDOFFRAME;

        omx_err =  OMX_EmptyThisBuffer(ilclient_get_handle(component),
                                       omx_buffer);
        exit_on_omx_error(omx_err, "Setting up video config failed %s\n");

    }
    return 1;
}

int send_audio_decoder_config(COMPONENT_T *component, FILE *out)
{
    WAVEFORMATEX wave_header =
        {.format_tag =  0x0001, //WAVE_FORMAT_PCM, // 32768,
         .channels = channels, .samples_per_sec = sample_rate,
         .avg_bytes_per_sec = 0, .block_align = block_align,
         .bits_per_sample = bits_per_sample,
         .cb_size = 0};
```

```
    OMX_ERRORTYPE omx_err   = OMX_ErrorNone;

    OMX_BUFFERHEADERTYPE *omx_buffer = ilclient_get_input_buffer(component,
                                                                  120,
                                                                  1 /* block */);

    if(omx_buffer == NULL) {
            fprintf(stderr, "%s - buffer error 0x%08x", __func__, omx_err);
            return 0;
    }

    omx_buffer->nOffset = 0;
    omx_buffer->nFilledLen = sizeof(wave_header);
    if(omx_buffer->nFilledLen > omx_buffer->nAllocLen) {
            fprintf(stderr, "%s - omx_buffer->nFilledLen > omx_buffer->nAllocLen",  __
            func__);
            return 0;
    }

    memset((unsigned char *)omx_buffer->pBuffer, 0x0,
           omx_buffer->nAllocLen);
    memcpy((unsigned char *)omx_buffer->pBuffer, &wave_header,
           sizeof(wave_header));
    omx_buffer->nFlags = OMX_BUFFERFLAG_CODECCONFIG | OMX_BUFFERFLAG_ENDOFFRAME;

    omx_err =  OMX_EmptyThisBuffer(ilclient_get_handle(component),
                                   omx_buffer);
    exit_on_omx_error(omx_err, "Error setting up audio config %s\n");

    return 1;
}

OMX_ERRORTYPE set_video_decoder_input_format(COMPONENT_T *component) {
    int err;

    // printf("Setting video decoder format\n");
    OMX_VIDEO_PARAM_PORTFORMATTYPE video_port_format;

    OMX_INIT_STRUCTURE(video_port_format);
    video_port_format.nPortIndex = 130;

    err = OMX_GetParameter(ilclient_get_handle(component),
                           OMX_IndexParamVideoPortFormat, &video_port_format);
    exit_on_omx_error(err, "Error getting video decoder format %s\n");

    video_port_format.nPortIndex = 130;
    video_port_format.nIndex = 0;
    video_port_format.eCompressionFormat = OMX_VIDEO_CodingAVC;
    video_port_format.eColorFormat = OMX_COLOR_FormatUnused;
    video_port_format.xFramerate = 0;
```

```
#if 1 // doesn't seem to make any difference!!!
    if (fpsscale > 0 && fpsrate > 0) {
        video_port_format.xFramerate =
            (long long)(1<<16)*fpsrate / fpsscale;
    } else {
        video_port_format.xFramerate = 25 * (1<<16);
    }
#endif
    err = OMX_SetParameter(ilclient_get_handle(component),
                        OMX_IndexParamVideoPortFormat, &video_port_format);
    exit_on_omx_error(err, "Error setting video decoder format %s\n");

    OMX_PARAM_PORTDEFINITIONTYPE port_param;
    OMX_INIT_STRUCTURE(port_param);
    port_param.nPortIndex = 130;

    err =  OMX_GetParameter(ilclient_get_handle(component),
                        OMX_IndexParamPortDefinition, &port_param);
    exit_on_omx_error(err, "Error getting video parameters %s\n");

    port_param.nPortIndex = 130;
    port_param.format.video.nFrameWidth  = img_width;
    port_param.format.video.nFrameHeight = img_height;

    err =  OMX_SetParameter(ilclient_get_handle(component),
                        OMX_IndexParamPortDefinition, &port_param);
    exit_on_omx_error(err, "Error setting video params %s\n");

    return OMX_ErrorNone;
}

int setup_demuxer(const char *filename) {
    // Register all formats and codecs
    av_register_all();
    if(avformat_open_input(&format_ctx, filename, NULL, NULL)!=0) {
        fprintf(stderr, "Can't get format\n");
        return -1; // Couldn't open file
    }
    // Retrieve stream information
    if (avformat_find_stream_info(format_ctx, NULL) < 0) {
        return -1; // Couldn't find stream information
    }
    av_dump_format(format_ctx, 0, filename, 0);

    int ret;
    ret = av_find_best_stream(format_ctx, AVMEDIA_TYPE_VIDEO, -1, -1, NULL, 0);
    if (ret >= 0) {
        video_stream_idx = ret;

        video_stream = format_ctx->streams[video_stream_idx];
        video_dec_ctx = video_stream->codec;
```

```
        img_width           = video_stream->codec->width;
        img_height          = video_stream->codec->height;
        extradata           = video_stream->codec->extradata;
        extradatasize       = video_stream->codec->extradata_size;
        fpsscale            = video_stream->r_frame_rate.den;
        fpsrate             = video_stream->r_frame_rate.num;
        vtime_base_num          = video_stream->time_base.num;
        vtime_base_den          = video_stream->time_base.den;

        AVCodec *codec = avcodec_find_decoder(video_stream->codec->codec_id);

        if (codec) {
            // printf("Codec name %s\n", codec->name);
        }
    }

    if ((ret = av_find_best_stream(format_ctx, AVMEDIA_TYPE_AUDIO, -1, -1, NULL, 0)) >= 0) {
        //AVCodecContext* codec_context;
        AVStream *audio_stream;

        audio_stream_idx = ret;
        fprintf(stderr, "Audio stream index is %d\n", ret);

        audio_stream = format_ctx->streams[audio_stream_idx];

        atime_base_num          = audio_stream->time_base.num;
        atime_base_den          = audio_stream->time_base.den;

        audio_dec_ctx = audio_stream->codec;

        sample_rate = audio_dec_ctx->sample_rate;
        channels = audio_dec_ctx->channels;
        block_align =  audio_dec_ctx->block_align;
        bits_per_sample =  audio_dec_ctx->bits_per_coded_sample;

        if (audio_dec_ctx->channel_layout == 0) {
            audio_dec_ctx->channel_layout =
                av_get_default_channel_layout(audio_dec_ctx->channels);
        }

        AVCodec *codec = avcodec_find_decoder(audio_stream->codec->codec_id);
        if (avcodec_open2(audio_dec_ctx, codec, NULL) < 0) {
            fprintf(stderr, "could not open codec\n");
            exit(1);
        }
        if (codec) {
            // printf("Codec name %s\n", codec->name);
        }
    }

    return 0;
}
```

```
void setup_video_decode_component(ILCLIENT_T  *handle,
                                  char *decode_component_name,
                                  COMPONENT_T **decode_component) {
    int err;

    err = ilclient_create_component(handle,
                                    decode_component,
                                    decode_component_name,
                                    ILCLIENT_DISABLE_ALL_PORTS
                                    |
                                    ILCLIENT_ENABLE_INPUT_BUFFERS
                                    |
                                    ILCLIENT_ENABLE_OUTPUT_BUFFERS
                                    );
    exit_on_il_error(err, "DecodeComponent create failed\n");

    err = ilclient_change_component_state(*decode_component,
                                          OMX_StateIdle);
    exit_on_il_error(err, "Couldn't change state to Idle\n");

    // must be done before we enable buffers
    set_video_decoder_input_format(*decode_component);
}

void setup_video_render_component(ILCLIENT_T  *handle,
                                  char *render_component_name,
                                  COMPONENT_T **render_component) {
    int err;

    err = ilclient_create_component(handle,
                                    render_component,
                                    render_component_name,
                                    ILCLIENT_DISABLE_ALL_PORTS
                                    |
                                    ILCLIENT_ENABLE_INPUT_BUFFERS
                                    );
    exit_on_il_error(err, "RenderComponent create failed\n");


    err = ilclient_change_component_state(*render_component,
                                          OMX_StateIdle);
    exit_on_il_error(err, "Couldn't change state to Idle\n");
}

/* For the RPi name can be "hdmi" or "local" */
void set_output_device(OMX_HANDLETYPE handle, const char *name) {
    OMX_ERRORTYPE err;
    OMX_CONFIG_BRCMAUDIODESTINATIONTYPE arDest;
```

```
    if (name && strlen(name) < sizeof(arDest.sName)) {
        memset(&arDest, 0, sizeof(OMX_CONFIG_BRCMAUDIODESTINATIONTYPE));
        arDest.nSize = sizeof(OMX_CONFIG_BRCMAUDIODESTINATIONTYPE);
        arDest.nVersion.nVersion = OMX_VERSION;

        strcpy((char *)arDest.sName, name);

        err = OMX_SetParameter(handle, OMX_IndexConfigBrcmAudioDestination,
                               &arDest);
        exit_on_omx_error(err, "Error on setting audio destination %s\n");
    }
}

void set_pcm_mode(OMX_HANDLETYPE handle, int startPortNumber) {
    OMX_AUDIO_PARAM_PCMMODETYPE pcm_mode;
    OMX_ERRORTYPE err;

    OMX_INIT_STRUCTURE(pcm_mode);
    pcm_mode.nPortIndex = startPortNumber;

    err = OMX_GetParameter(handle, OMX_IndexParamAudioPcm, &pcm_mode);
    exit_on_omx_error(err, "Error getting audio params %s\n");

    pcm_mode.nSamplingRate = sample_rate;
    pcm_mode.nChannels = channels;
    pcm_mode.nBitPerSample = bits_per_sample;

    err = OMX_SetParameter(handle, OMX_IndexParamAudioPcm, &pcm_mode);
    exit_on_omx_error(err, "PCM mode unsupported %s\n");
}

 void set_audio_render_input_format(COMPONENT_T *component) {
    OMX_AUDIO_PARAM_PORTFORMATTYPE audio_port_format;

    OMX_INIT_STRUCTURE(audio_port_format);

#if AUDIO_DECODE
    int port_index = 120;
#else
    int port_index = 100;
#endif

    audio_port_format.nPortIndex = port_index;

    OMX_GetParameter(ilclient_get_handle(component),
                     OMX_IndexParamAudioPortFormat, &audio_port_format);


    audio_port_format.eEncoding = OMX_AUDIO_CodingPCM;
    OMX_SetParameter(ilclient_get_handle(component),
                     OMX_IndexParamAudioPortFormat, &audio_port_format);
```

```
#if AUDIO_DECODE

#else
    set_pcm_mode(ilclient_get_handle(component), 100);
#endif
}

void setup_resample_context() {
    swr = avresample_alloc_context();


    av_opt_set_int(swr, "in_channel_layout",
                   av_get_default_channel_layout(audio_dec_ctx->channels) , 0);
    av_opt_set_int(swr, "out_channel_layout", AV_CH_LAYOUT_STEREO,  0);
    av_opt_set_int(swr, "in_sample_rate",     audio_dec_ctx->sample_rate, 0);
    av_opt_set_int(swr, "out_sample_rate",    audio_dec_ctx->sample_rate, 0);
    av_opt_set_int(swr, "in_sample_fmt",  audio_dec_ctx->sample_fmt, 0);
    av_opt_set_int(swr, "out_sample_fmt", AV_SAMPLE_FMT_S16,  0);
    avresample_open(swr);
}

#if AUDIO_DECODE
void setup_audio_components(ILCLIENT_T  *handle,
                            char *decode_component_name,
                            COMPONENT_T **decode_component,
                            char *render_component_name,
                            COMPONENT_T **render_component) {
    int err;

    err = ilclient_create_component(handle,
                                    decode_component,
                                    decode_component_name,
                                    ILCLIENT_DISABLE_ALL_PORTS
                                    |
                                    ILCLIENT_ENABLE_INPUT_BUFFERS
                                    );
    exit_on_il_error(err, "DecodeComponent create failed\n");

    err = ilclient_change_component_state(*decode_component,
                                          OMX_StateIdle);
    exit_on_il_error(err,  "Couldn't change state to Idle\n");

    set_audio_render_input_format(*decode_component);

    err = ilclient_create_component(handle,
                                    render_component,
                                    render_component_name,
                                    ILCLIENT_DISABLE_ALL_PORTS
                                    |
                                    ILCLIENT_ENABLE_INPUT_BUFFERS
                                    );
```

```
    exit_on_il_error(err, "RenderComponent create failed\n");

    err = ilclient_change_component_state(*render_component,
                                          OMX_StateIdle);
    exit_on_il_error(err,  "Couldn't change state to Idle\n");


    // "local" or "hdmi"
    set_output_device(ilclient_get_handle(*render_component), "local");

#if 0 // omxplayer uses this sometimes
    OMX_CONFIG_BOOLEANTYPE bool_type;
    OMX_INIT_STRUCTURE(bool_type);

    bool_type.bEnabled = OMX_TRUE;
    OMX_SetParameter(ilclient_get_handle(*decode_component),
                     OMX_IndexParamBrcmDecoderPassThrough, &bool_type);
#endif
}
#else
void setup_audio_render_component(ILCLIENT_T  *handle,
                                  char *render_component_name,
                                  COMPONENT_T **render_component) {
    int err;

    err = ilclient_create_component(handle,
                                    render_component,
                                    render_component_name,
                                    ILCLIENT_DISABLE_ALL_PORTS
                                    |
                                    ILCLIENT_ENABLE_INPUT_BUFFERS
                                    );
    exit_on_il_error(err, "RenderComponent create failed\n");

    err = ilclient_change_component_state(*render_component,
                                          OMX_StateIdle);
    exit_on_il_error(err, "Couldn't change state to Idle\n");

    set_audio_render_input_format(*render_component);


    // "local" is 3.5 mm audio, "hdmi" is ... hdmi
    set_output_device(ilclient_get_handle(*render_component), "local");
}
#endif

void setup_scheduler_component(ILCLIENT_T  *handle,
                               char *scheduler_component_name,
                               COMPONENT_T **scheduler_component) {
    int err;
```

```
    err = ilclient_create_component(handle,
                                    scheduler_component,
                                    scheduler_component_name,
                                    ILCLIENT_DISABLE_ALL_PORTS
                                    |
                                    ILCLIENT_ENABLE_INPUT_BUFFERS
                                    );
    exit_on_il_error(err, "SchedulerComponent create failed\n");


    err = ilclient_change_component_state(*scheduler_component,
                                          OMX_StateIdle);
    exit_on_il_error(err,  "Couldn't change state to Idle\n");
}

void setup_clock_component(ILCLIENT_T  *handle,
                           char *clock_component_name,
                           COMPONENT_T **clock_component) {
    int err;


    err = ilclient_create_component(handle,
                                    clock_component,
                                    clock_component_name,
                                    ILCLIENT_DISABLE_ALL_PORTS
                                    );
    exit_on_il_error(err,  "ClockComponent create failed\n");

    err = ilclient_change_component_state(*clock_component,
                                          OMX_StateIdle);
    exit_on_il_error(err,  "Couldn't change state to Idle\n");

    OMX_TIME_CONFIG_ACTIVEREFCLOCKTYPE reference_clock;
    OMX_INIT_STRUCTURE(reference_clock);
    reference_clock.eClock = OMX_TIME_RefClockAudio;

    err = OMX_SetConfig(ilclient_get_handle(*clock_component),
                        OMX_IndexConfigTimeActiveRefClock, &reference_clock);
    exit_on_omx_error(err, "Set config on clock failed %s\n");

    OMX_TIME_CONFIG_SCALETYPE scale_type;
    OMX_INIT_STRUCTURE(scale_type);
    scale_type.xScale = 0x00010000;s

    err = OMX_SetConfig(ilclient_get_handle(*clock_component),
                        OMX_IndexConfigTimeScale, &scale_type);
    exit_on_omx_error(err, "Set config failed on clock %s\n");
}
```

```
void setup_components() {
    setup_video_decode_component(handle, video_decode_component_name,
                                 &video_decode_component);
    setup_video_render_component(handle, video_render_component_name,
                                 &video_render_component);
#if AUDIO_DECODE
    setup_audio_components(handle,
                           audio_decode_component_name,
                           &audio_decode_component,
                           audio_render_component_name,
                           &audio_render_component);
#else
    setup_audio_render_component(handle, audio_render_component_name,
                                 &audio_render_component);
#endif
    setup_scheduler_component(handle, scheduler_component_name,
                              &scheduler_component);
    setup_clock_component(handle, clock_component_name,
                          &clock_component);
}

void setup_video_tunnel(COMPONENT_T *decode_component,
                        COMPONENT_T *render_component,
                        COMPONENT_T *scheduler_component,
                        COMPONENT_T *clock_component) {
    int err;

    TUNNEL_T decode_tunnel;
    set_tunnel(&decode_tunnel, decode_component, 131, scheduler_component, 10);
    err = ilclient_setup_tunnel(&decode_tunnel, 0, 0);
    exit_on_il_error(err, "Error setting up decode tunnel\n");

    TUNNEL_T scheduler_tunnel;
    set_tunnel(&scheduler_tunnel, scheduler_component, 11, render_component, 90);
    err = ilclient_setup_tunnel(&scheduler_tunnel, 0, 0);
    exit_on_il_error(err, "Error setting up scheduler tunnel\n");

    TUNNEL_T clock_tunnel;
    set_tunnel(&clock_tunnel, clock_component, 80, scheduler_component, 12);
    err = ilclient_setup_tunnel(&clock_tunnel, 0, 0);
    exit_on_il_error(err, "Error setting up clock tunnel\n");

    start_clock(clock_component);
    print_clock_state(clock_component);

    // Okay to go back to processing data
    // enable the decode output ports
    OMX_SendCommand(ilclient_get_handle(decode_component),
                    OMX_CommandPortEnable, 131, NULL);
    ilclient_enable_port(decode_component, 131);
```

```
    // enable the clock output ports
    OMX_SendCommand(ilclient_get_handle(clock_component),
                    OMX_CommandPortEnable, 80, NULL);
    ilclient_enable_port(clock_component, 80);

    // enable the scheduler ports
    OMX_SendCommand(ilclient_get_handle(scheduler_component),
                    OMX_CommandPortEnable, 10, NULL);
    ilclient_enable_port(scheduler_component, 10);

    OMX_SendCommand(ilclient_get_handle(scheduler_component),
                    OMX_CommandPortEnable, 11, NULL);
    ilclient_enable_port(scheduler_component, 11);

    OMX_SendCommand(ilclient_get_handle(scheduler_component),
                    OMX_CommandPortEnable, 12, NULL);
    ilclient_enable_port(scheduler_component, 12);

    // enable the render input ports

    OMX_SendCommand(ilclient_get_handle(render_component),
                    OMX_CommandPortEnable, 90, NULL);
    ilclient_enable_port(render_component, 90);


    // set both components to executing state
    err = ilclient_change_component_state(decode_component,
                                          OMX_StateExecuting);
    exit_on_il_error(err, "Couldn't change state to Executing\n");

    err = ilclient_change_component_state(render_component,
                                          OMX_StateExecuting);
    exit_on_il_error(err, "Couldn't change state to Executing\n");

    err = ilclient_change_component_state(scheduler_component,
                                          OMX_StateExecuting);
    exit_on_il_error(err, "Couldn't change state to Executing\n");

    err = ilclient_change_component_state(clock_component,
                                          OMX_StateExecuting);
    exit_on_il_error(err, "Couldn't change state to Executing\n");
}

void setup_audio_tunnel(COMPONENT_T *audio_decode_component,
                        COMPONENT_T *audio_render_component,
                        COMPONENT_T *scheduler_component,
                        COMPONENT_T *clock_component) {
    int err;
```

```
    // for debugging
    OMX_AUDIO_PARAM_PORTFORMATTYPE audio_port_format;
    OMX_INIT_STRUCTURE(audio_port_format);
#if AUDIO_DECODE
    int port_index = 121;
#else
    int port_index = 100;
#endif
    audio_port_format.nPortIndex = port_index;

    OMX_GetParameter(ilclient_get_handle(audio_decode_component),
                     OMX_IndexParamAudioPortFormat, &audio_port_format);

    OMX_AUDIO_PARAM_PCMMODETYPE pcm_mode;

    OMX_INIT_STRUCTURE(pcm_mode);
    pcm_mode.nPortIndex = port_index;

    OMX_GetParameter(ilclient_get_handle(audio_decode_component),
                     OMX_IndexParamAudioPcm, &pcm_mode);
    printf("Sampling rate %d, channels %d\n",
           pcm_mode.nSamplingRate,
           pcm_mode.nChannels);
    // end debugging

    TUNNEL_T audio_decode_tunnel;
    set_tunnel(&audio_decode_tunnel, audio_decode_component, 121,
               audio_render_component, 100);
    err = ilclient_setup_tunnel(&audio_decode_tunnel, 0, 0);
    exit_on_il_error(err, "Error setting up decode tunnel\n");

    TUNNEL_T clock_tunnel;
    set_tunnel(&clock_tunnel, clock_component, 81, audio_render_component, 101);
    err = ilclient_setup_tunnel(&clock_tunnel, 0, 0);
    exit_on_il_error(err, "Error setting up clock tunnel\n");

    OMX_SendCommand(ilclient_get_handle(audio_decode_component),
                    OMX_CommandPortEnable, 121, NULL);
    ilclient_enable_port(audio_decode_component, 121);

    err = ilclient_change_component_state(audio_render_component,
                                          OMX_StateExecuting);
    exit_on_il_error(err, "Couldn't change state to  Executing\n");
}

void handle_video_packet(AVPacket *pkt) {
    int err;

    pkt->dts = convert_timestamp(pkt->dts, vtime_base_num, vtime_base_den);
    pkt->pts = convert_timestamp(pkt->pts, vtime_base_num, vtime_base_den);
```

```
    copy_video_into_buffer_and_empty(pkt,
                                     video_decode_component //,
                                     //buff_header
                                     );
    err = ilclient_wait_for_event(video_decode_component,
                                  OMX_EventPortSettingsChanged,
                                  131, 0, 0, 1,
                                  ILCLIENT_EVENT_ERROR | ILCLIENT_PARAMETER_CHANGED,
                                  0 /* no block */);
    if (err < 0) {
        //printf("No port settings change\n");
    } else {
        printf("Video port settings changed\n");
        setup_video_tunnel(video_decode_component,
                           video_render_component,
                           scheduler_component,
                           clock_component);
    }
}

void handle_audio_packet(AVPacket *pkt) {
    int err;
    int got_frame = 0;
    int len;
    AVFrame *decoded_frame = NULL;

    if (!decoded_frame) {
        if (!(decoded_frame = avcodec_alloc_frame())) {
            fprintf(stderr, "out of memory\n");
            exit(1);
        }
    }

    pkt->dts = convert_timestamp(pkt->dts, atime_base_num, atime_base_den);
    pkt->pts = convert_timestamp(pkt->pts, atime_base_num, atime_base_den);

    len = avcodec_decode_audio4(audio_dec_ctx,
                                decoded_frame, &got_frame, pkt);
    if (len < 0) {
        fprintf(stderr, "Error while decoding\n");
        exit(1);
    }
    if (got_frame) {
        /* if a frame has been decoded, we want to send it to OpenMAX */
        int data_size =
            av_samples_get_buffer_size(NULL, audio_dec_ctx->channels,
                                       decoded_frame->nb_samples,
                                       audio_dec_ctx->sample_fmt, 1);

        // Empty into decode/render_audio input buffers
```

```
#if AUDIO_DECODE
        read_audio_into_buffer_and_empty(decoded_frame,
                                         audio_decode_component,
                                         data_size
                                         );
#else
        read_audio_into_buffer_and_empty(decoded_frame,
                                         audio_render_component,
                                         data_size
                                         );
#endif
    }
#if AUDIO_DECODE
    err = ilclient_wait_for_event(audio_decode_component,
                                  OMX_EventPortSettingsChanged,
                                  121, 0, 0, 1,
                                  ILCLIENT_EVENT_ERROR | ILCLIENT_PARAMETER_CHANGED,
                                  0 /* no block */);
    if (err < 0) {
        //printf("No port settings change\n");
    } else {
        printf("Audio port settings changed\n");
        setup_audio_tunnel(audio_decode_component,
                           audio_render_component,
                           scheduler_component,
                           clock_component);
    }
#endif
}

int main(int argc, char** argv) {
    int err;

    // override default file
    if (argc > 1) {
        fname = argv[1];
    }

    setup_demuxer(fname);

    bcm_host_init();
    // vcos_log_set_level(VCOS_LOG_CATEGORY, VCOS_LOG_TRACE);

    handle = ilclient_init();
    if (handle == NULL) {
        fprintf(stderr, "IL client init failed\n");
        exit(1);
    }
```

```
    if (OMX_Init() != OMX_ErrorNone) {
        ilclient_destroy(handle);
        fprintf(stderr, "OMX init failed\n");
        exit(1);
    }

    setup_callbacks();
    setup_components();
    setup_resample_context();

    start_clock(clock_component);
    print_clock_state(clock_component);

    // both components now in Idle state, no buffers, ports disabled

    // video input port
    err = ilclient_enable_port_buffers(video_decode_component, 130,
                                       NULL, NULL, NULL);
    exit_on_il_error(err, "Couldn't enable buffers\n");

    ilclient_enable_port(video_decode_component, 130);

    err = ilclient_change_component_state(video_decode_component,
                                          OMX_StateExecuting);
    exit_on_il_error(err, "Couldn't change state to  Executing\n");

    // audio render
#if AUDIO_DECODE
    ilclient_enable_port_buffers(audio_decode_component, 120,
                                 NULL, NULL, NULL);
    ilclient_enable_port(audio_decode_component, 120);

    err = ilclient_change_component_state(audio_decode_component,
                                          OMX_StateExecuting);
    send_audio_decoder_config(audio_decode_component, NULL);
#else
    ilclient_enable_port_buffers(audio_render_component, 100,
                                 NULL, NULL, NULL);
    ilclient_enable_port(audio_render_component, 100);

    err = ilclient_change_component_state(audio_render_component,
                                          OMX_StateExecuting);
#endif
    exit_on_il_error(err, "Couldn't change state to  Executing\n");

    send_video_decoder_config(video_decode_component);

    /* read frames from the file */
    while (av_read_frame(format_ctx, &pkt) >= 0) {
        AVPacket orig_pkt = pkt;
```

```
        if (pkt.stream_index == video_stream_idx) {
            handle_video_packet(&pkt);
        } else if (pkt.stream_index == audio_stream_idx) {
            handle_audio_packet(&pkt);
        }
        av_free_packet(&orig_pkt);
    }

    ilclient_wait_for_event(video_render_component,
                            OMX_EventBufferFlag,
                            90, 0, OMX_BUFFERFLAG_EOS, 0,
                            ILCLIENT_BUFFER_FLAG_EOS, 10000);
    printf("EOS on render\n");

    exit(0);
}
```

Conclusion

This chapter looked at playing files containing both audio and video streams. You have considered only a limited number of formats: MP4 files containing H.264 video and AAC audio. The player omxplayer deals with far more cases and hence has increased complexity. Nevertheless, the concepts discussed and demonstrated here are applicable in those cases too.

Resources

- *Proposal for OMX.broadcom.read_media*: www.raspberrypi.org/forums/viewtopic.php?t=32601&p=280735
- *H.264 extradata (partially) explained - for dummies by Aviad Rozenhek*: http://aviadr1.blogspot.com/2010/05/h264-extradata-partially-explained-for.html
- *Possible Locations for Sequence/Picture Parameter Set(s) for H.264 Stream byszatmary*: http://stackoverflow.com/questions/24884827/possible-locations-for-sequence-picture-parameter-sets-for-h-264-stream
- *Tomaka's blog: libavcodec/libavformat tutorial*: http://blog.tomaka17.com/2012/03/libavcodeclibavformat-tutorial/
- *An ffmpeg and SDL Tutorial: Tutorial 05: Synching Video by dranger*: http://dranger.com/ffmpeg/tutorial05.html

CHAPTER 17

Basic OpenVG on the Raspberry Pi

OpenVG is an application programming interface (API) for hardware-accelerated two-dimensional vector and raster graphics. It has not apparently proven immensely popular, but I will need it for drawing text overlays on video images in a later chapter. This is a minimal chapter on OpenVG.

Building Programs

You can build the programs in this chapter using the following `Makefile`:

```
DMX_INC =  -I/opt/vc/include/ -I /opt/vc/include/interface/vmcs_host/ -I/opt/vc/include/
interface/vcos/pthreads -I/opt/vc/include/interface/vmcs_host/linux
EGL_INC =
OPENVG_INC =
INCLUDES = $(DMX_INC) $(EGL_INC) $(OPENVG_INC)

CFLAGS= $(INCLUDES)
CPPFLAGS =

DMX_LIBS =  -L/opt/vc/lib/ -lbcm_host -lvcos -lvchiq_arm -lpthread
EGL_LIBS = -L/opt/vc/lib/ -lEGL -lGLESv2
OPENVG_LIBS = -lOpenVG

LDFLAGS =  $(DMX_LIBS) $(EGL_LIBS) $(OPENVG_LIBS)

SRC = ellipse.c simple_shape.c window.c image.c

all: ellipse simple_shape window image
```

Introduction

The OpenVG API is for hardware-accelerated two-dimensional vector and raster graphics. According to the OpenVG specification, it is intended to provide the following:

> *SVG and Adobe Flash viewers*: OpenVG can provide the drawing functionality required for high-performance SVG document viewers that are conformant with version 1.2 of the SVG Tiny profile.

J. Newmarch, *Raspberry Pi GPU Audio Video Programming*, DOI 10.1007/978-1-4842-2472-4_17

Portable mapping applications: OpenVG can provide dynamic features for map display that would be difficult or impossible to do with an SVG or Flash viewer alone, such as dynamic placement and sizing of street names and markers and efficient viewport culling.

E-book readers: The OpenVG API can provide fast rendering of readable text in Western, Asian, and other scripts.

Games: The OpenVG API is useful for defining sprites, backgrounds, and textures for use in both 2D and 3D games. It can provide two-dimensional overlays (for example, for maps or scores) on top of 3D content.

Scalable user interfaces: OpenVG can be used to render scalable user interfaces, particularly for applications that want to present users with a unique look and feel that is consistent across different screen resolutions.

Low-level graphics device interface: OpenVG can be used as a low-level graphics device interface. Other graphical toolkits, such as windowing systems, may be implemented above OpenVG.

OpenVG hasn't taken off in general. Mesa is the main open source supplier for many graphical toolkits such as OpenGL. While it previously supported OpenVG, this has recently been removed, according to `http://cgit.freedesktop.org/mesa/mesa/commit/?id=3acd7a34ab05b87521b74f626ec637e7fdcc6595`.

OpenVG API seems to have dwindled away. The code would still be interesting if we wanted to implement NV_path_rendering, but given the trend of the next-gen graphics APIs, it seems unlikely that this becomes ARB or core.

The "next-gen" graphics API this is referring to is probably Vulkan. On the other hand, there are now more than 10 million RPi computers out there capable of running OpenVG. Will that make a difference?

Dispmanx and EGL

Just like OpenGL, OpenVG depends on an EGL surface, which on the RPi is created using Dispmanx. The bare-minimum program is just like `window.c` from Chapter 5 with these two changes:

- The API is bound to OpenVG rather than OpenGL ES with the following:

```
eglBindAPI(EGL_OPENVG_API);
```

- Drawing is done using OpenVG calls, where the minimum to clear and draw against a white background is as follows:

```
float c = 1.0;
float clearColor[4] = {c, c, c, 0.5}; // white, semi transparent
vgSetfv(VG_CLEAR_COLOR, 4, clearColor);
vgClear(0, 0, 1920, 1080); // window_width(), window_height());

// do lots of drawing in here

vgFlush();
```

The program is `window.c`.

```
/*
 * code adapted from openGL-RPi-tutorial-master/encode_OGL/
 */

#include <stdio.h>
#include <assert.h>
#include <math.h>
#include <stdlib.h>

#include <EGL/egl.h>
#include <EGL/eglext.h>
#include <GLES2/gl2.h>
#include <VG/openvg.h>
#include <VG/vgu.h>

typedef struct
{
    EGLDisplay display;
    EGLSurface surface;
    EGLContext context;
    EGLConfig config;
} EGL_STATE_T;

struct egl_manager {
    EGLNativeDisplayType xdpy;
    EGLNativeWindowType xwin;
    EGLNativePixmapType xpix;

    EGLDisplay dpy;
    EGLConfig conf;
    EGLContext ctx;

    EGLSurface win;
    EGLSurface pix;
    EGLSurface pbuf;
    EGLImageKHR image;

    EGLBoolean verbose;
    EGLint major, minor;
};

EGL_STATE_T state, *p_state = &state;

void init_egl(EGL_STATE_T *state)
{
    EGLint num_configs;
    EGLBoolean result;

    //bcm_host_init();
```

```
    static const EGLint attribute_list[] =
        {
            EGL_RED_SIZE, 8,
            EGL_GREEN_SIZE, 8,
            EGL_BLUE_SIZE, 8,
            EGL_ALPHA_SIZE, 8,
            EGL_SURFACE_TYPE, EGL_WINDOW_BIT,
            EGL_RENDERABLE_TYPE, EGL_OPENVG_BIT,
            EGL_NONE
        };

    static const EGLint context_attributes[] =
        {
            EGL_CONTEXT_CLIENT_VERSION, 2,
            EGL_NONE
        };

    // get an EGL display connection
    state->display = eglGetDisplay(EGL_DEFAULT_DISPLAY);

    // initialize the EGL display connection
    result = eglInitialize(state->display, NULL, NULL);

    // get an appropriate EGL frame buffer configuration
    result = eglChooseConfig(state->display, attribute_list, &state->config, 1,
            &num_configs);
    assert(EGL_FALSE != result);

    //result = eglBindAPI(EGL_OPENGL_ES_API);
    result = eglBindAPI(EGL_OPENVG_API);
    assert(EGL_FALSE != result);

    // create an EGL rendering context
    state->context = eglCreateContext(state->display,
                                      state->config, EGL_NO_CONTEXT,
                                      NULL);
                                      // breaks if we use this: context_attributes);
    assert(state->context!=EGL_NO_CONTEXT);
}

void init_dispmanx(EGL_DISPMANX_WINDOW_T *nativewindow) {
    int32_t success = 0;
    uint32_t screen_width;
    uint32_t screen_height;

    DISPMANX_ELEMENT_HANDLE_T dispman_element;
    DISPMANX_DISPLAY_HANDLE_T dispman_display;
    DISPMANX_UPDATE_HANDLE_T dispman_update;
    VC_RECT_T dst_rect;
    VC_RECT_T src_rect;

    bcm_host_init();
```

```
    // create an EGL window surface
    success = graphics_get_display_size(0 /* LCD */,
                                        &screen_width,
                                        &screen_height);
    assert( success >= 0 );

    dst_rect.x = 0;
    dst_rect.y = 0;
    dst_rect.width = screen_width;
    dst_rect.height = screen_height;

    src_rect.x = 0;
    src_rect.y = 0;
    src_rect.width = screen_width << 16;
    src_rect.height = screen_height << 16;

    dispman_display = vc_dispmanx_display_open( 0 /* LCD */);
    dispman_update = vc_dispmanx_update_start( 0 );

    dispman_element =
        vc_dispmanx_element_add(dispman_update, dispman_display,
                                0/*layer*/, &dst_rect, 0/*src*/,
                                &src_rect, DISPMANX_PROTECTION_NONE,
                                0 /*alpha*/, 0/*clamp*/, 0/*transform*/);

    // Build an EGL_DISPMANX_WINDOW_T from the Dispmanx window
    nativewindow->element = dispman_element;
    nativewindow->width = screen_width;
    nativewindow->height = screen_height;
    vc_dispmanx_update_submit_sync(dispman_update);

    printf("Got a Dispmanx window\n");
}

void egl_from_dispmanx(EGL_STATE_T *state,
                       EGL_DISPMANX_WINDOW_T *nativewindow) {
    EGLBoolean result;

    state->surface = eglCreateWindowSurface(state->display,
                                            state->config,
                                            nativewindow, NULL );
    assert(state->surface != EGL_NO_SURFACE);

    // connect the context to the surface
    result = eglMakeCurrent(state->display, state->surface, state->surface, state->context);
    assert(EGL_FALSE != result);
}

void draw(){
    float c = 1.0;
    float clearColor[4] = {c, c, c, 0.5};  // white, no transparency
    vgSetfv(VG_CLEAR_COLOR, 4, clearColor);
```

```
    vgClear(0, 0, 1920, 1080); // window_width(), window_height());

    // do lots of drawing in here

   vgFlush();
}

int
main(int argc, char *argv[])
{
    EGL_DISPMANX_WINDOW_T nativewindow;

    init_egl(p_state);
    init_dispmanx(&nativewindow);
    egl_from_dispmanx(p_state, &nativewindow);

    draw();
    eglSwapBuffers(p_state->display, p_state->surface);

    sleep(10);
    eglTerminate(p_state->display);

    exit(0);
}
```

This is run by `window` and just clears the screen and shows a white background.

OpenVG Pipeline

OpenVG, like any graphics system, has a multistage pipeline to process drawing. The OpenVG specification shows these for drawing a colored dashed line on a scene of images and various shapes.

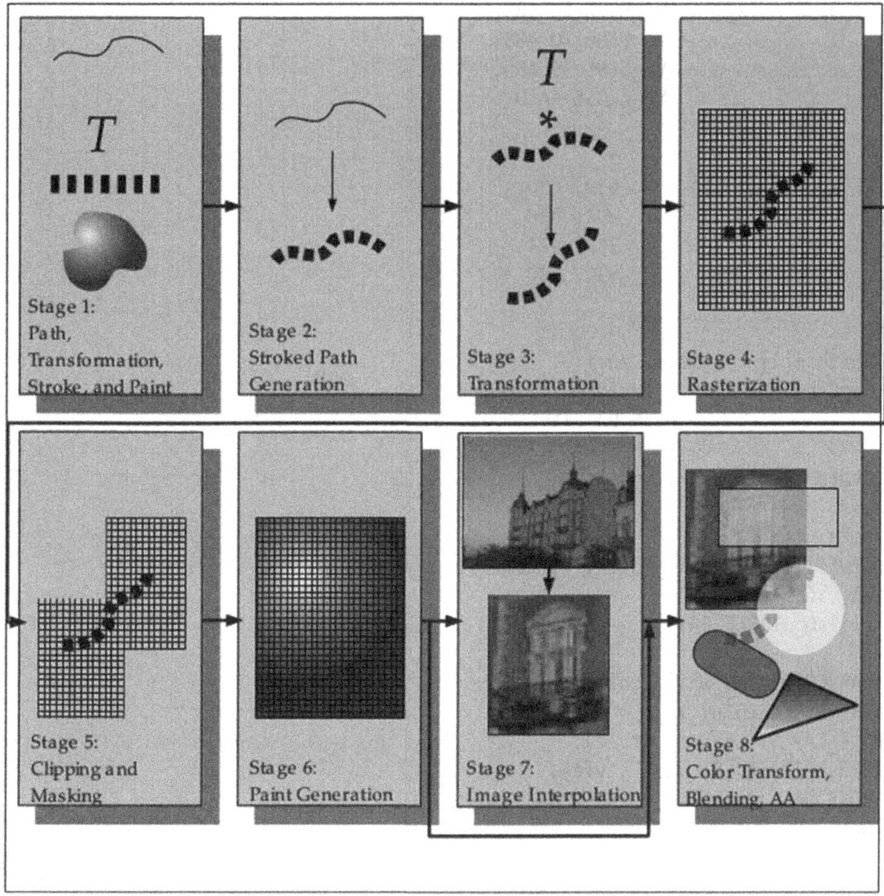

To actually draw anything means specifying what might happen at each of these stages.

Drawing a Pink Triangle

In OpenVG you draw and fill *paths*. In the simplest cases, the paths are made up of lines. These can be solid or made up of various dots and dashes. They can be straight or curved in various ways. Where lines join, there are several ways the joins can be managed. The path can be filled with a color, and the lines themselves can be colored. These can all be set by calls to vgSetxxx, where xxx is used to tell the data type being set. For example, vgSetfv means the parameter is a vector of four float values. At the conclusion of setting a path, the path is drawn by vgDrawPath.

A pink dashed triangle is drawn with the following:

```
VGint cap_style = VG_CAP_BUTT;
VGint join_style = VG_JOIN_MITER;
VGfloat color[4] = {1.0, 0.1, 1.0, 0.2}; // pink
VGPaint fill;
```

```
    static const VGubyte cmds[] = {VG_MOVE_TO_ABS,
                                   VG_LINE_TO_ABS,
                                   VG_LINE_TO_ABS,
                                   VG_CLOSE_PATH
    };

    static const VGfloat coords[]   = {630, 630,
                                       902, 630,
                                       750, 924,
                                       630, 630
    };

    VGfloat dash_pattern[2] = { 20.f, 20.f };
    VGPath path = vgCreatePath(VG_PATH_FORMAT_STANDARD,
                               VG_PATH_DATATYPE_F, 1, 0, 0, 0,
                               VG_PATH_CAPABILITY_APPEND_TO);
    vgAppendPathData(path, 4, cmds, coords);

    fill = vgCreatePaint();
    vgSetParameterfv(fill, VG_PAINT_COLOR, 4, color);
    vgSetPaint(fill, VG_FILL_PATH);
    vgSetPaint(fill, VG_STROKE_PATH);

    vgSetfv(VG_CLEAR_COLOR, 4, white_color);
    vgSetf(VG_STROKE_LINE_WIDTH, 20);
    vgSeti(VG_STROKE_CAP_STYLE, cap_style);
    vgSeti(VG_STROKE_JOIN_STYLE, join_style);
    vgSetfv(VG_STROKE_DASH_PATTERN, 2, dash_pattern);
    vgSetf(VG_STROKE_DASH_PHASE, 0.0f);

    vgDrawPath(path, VG_STROKE_PATH|VG_FILL_PATH);
```

The program is `simple_shape.c`.

```
/*
 * code adapted from openGL-RPi-tutorial-master/encode_OGL/
 */

#include <stdio.h>
#include <assert.h>
#include <math.h>
#include <stdlib.h>

#include <EGL/egl.h>
#include <EGL/eglext.h>
#include <GLES2/gl2.h>
#include <VG/openvg.h>
#include <VG/vgu.h>
```

```
typedef struct
{
    EGLDisplay display;
    EGLSurface surface;
    EGLContext context;
    EGLConfig config;
} EGL_STATE_T;

struct egl_manager {
    EGLNativeDisplayType xdpy;
    EGLNativeWindowType xwin;
    EGLNativePixmapType xpix;

    EGLDisplay dpy;
    EGLConfig conf;
    EGLContext ctx;

    EGLSurface win;
    EGLSurface pix;
    EGLSurface pbuf;
    EGLImageKHR image;

    EGLBoolean verbose;
    EGLint major, minor;
};

EGL_STATE_T state, *p_state = &state;

void init_egl(EGL_STATE_T *state)
{
    EGLint num_configs;
    EGLBoolean result;

    //bcm_host_init();

    static const EGLint attribute_list[] =
        {
            EGL_RED_SIZE, 8,
            EGL_GREEN_SIZE, 8,
            EGL_BLUE_SIZE, 8,
            EGL_ALPHA_SIZE, 8,
            EGL_SURFACE_TYPE, EGL_WINDOW_BIT,
            EGL_RENDERABLE_TYPE, EGL_OPENVG_BIT,
            EGL_NONE
        };

    static const EGLint context_attributes[] =
        {
            EGL_CONTEXT_CLIENT_VERSION, 2,
            EGL_NONE
        };
```

```
    // get an EGL display connection
    state->display = eglGetDisplay(EGL_DEFAULT_DISPLAY);

    // initialize the EGL display connection
    result = eglInitialize(state->display, NULL, NULL);

    // get an appropriate EGL frame buffer configuration
    result = eglChooseConfig(state->display, attribute_list, &state->config, 1,
            &num_configs);
    assert(EGL_FALSE != result);

    //result = eglBindAPI(EGL_OPENGL_ES_API);
    result = eglBindAPI(EGL_OPENVG_API);
    assert(EGL_FALSE != result);

    // create an EGL rendering context
    state->context = eglCreateContext(state->display,
                                      state->config, EGL_NO_CONTEXT,
                                      NULL);
                                      // breaks if we use this: context_attributes);
    assert(state->context!=EGL_NO_CONTEXT);
}

void init_dispmanx(EGL_DISPMANX_WINDOW_T *nativewindow) {
    int32_t success = 0;
    uint32_t screen_width;
    uint32_t screen_height;

    DISPMANX_ELEMENT_HANDLE_T dispman_element;
    DISPMANX_DISPLAY_HANDLE_T dispman_display;
    DISPMANX_UPDATE_HANDLE_T dispman_update;
    VC_RECT_T dst_rect;
    VC_RECT_T src_rect;

    bcm_host_init();

    // create an EGL window surface
    success = graphics_get_display_size(0 /* LCD */,
                                        &screen_width,
                                        &screen_height);
    assert( success >= 0 );

    dst_rect.x = 0;
    dst_rect.y = 0;
    dst_rect.width = screen_width;
    dst_rect.height = screen_height;

    src_rect.x = 0;
    src_rect.y = 0;
    src_rect.width = screen_width << 16;
    src_rect.height = screen_height << 16;
```

```
    dispman_display = vc_dispmanx_display_open( 0 /* LCD */);
    dispman_update = vc_dispmanx_update_start( 0 );

    dispman_element =
        vc_dispmanx_element_add(dispman_update, dispman_display,
                                0/*layer*/, &dst_rect, 0/*src*/,
                                &src_rect, DISPMANX_PROTECTION_NONE,
                                0 /*alpha*/, 0/*clamp*/, 0/*transform*/);

    // Build an EGL_DISPMANX_WINDOW_T from the Dispmanx window
    nativewindow->element = dispman_element;
    nativewindow->width = screen_width;
    nativewindow->height = screen_height;
    vc_dispmanx_update_submit_sync(dispman_update);

    printf("Got a Dispmanx window\n");
}

void egl_from_dispmanx(EGL_STATE_T *state,
                       EGL_DISPMANX_WINDOW_T *nativewindow) {
    EGLBoolean result;

    state->surface = eglCreateWindowSurface(state->display,
                                            state->config,
                                            nativewindow, NULL );
    assert(state->surface != EGL_NO_SURFACE);

    // connect the context to the surface
    result = eglMakeCurrent(state->display, state->surface, state->surface, state->context);
    assert(EGL_FALSE != result);
}

void simple_shape() {
    VGint cap_style = VG_CAP_BUTT;
    VGint join_style = VG_JOIN_MITER;
    VGfloat color[4] = {1.0, 0.1, 1.0, 0.2};
    VGfloat white_color[4] = {1.0, 1.0, 1.0, 0.0}; //1.0};
    VGPaint fill;

    static const VGubyte cmds[] = {VG_MOVE_TO_ABS,
                                   VG_LINE_TO_ABS,
                                   VG_LINE_TO_ABS,
                                   VG_CLOSE_PATH
    };

    static const VGfloat coords[]   = {630, 630,
                                       902, 630,
                                       750, 924
    };
```

```
    VGfloat dash_pattern[2] = { 20.f, 20.f };
    VGPath path = vgCreatePath(VG_PATH_FORMAT_STANDARD,
                               VG_PATH_DATATYPE_F, 1, 0, 0, 0,
                               VG_PATH_CAPABILITY_APPEND_TO);
    vgAppendPathData(path, 4, cmds, coords);

    fill = vgCreatePaint();
    vgSetParameterfv(fill, VG_PAINT_COLOR, 4, color);
    vgSetPaint(fill, VG_FILL_PATH);
    vgSetPaint(fill, VG_STROKE_PATH);

    vgSetfv(VG_CLEAR_COLOR, 4, white_color);
    vgSetf(VG_STROKE_LINE_WIDTH, 20);
    vgSeti(VG_STROKE_CAP_STYLE, cap_style);
    vgSeti(VG_STROKE_JOIN_STYLE, join_style);
    vgSetfv(VG_STROKE_DASH_PATTERN, 2, dash_pattern);
    vgSetf(VG_STROKE_DASH_PHASE, 0.0f);

    vgDrawPath(path, VG_STROKE_PATH|VG_FILL_PATH);
}

void draw(){
    float c = 1.0;
    float clearColor[4] = {c, c, c, c};  // white, no transparency
    vgSetfv(VG_CLEAR_COLOR, 4, clearColor);

    vgClear(0, 0, 1920, 1080);

    simple_shape();

    vgFlush();
}

int
main(int argc, char *argv[])
{
    EGL_DISPMANX_WINDOW_T nativewindow;

    init_egl(p_state);
    init_dispmanx(&nativewindow);
    egl_from_dispmanx(p_state, &nativewindow);

    draw();
    eglSwapBuffers(p_state->display, p_state->surface);

    sleep(100);
    eglTerminate(p_state->display);

    exit(0);
}
```

This is run with `simple_shape`.

The figure drawn is as follows (image captured by raspi2png):

Drawing Standard Shapes

The VGU utility library is an optional library that provides a set of common shapes such as polygons, rectangles, and ellipses. The shapes still need to have the stroke and fill parameters set, and so on. The program `ellipse.c` draws a purple rectangle that decreases in height on each display.

```
#include <stdio.h>
#include <stdlib.h>
//#include <sys/stat.h>
#include <signal.h>

#include <assert.h>
#include <vc_dispmanx.h>
#include <bcm_host.h>
#include <EGL/egl.h>
#include <EGL/eglext.h>
#include <GLES2/gl2.h>
#include <VG/openvg.h>
#include <VG/vgu.h>

typedef struct
{
    EGLDisplay display;
    EGLSurface surface;
    EGLContext context;
    EGLConfig config;
} EGL_STATE_T;

EGL_STATE_T state, *p_state = &state;

void init_egl(EGL_STATE_T *state)
{
    EGLint num_configs;
    EGLBoolean result;
```

```
    //bcm_host_init();

    static const EGLint attribute_list[] =
        {
            EGL_RED_SIZE, 8,
            EGL_GREEN_SIZE, 8,
            EGL_BLUE_SIZE, 8,
            EGL_ALPHA_SIZE, 8,
            EGL_SURFACE_TYPE, EGL_WINDOW_BIT,
            EGL_SAMPLES, 1,
            EGL_NONE
        };

    static const EGLint context_attributes[] =
        {
            EGL_CONTEXT_CLIENT_VERSION, 2,
            EGL_NONE
        };

    // get an EGL display connection
    state->display = eglGetDisplay(EGL_DEFAULT_DISPLAY);

    // initialize the EGL display connection
    result = eglInitialize(state->display, NULL, NULL);

    // get an appropriate EGL frame buffer configuration
    result = eglChooseConfig(state->display, attribute_list, &state->config, 1,
            &num_configs);
    assert(EGL_FALSE != result);

    // Choose the OpenVG API
    result = eglBindAPI(EGL_OPENVG_API);
    assert(EGL_FALSE != result);

    // create an EGL rendering context
    state->context = eglCreateContext(state->display,
                                      state->config,
                                      NULL, // EGL_NO_CONTEXT,
                                      NULL);
                                      // breaks if we use this: context_attributes);
    assert(state->context!=EGL_NO_CONTEXT);
}

void init_dispmanx(EGL_DISPMANX_WINDOW_T *nativewindow) {
    int32_t success = 0;
    uint32_t screen_width;
    uint32_t screen_height;

    DISPMANX_ELEMENT_HANDLE_T dispman_element;
    DISPMANX_DISPLAY_HANDLE_T dispman_display;
    DISPMANX_UPDATE_HANDLE_T dispman_update;
    VC_RECT_T dst_rect;
    VC_RECT_T src_rect;
```

```
    bcm_host_init();

    // create an EGL window surface
    success = graphics_get_display_size(0 /* LCD */,
                                        &screen_width,
                                        &screen_height);
    assert( success >= 0 );

    dst_rect.x = 0;
    dst_rect.y = 0;
    dst_rect.width = screen_width;
    dst_rect.height = screen_height;

    src_rect.x = 0;
    src_rect.y = 0;
    src_rect.width = screen_width << 16;
    src_rect.height = screen_height << 16;

    dispman_display = vc_dispmanx_display_open( 0 /* LCD */);
    dispman_update = vc_dispmanx_update_start( 0 );

    dispman_element =
        vc_dispmanx_element_add(dispman_update, dispman_display,
                                0 /*layer*/, &dst_rect, 0 /*src*/,
                                &src_rect, DISPMANX_PROTECTION_NONE,
                                0 /*alpha*/, 0 /*clamp*/, 0 /*transform*/);

    // Build an EGL_DISPMANX_WINDOW_T from the Dispmanx window
    nativewindow->element = dispman_element;
    nativewindow->width = screen_width;
    nativewindow->height = screen_height;
    vc_dispmanx_update_submit_sync(dispman_update);

    printf("Got a Dispmanx window\n");
}

void egl_from_dispmanx(EGL_STATE_T *state,
                       EGL_DISPMANX_WINDOW_T *nativewindow) {
    EGLBoolean result;
    static const EGLint attribute_list[] =
        {
            EGL_RENDER_BUFFER, EGL_SINGLE_BUFFER,
            EGL_NONE
        };

    state->surface = eglCreateWindowSurface(state->display,
                                            state->config,
                                            nativewindow,
                                            NULL );
                                            //attribute_list);
    assert(state->surface != EGL_NO_SURFACE);
```

```
    // connect the context to the surface
    result = eglMakeCurrent(state->display, state->surface, state->surface, state->context);
    assert(EGL_FALSE != result);
}

// setfill sets the fill color
void setfill(float color[4]) {
    VGPaint fillPaint = vgCreatePaint();
    vgSetParameteri(fillPaint, VG_PAINT_TYPE, VG_PAINT_TYPE_COLOR);
    vgSetParameterfv(fillPaint, VG_PAINT_COLOR, 4, color);
    vgSetPaint(fillPaint, VG_FILL_PATH);
    vgDestroyPaint(fillPaint);
}

// setstroke sets the stroke color and width
void setstroke(float color[4], float width) {
    VGPaint strokePaint = vgCreatePaint();
    vgSetParameteri(strokePaint, VG_PAINT_TYPE, VG_PAINT_TYPE_COLOR);
    vgSetParameterfv(strokePaint, VG_PAINT_COLOR, 4, color);
    vgSetPaint(strokePaint, VG_STROKE_PATH);
    vgSetf(VG_STROKE_LINE_WIDTH, width);
    vgSeti(VG_STROKE_CAP_STYLE, VG_CAP_BUTT);
    vgSeti(VG_STROKE_JOIN_STYLE, VG_JOIN_MITER);
    vgDestroyPaint(strokePaint);
}

// Ellipse makes an ellipse at the specified location and dimensions, applying style
void Ellipse(float x, float y, float w, float h, float sw, float fill[4], float stroke[4]) {
    VGPath path = vgCreatePath(VG_PATH_FORMAT_STANDARD, VG_PATH_DATATYPE_F, 1.0f, 0.0f, 0,
                  0, VG_PATH_CAPABILITY_ALL);
    vguEllipse(path, x, y, w, h);
    setfill(fill);
    setstroke(stroke, sw);
    vgDrawPath(path, VG_FILL_PATH | VG_STROKE_PATH);
    vgDestroyPath(path);
}

void draw() {
    EGL_DISPMANX_WINDOW_T nativewindow;

    init_egl(p_state);
    init_dispmanx(&nativewindow);

    egl_from_dispmanx(p_state, &nativewindow);

    vgClear(0, 0, 1920, 1080);

    VGfloat color[4] = {0.4, 0.1, 1.0, 1.0}; // purple
    float c = 1.0;
    float clearColor[4] = {c, c, c, c}; // white non-transparent
    vgSetfv(VG_CLEAR_COLOR, 4, clearColor);
    vgClear(0, 0, 1920, 1080);
```

```
    vgSeti(VG_BLEND_MODE, VG_BLEND_SRC_OVER);
    static float ht = 200;
    while (1) {
        vgClear(0, 0, 1920, 1080);
        Ellipse(1920/2, 1080/2, 400, ht--, 0, color, color);
        if (ht <= 0) {
            ht = 200;
        }
        eglSwapBuffers(p_state->display, p_state->surface);
    }
    assert(vgGetError() == VG_NO_ERROR);

    vgFlush();

    eglSwapBuffers(p_state->display, p_state->surface);
}

void sig_handler(int sig) {
    eglTerminate(p_state->display);
    exit(1);
}

int main(int argc, char** argv) {
    signal(SIGINT, sig_handler);

    draw();

    exit(0);
}
```

This is run by **ellipse**.
It looks like this:

Images

OpenVG has the type VGImage to represent images. An image has a format such as RGB, and each pixel takes a number of bytes. For example, the format VG_sXRGB_8888 takes up 4 bytes. Each image consists of rows of pixels, and some images round up the number of bytes per row to, for example, a multiple of 32 bytes. This rounded-up number is called the *stride*. For example, 100 pixels in VG_sXRGB_8888 format would take up 400 bytes, but this might be rounded up to a stride of 416 bytes.

Images can come from a number of sources.

- An external file may hold an image in, for example, PNG format. This will have to be loaded into memory and decoded into the correct format for a `VGImage`.
- A program may generate an image in memory in the format required for a `VGImage`.
- An image may be created as an off-screen buffer, and VG drawing calls can be made into it.

The resulting image can be drawn to the current OpenVG surface, either in total or in part.

Creating a VGImage from Memory Data

You can create a grayscale image where each pixel takes 1 byte by creating an image of type `VG_sL_8`. You can choose the size as 256 × 256. The data for this will then be stored in a 256 × 256 unsigned `char` array. For simplicity, you just fill it with increasing values (overflowing when necessary).

```
VGImage img;
img = vgCreateImage(VG_sL_8,
                    256, 256,
                    VG_IMAGE_QUALITY_NONANTIALIASED);
if (img == VG_INVALID_HANDLE) {
    fprintf(stderr, "Can't create simple image\n");
    fprintf(stderr, "Error code %x\n", vgGetError());
    exit(2);
}
unsigned char val;
unsigned char data[256*256];
int n;

for (n = 0; n < 256*256; n++) {
    val = (unsigned char) n;
    data[n] = val;
}
```

To load an array into an image, you use the function `vgImageSubData`. This takes the image, the data array, the image format, the stride, and the image coordinates.

```
vgImageSubData(img, data,
               256, VG_sL_8,
               0, 0, 256, 256);
```

Drawing an Image

To draw the image, you must set up an "image-user-to-surface" transformation. This requires setting up a matrix to transform the image to user coordinates, such as a projective transformation. You just use an identity matrix. The resulting image may be subjected to operations such as translation to a different origin. The image can then be drawn as follows:

```
vgSeti(VG_MATRIX_MODE, VG_MATRIX_IMAGE_USER_TO_SURFACE);
vgLoadIdentity();
vgTranslate(200, 30);

vgDrawImage(img);
```

A function to create and draw an image is then as follows:

```
void simple_image() {
    VGImage img;
    img = vgCreateImage(VG_sL_8,
                        256, 256,
                        VG_IMAGE_QUALITY_NONANTIALIASED);
    if (img == VG_INVALID_HANDLE) {
        fprintf(stderr, "Can't create simple image\n");
        fprintf(stderr, "Error code %x\n", vgGetError());
        exit(2);
    }
    unsigned char val;
    unsigned char data[256*256];
    int n;

    for (n = 0; n < 256*256; n++) {
        val = (unsigned char) n;
        data[n] = val;
    }
    vgImageSubData(img, data,
                   256, VG_sL_8,
                   0, 0, 256, 256);
    vgSeti(VG_MATRIX_MODE, VG_MATRIX_IMAGE_USER_TO_SURFACE);
    vgLoadIdentity();
    vgTranslate(200, 30);

    vgDrawImage(img);

    vgDestroyImage(img);
}
```

The image looks like this:

Drawing an Image Using OpenVG Calls

OpenVG can also be used to draw into an off-screen buffer image. This is much more complicated because an image has to be created that is compatible with the display, and then an off-screen surface using a buffer has to be created from this image. The surface must then be made the current drawing surface, and then OpenVG calls can be made to it. The original drawing surface must be restored, and then the image can be drawn to it, as shown earlier.

In setting up the EGL surfaces, you had to create an EGL context. This used the `EGLDisplay` and also an `EGLConfig`. The configuration used a list of attributes that typically specify the RGBA characteristics required. This is done by passing to `eglChooseConfig` an attribute list such as the following:

```
EGLint attribute_list[] =
       {
           EGL_RED_SIZE, 8,
           EGL_GREEN_SIZE, 8,
           EGL_BLUE_SIZE, 8,
           EGL_ALPHA_SIZE, 8,
           ...
           EGL_NONE
       };
```

By experimenting I have found that for the RPi, both `RGBA_8888` and `RGB_565` but not much else seems to work for a window configuration. I couldn't get a Luminance setting (grayscale) to work.

An off-screen image seems to be even fussier, and I have only been able to get a setting of `RGBA_8888` to work. You choose a configuration and build context for the off-screen image using a *pbuffer*, as follows:

```
EGLContext context;

static const EGLint attribute_list[] =
    {
        EGL_RED_SIZE, 8,
        EGL_GREEN_SIZE, 8,
        EGL_BLUE_SIZE, 8,
        EGL_ALPHA_SIZE, 8,
        EGL_SURFACE_TYPE, EGL_PBUFFER_BIT,
        EGL_RENDERABLE_TYPE, EGL_OPENVG_BIT,
        EGL_NONE
    };
EGLConfig config;

drawn_image = vgCreateImage(VG_sRGBA_8888,
                            256, 256,
                            VG_IMAGE_QUALITY_NONANTIALIASED);
if (drawn_image == VG_INVALID_HANDLE) {
    fprintf(stderr, "Can't create drawn image\n");
    fprintf(stderr, "Error code %x\n", vgGetError());
    exit(2);
}
vgClearImage(drawn_image, 0, 0, 256, 256);
```

```
int result, num_configs;
result = eglChooseConfig(p_state->display,
                         attribute_list,
                         &config,
                         1,
                         &num_configs);
assert(EGL_FALSE != result);
context = eglCreateContext(p_state->display,
                           config, EGL_NO_CONTEXT,
                           NULL);
```

The next step is to create an off-screen surface using the call eglCreatePbufferFromClientBuffer. This takes the image coerced to type EGLClientBuffer.

```
img_surface = eglCreatePbufferFromClientBuffer(p_state->display,
                                               EGL_OPENVG_IMAGE,
                                               (EGLClientBuffer) drawn_image,
                                               config,
                                               NULL);
```

This needs to be made into the current surface with the following:

```
eglMakeCurrent(p_state->display, img_surface, img_surface,
               context);
```

OpenVG calls can then be made: create a path, set objects into the path, set fill parameters, and so on, and then finally draw the path.

```
VGPath path = vgCreatePath(VG_PATH_FORMAT_STANDARD,
                           VG_PATH_DATATYPE_F, 1.0f,
                           0.0f, 0, 0, VG_PATH_CAPABILITY_ALL);

float height = 40.0;
float arcW = 15.0;
float arcH = 15.0;

vguRoundRect(path, 28, 10,
             200, 60, arcW, arcH);
vguEllipse(path, 128, 200, 60, 40);

setfill(drawn_color);
vgDrawPath(path, VG_FILL_PATH);
```

The final steps are to reset the drawing surface and then draw the image onto the surface.

```
eglMakeCurrent(p_state->display, p_state->surface,
               p_state->surface,
               p_state->context);

vgSeti(VG_MATRIX_MODE, VG_MATRIX_IMAGE_USER_TO_SURFACE);
vgLoadIdentity();
vgTranslate(800, 300);
```

```
vgDrawImage(drawn_image);

vgDestroyImage(drawn_image);
```

The final function to do all this is as follows:

```
void drawn_image() {
    EGLSurface img_surface;
    VGImage drawn_image;
    EGLContext context;

    static const EGLint attribute_list[] =
        {
            //EGL_COLOR_BUFFER_TYPE, EGL_LUMINANCE_BUFFER,
            // EGL_LUMINANCE_SIZE, 8,
            EGL_RED_SIZE, 8,
            EGL_GREEN_SIZE, 8,
            EGL_BLUE_SIZE, 8,
            EGL_ALPHA_SIZE, 8,
            EGL_SURFACE_TYPE, EGL_PBUFFER_BIT,
            EGL_RENDERABLE_TYPE, EGL_OPENVG_BIT,
            EGL_NONE
        };
    EGLConfig config;

    drawn_image = vgCreateImage(VG_sRGBA_8888,
                                256, 256,
                                VG_IMAGE_QUALITY_NONANTIALIASED);
    if (drawn_image == VG_INVALID_HANDLE) {
        fprintf(stderr, "Can't create drawn image\n");
        fprintf(stderr, "Error code %x\n", vgGetError());
        exit(2);
    }
    vgClearImage(drawn_image, 0, 0, 256, 256);

    int result, num_configs;
    result = eglChooseConfig(p_state->display,
                             attribute_list,
                             &config,
                             1,
                             &num_configs);
    assert(EGL_FALSE != result);gt;display,
    context = eglCreateContext(p_state->display,
                               config, EGL_NO_CONTEXT,
                               NULL);

    img_surface = eglCreatePbufferFromClientBuffer(p_state->display,
                                                   EGL_OPENVG_IMAGE,
                                                   (EGLClientBuffer) drawn_image,
                                                   config,
                                                   NULL);
```

```
    if (img_surface == EGL_NO_SURFACE) {
        fprintf(stderr, "Couldn't create pbuffer\n");
        fprintf(stderr, "Error code %x\n", eglGetError());
        exit(1);
    }
    eglMakeCurrent(p_state->display, img_surface, img_surface,
                   context); //p_state->context);

    VGPath path = vgCreatePath(VG_PATH_FORMAT_STANDARD,
                               VG_PATH_DATATYPE_F, 1.0f,
                               0.0f, 0, 0, VG_PATH_CAPABILITY_ALL);

    float height = 40.0;
    float arcW = 15.0;
    float arcH = 15.0;

    vguRoundRect(path, 28, 10,
                 200, 60, arcW, arcH);
    vguEllipse(path, 128, 200, 60, 40);

    setfill(drawn_color);
    vgDrawPath(path, VG_FILL_PATH);

    eglMakeCurrent(p_state->display, p_state->surface,
                   p_state->surface,
                   p_state->context);

    vgSeti(VG_MATRIX_MODE, VG_MATRIX_IMAGE_USER_TO_SURFACE);
    vgLoadIdentity();
    vgTranslate(800, 300);

    vgDrawImage(drawn_image);

    vgDestroyImage(drawn_image);
}
```

The image looks like this:

The complete program to draw the simple images is `image.c`.

Conclusion

This chapter showed you how to draw OpenVG surfaces on a Dispmanx window and introduced you to the OpenVG API.

Resources

- *Khronos Group: OpenVG - The Standard for Vector Graphics Acceleration*: `https://www.khronos.org/openvg/`
- *OpenVG Specification*: `https://www.khronos.org/registry/vg/specs/openvg-1.1.pdf`
- *Huang Loren OpenVG Graphics Solutions Tutorial A very good overview of OpenVG capabilities*: `https://www.yumpu.com/en/document/view/4775633/openvg-graphics-solutions-tutorial`
- *Mindchunk: OpenVG on the Raspberry Pi describes a simple library to program OpenVG on the RPi*: `https://github.com/ajstarks/openvg`
- *Why have hardware-accelerated vector graphics not taken off?*: `http://programmers.stackexchange.com/questions/191472/why-have-hardware-accelerated-vector-graphics-not-taken-off`

CHAPTER 18

Text Processing in OpenVG on the Raspberry Pi

Displaying text is an important function of any application. *Drawing* the text should be a low-level operation that the application should not be concerned with. It has to be done explicitly in OpenVG, and this chapter discusses various techniques of doing that.

Building Programs

The programs in this chapter require extra libraries in addition to the OpenVG libraries. These need to be installed with the following:

```
sudo apt-get install libcairo2-dev libpango1.0-dev
```

The FreeType 2 packages should already be installed, but if they are not, you will need to get the package freetype2.

You can then build the programs in this chapter using the following Makefile:

```
DMX_INC =  -I/opt/vc/include/ -I /opt/vc/include/interface/vmcs_host/ -I/opt/vc/include/
interface/vcos/pthreads -I/opt/vc/include/interface/vmcs_host/linux
EGL_INC =
OPENVG_INC =
PANGO_CAIRO_INC = $(shell pkg-config --cflags pangocairo)
FREETYPE_INC = $(shell pkg-config --cflags freetype2)
INCLUDES = $(DMX_INC) $(EGL_INC) $(OPENVG_INC) $(PANGO_CAIRO_INC) $(FREETYPE_INC)

CFLAGS= $(INCLUDES)
CPPFLAGS =

DMX_LIBS =  -L/opt/vc/lib/ -lbcm_host -lvcos -lvchiq_arm -lpthread
EGL_LIBS = -L/opt/vc/lib/ -lEGL -lGLESv2
OPENVG_LIBS = -lOpenVG
PANGO_CAIRO_LIBS = $(shell pkg-config --libs pangocairo)
FREETYPE_LIBS = $(shell pkg-config --libs freetype2)
LDFLAGS =  $(DMX_LIBS) $(EGL_LIBS) $(OPENVG_LIBS)  $(PANGO_CAIRO_LIBS) \
   $(FREETYPE_LIBS)

all: cairo  pango  text-bitmap  text-font  text-outline
```


J. Newmarch, *Raspberry Pi GPU Audio Video Programming*, DOI 10.1007/978-1-4842-2472-4_18

Drawing Text

Drawing text is an immensely difficult area. Even in English there are the following issues:

- Letters such as *g* reach below the baseline and need to have space allocated for this.
- Letter combinations such as *ff* are often rendered with a different (ligature) format such as *ﬀ*.
- Letter pairs such as *AW* are often rendered with less space between the characters to avoid an ugly space between the letters.
- In cursive writing, the shapes of the letters may change to join up with the following or preceding letters.

Once you switch to other languages, the complexities multiply.

- Some languages have right-to-left writing such as Arabic or (often) top to bottom in Chinese.
- Languages such as Chinese have hieroglyphs typically taking twice the width of Latin languages.
- Joining letters in Arabic is an extremely complex issue.

Displaying text involves a number of issues.

- For each character, a *glyph* has to be chosen. A glyph is the visual representation of a character and can vary. In English, varieties can include bold, italic, cursive, and so on, and are often associated with fonts.
- Glyphs have to be laid out on the screen using complex rules for spacing, direction, and so on.

OpenVG has the capability of finding fonts and laying them out at locations on the rendering surface. The *rules* for layout are not part of OpenVG. The simplest rule is just to lay them out one after another, but this will often get them wrong or not visually satisfying.

In this chapter, I will look at a variety of ways of drawing text using OpenVG, trading off ease of use versus complexity of programming and CPU usage on the RPi. Of necessity, this will involve using other libraries such as Cairo, Pango, and the freetype libraries.

Drawing Text Using Cairo

Cairo is a popular 2D graphics library that includes support for "toy" text and complex text processing. The toy library is suitable for simple text layout rules, particularly for Latin languages.

Cairo draws 2D shapes onto surfaces of various kinds. A good tutorial is at `http://zetcode.com/gfx/cairo/`. Drawing involves creating a surface, drawing into it, and then getting the data out of the surface to display it or otherwise process it. The link with OpenVG from Cairo is provided by a Cairo function and an OpenVG function.

```
unsigned char* data = cairo_image_surface_get_data(surface);

vgWritePixels(data, tex_s,
              VG_lBGRA_8888, left, 300, tex_w, tex_h );
```

The text parameters are from Cairo.

```
    int tex_w = cairo_image_surface_get_width(surface);
    int tex_h = cairo_image_surface_get_height(surface);
    int tex_s = cairo_image_surface_get_stride(surface);
    int left = 1920/2 - extents.width/2;
```

A program to display a set of lines progressively using the Cairo “toy” text interface is `cairo.c`.

```
#include <stdio.h>
#include <stdlib.h>
#include <sys/stat.h>

#include <assert.h>

#include <EGL/egl.h>
#include <EGL/eglext.h>
#include <GLES2/gl2.h>
#include <VG/openvg.h>
#include <VG/vgu.h>

#include <bcm_host.h>

#include <cairo/cairo.h>

char *lines[] = {"hello", "world", "AWAKE"};
int num_lines = 3;

typedef struct
{
    EGLDisplay display;
    EGLSurface surface;
    EGLContext context;
    EGLConfig config;
} EGL_STATE_T;

EGL_STATE_T state, *p_state = &state;

void init_egl(EGL_STATE_T *state)
{
    EGLint num_configs;
    EGLBoolean result;

    //bcm_host_init();

    static const EGLint attribute_list[] =
        {
            EGL_RED_SIZE, 8,
            EGL_GREEN_SIZE, 8,
            EGL_BLUE_SIZE, 8,
            EGL_ALPHA_SIZE, 8,
            EGL_SURFACE_TYPE, EGL_WINDOW_BIT,
            EGL_NONE
        };
```

```
    // get an EGL display connection
    state->display = eglGetDisplay(EGL_DEFAULT_DISPLAY);

    // initialize the EGL display connection
    result = eglInitialize(state->display, NULL, NULL);

    // get an appropriate EGL frame buffer configuration
    result = eglChooseConfig(state->display, attribute_list, &state->config, 1,
    &num_configs);
    assert(EGL_FALSE != result);

    // Choose the OpenVG API
    result = eglBindAPI(EGL_OPENVG_API);
    assert(EGL_FALSE != result);

    // create an EGL rendering context
    state->context = eglCreateContext(state->display,
                                      state->config, EGL_NO_CONTEXT,
                                      NULL);

    assert(state->context!=EGL_NO_CONTEXT);
}

void init_dispmanx(EGL_DISPMANX_WINDOW_T *nativewindow) {
    int32_t success = 0;
    uint32_t screen_width;
    uint32_t screen_height;

    DISPMANX_ELEMENT_HANDLE_T dispman_element; // | FT_LOAD_NO_SCALE);
    DISPMANX_DISPLAY_HANDLE_T dispman_display;
    DISPMANX_UPDATE_HANDLE_T dispman_update;
    VC_RECT_T dst_rect;
    VC_RECT_T src_rect;

    bcm_host_init();

    // create an EGL window surface
    success = graphics_get_display_size(0 /* LCD */,
                                        &screen_width,
                                        &screen_height);
    assert( success >= 0 );

    dst_rect.x = 0;
    dst_rect.y = 0;
    dst_rect.width = screen_width;
    dst_rect.height = screen_height;

    src_rect.x = 0;
    src_rect.y = 0;
    src_rect.width = screen_width << 16;
    src_rect.height = screen_height << 16;
```

```
    dispman_display = vc_dispmanx_display_open( 0 /* LCD */);
    dispman_update = vc_dispmanx_update_start( 0 );

    dispman_element =
        vc_dispmanx_element_add(dispman_update, dispman_display,
                                0 /*layer*/, &dst_rect, 0 /*src*/,
                                &src_rect, DISPMANX_PROTECTION_NONE,
                                0 /*alpha*/, 0 /*clamp*/, 0 /*transform*/);

    // Build an EGL_DISPMANX_WINDOW_T from the Dispmanx window
    nativewindow->element = dispman_element;
    nativewindow->width = screen_width;
    nativewindow->height = screen_height;
    vc_dispmanx_update_submit_sync(dispman_update);

    printf("Got a Dispmanx window\n");
}

void egl_from_dispmanx(EGL_STATE_T *state,
                       EGL_DISPMANX_WINDOW_T *nativewindow) {
    EGLBoolean result;

    state->surface = eglCreateWindowSurface(state->display,
                                            state->config,
                                            nativewindow, NULL );
    assert(state->surface != EGL_NO_SURFACE);

    // connect the context to the surface
    result = eglMakeCurrent(state->display, state->surface, state->surface, state->context);
    assert(EGL_FALSE != result);
}

void cairo(char *text) {
    int width = 500;
    int height = 200;

    cairo_surface_t *surface = cairo_image_surface_create (CAIRO_FORMAT_ARGB32,
                                                           width, height);
    cairo_t *cr = cairo_create(surface);

    cairo_rectangle(cr, 0, 0, width, height);
    cairo_set_source_rgb(cr, 0.0, 0.0, 0.5);
    cairo_fill(cr);

    // draw some white text on top
    cairo_set_source_rgb(cr, 1.0, 1.0, 1.0);
    // this is a standard font for Cairo
    cairo_select_font_face (cr, "cairo:serif",
                            CAIRO_FONT_SLANT_NORMAL,
                            CAIRO_FONT_WEIGHT_BOLD);
    cairo_set_font_size (cr, 36);
    cairo_move_to(cr, 10.0, 50.0);
```

```
    cairo_scale(cr, 1.0f, -1.0f);
    cairo_translate(cr, 0.0f, -50);

    cairo_text_extents_t extents;
    cairo_text_extents(cr, text, &extents);

    cairo_show_text (cr, text);

    int tex_w = cairo_image_surface_get_width(surface);
    int tex_h = cairo_image_surface_get_height(surface);
    int tex_s = cairo_image_surface_get_stride(surface);
    unsigned char* data = cairo_image_surface_get_data(surface);

    int left = 1920/2 - extents.width/2;
    vgWritePixels(data, tex_s,
                  VG_lBGRA_8888, left, 300, tex_w, tex_h );
}

void draw() {
    EGL_DISPMANX_WINDOW_T nativewindow;

    init_egl(p_state);
    init_dispmanx(&nativewindow);

    egl_from_dispmanx(p_state, &nativewindow);

    int n = 0;
    while (1) {
        cairo(lines[n++]);
        n %= num_lines;

        eglSwapBuffers(p_state->display, p_state->surface);
        sleep(3);
    }
}

int main(int argc, char** argv) {

    draw();

    exit(0);
}
```

Run the program with cairo.

This draws each line of text in white on a blue background, as shown here:

Drawing Text Using Pango

While Cairo can draw any form of text, the functions such as `cairo_show_text` do not have much flexibility. Drawing in, say, multiple colors will involve a lot of work. Pango is a library for handling all aspects of text. There is a Pango Reference Manual (`https://developer.gnome.org/pango/stable/`), and you can find a good tutorial at `www.ibm.com/developerworks/library/l-u-pango2/`.

The simplest way of coloring text (and some other effects) is to create the text marked up with HTML, as follows:

```
gchar *markup_text = "<span foreground=\"red\">hello </span><span
foreground=\"black\">world</span>";
```

This has *hello* in red and *world* in black. This is then parsed into the text itself, *red black*, and a set of attribute markups.

```
gchar *markup_text = "<span foreground=\"red\">hello </span><span
foreground=\"black\">world</span>";
PangoAttrList *attrs;
gchar *text;

pango_parse_markup (markup_text, -1,0, &attrs, &text, NULL, NULL);
```

This can be rendered into a Cairo context by creating a `PangoLayout` from the Cairo context, laying out the text with its attributes in the Pango layout, and then showing this layout in the Cairo context.

```
PangoLayout *layout;
PangoFontDescription *desc;
cairo_move_to(cr, 300.0, 50.0);
layout = pango_cairo_create_layout (cr);
pango_layout_set_text (layout, text, -1);
pango_layout_set_attributes(layout, attrs);
pango_cairo_update_layout (cr, layout);
pango_cairo_show_layout (cr, layout);
```

(Yes, there is a lot of jumping around between libraries in all of this!)

Once the Cairo surface has been drawn, the data can be extracted as in the Cairo example and written into the OpenVG surface. The program is pango.c and is as follows:

```
#include <stdio.h>
#include <stdlib.h>
#include <sys/stat.h>

#include <assert.h>

#include <EGL/egl.h>
#include <EGL/eglext.h>
#include <GLES2/gl2.h>
#include <VG/openvg.h>
#include <VG/vgu.h>

#include <bcm_host.h>

#include <pango/pangocairo.h>

typedef struct
{
    EGLDisplay display;
    EGLSurface surface;
    EGLContext context;
    EGLConfig config;
} EGL_STATE_T;

EGL_STATE_T state, *p_state = &state;

void init_egl(EGL_STATE_T *state)
{
    EGLint num_configs;
    EGLBoolean result;

    //bcm_host_init();

    static const EGLint attribute_list[] =
        {
            EGL_RED_SIZE, 8,
            EGL_GREEN_SIZE, 8,
            EGL_BLUE_SIZE, 8,
            EGL_ALPHA_SIZE, 8,
            EGL_SURFACE_TYPE, EGL_WINDOW_BIT,
            EGL_NONE
        };

    // get an EGL display connection
    state->display = eglGetDisplay(EGL_DEFAULT_DISPLAY);

    // initialize the EGL display connection
    result = eglInitialize(state->display, NULL, NULL);
```

```
    // get an appropriate EGL frame buffer configuration
    result = eglChooseConfig(state->display, attribute_list, &state->config, 1,
             &num_configs);
    assert(EGL_FALSE != result);

    // Choose the OpenVG API
    result = eglBindAPI(EGL_OPENVG_API);
    assert(EGL_FALSE != result);

    // create an EGL rendering context
    state->context = eglCreateContext(state->display,
                                      state->config, EGL_NO_CONTEXT,
                                      NULL);

    assert(state->context!=EGL_NO_CONTEXT);
}

void init_dispmanx(EGL_DISPMANX_WINDOW_T *nativewindow) {
    int32_t success = 0;
    uint32_t screen_width;
    uint32_t screen_height;

    DISPMANX_ELEMENT_HANDLE_T dispman_element;
    DISPMANX_DISPLAY_HANDLE_T dispman_display;
    DISPMANX_UPDATE_HANDLE_T dispman_update;
    VC_RECT_T dst_rect;
    VC_RECT_T src_rect;

    bcm_host_init();

    // create an EGL window surface
    success = graphics_get_display_size(0 /* LCD */,
                                        &screen_width,
                                        &screen_height);
    assert( success >= 0 );

    dst_rect.x = 0;
    dst_rect.y = 0;
    dst_rect.width = screen_width;
    dst_rect.height = screen_height;

    src_rect.x = 0;
    src_rect.y = 0;
    src_rect.width = screen_width << 16;
    src_rect.height = screen_height << 16;

    dispman_display = vc_dispmanx_display_open( 0 /* LCD */);
    dispman_update = vc_dispmanx_update_start( 0 );

    dispman_element =
```

```
        vc_dispmanx_element_add(dispman_update, dispman_display,
                                0/*layer*/, &dst_rect, 0/*src*/,
                                &src_rect, DISPMANX_PROTECTION_NONE,
                                0 /*alpha*/, 0/*clamp*/, 0/*transform*/);

    // Build an EGL_DISPMANX_WINDOW_T from the Dispmanx window
    nativewindow->element = dispman_element;
    nativewindow->width = screen_width;
    nativewindow->height = screen_height;
    vc_dispmanx_update_submit_sync(dispman_update);

    printf("Got a Dispmanx window\n");
}

void egl_from_dispmanx(EGL_STATE_T *state,
                       EGL_DISPMANX_WINDOW_T *nativewindow) {
    EGLBoolean result;

    state->surface = eglCreateWindowSurface(state->display,
                                            state->config,
                                            nativewindow, NULL );
    assert(state->surface != EGL_NO_SURFACE);

    // connect the context to the surface
    result = eglMakeCurrent(state->display, state->surface, state->surface, state->context);
    assert(EGL_FALSE != result);
}

// setfill sets the fill color
void setfill(float color[4]) {
    VGPaint fillPaint = vgCreatePaint();
    vgSetParameteri(fillPaint, VG_PAINT_TYPE, VG_PAINT_TYPE_COLOR);
    vgSetParameterfv(fillPaint, VG_PAINT_COLOR, 4, color);
    vgSetPaint(fillPaint, VG_FILL_PATH);
    vgDestroyPaint(fillPaint);
}

// setstroke sets the stroke color and width
void setstroke(float color[4], float width) {
    VGPaint strokePaint = vgCreatePaint();
    vgSetParameteri(strokePaint, VG_PAINT_TYPE, VG_PAINT_TYPE_COLOR);
    vgSetParameterfv(strokePaint, VG_PAINT_COLOR, 4, color);
    vgSetPaint(strokePaint, VG_STROKE_PATH);
    vgSetf(VG_STROKE_LINE_WIDTH, width);
    vgSeti(VG_STROKE_CAP_STYLE, VG_CAP_BUTT);
    vgSeti(VG_STROKE_JOIN_STYLE, VG_JOIN_MITER);
    vgDestroyPaint(strokePaint);
}

// Ellipse makes an ellipse at the specified location and dimensions, applying style
void Ellipse(float x, float y, float w, float h, float sw, float fill[4], float stroke[4]) {
```

```
    VGPath path = vgCreatePath(VG_PATH_FORMAT_STANDARD, VG_PATH_DATATYPE_F, 1.0f, 0.0f, 0,
    0, VG_PATH_CAPABILITY_ALL);
    vguEllipse(path, x, y, w, h);
    setfill(fill);
    setstroke(stroke, sw);
    vgDrawPath(path, VG_FILL_PATH | VG_STROKE_PATH);
    vgDestroyPath(path);
}

void pango() {
    int width = 500;
    int height = 200;

    cairo_surface_t *surface = cairo_image_surface_create (CAIRO_FORMAT_ARGB32,
                                                           width, height);
    cairo_t *cr = cairo_create(surface);

    cairo_rectangle(cr, 0, 0, width, height);
    cairo_set_source_rgb(cr, 0.5, 0.5, 0.5);
    cairo_fill(cr);

    // Pango marked up text, half red, half black
    gchar *markup_text = "<span foreground=\"red\" font='48'>hello</span>\
<span foreground=\"black\" font='48'>AWA狴犬</span>";
    PangoAttrList *attrs;
    gchar *text;

    pango_parse_markup (markup_text, -1, 0, &attrs, &text, NULL, NULL);

    // draw Pango text
    PangoLayout *layout;
    PangoFontDescription *desc;

    cairo_move_to(cr, 30.0, 150.0);
    cairo_scale(cr, 1.0f, -1.0f);
    layout = pango_cairo_create_layout (cr);
    pango_layout_set_text (layout, text, -1);
    pango_layout_set_attributes(layout, attrs);
    pango_cairo_update_layout (cr, layout);
    pango_cairo_show_layout (cr, layout);

    int tex_w = cairo_image_surface_get_width(surface);
    int tex_h = cairo_image_surface_get_height(surface);
    int tex_s = cairo_image_surface_get_stride(surface);
    cairo_surface_flush(surface);
    cairo_format_t type =
        cairo_image_surface_get_format (surface);
    printf("Format is (0 is ARGB32) %d\n", type);
    unsigned char* data = cairo_image_surface_get_data(surface);

    PangoLayoutLine *layout_line =
        pango_layout_get_line(layout, 0);
```

```
    PangoRectangle ink_rect;
    PangoRectangle logical_rect;
    pango_layout_line_get_pixel_extents (layout_line,
                                          &ink_rect,
                                          &logical_rect);
    printf("Layout line rectangle is %d, %d, %d, %d\n",
           ink_rect.x, ink_rect.y, ink_rect.width, ink_rect.height);

    int left = 1920/2 - ink_rect.width/2;
    vgWritePixels(data, tex_s,
                  //VG_lBGRA_8888,
                  VG_sARGB_8888 ,
                  left, 300, tex_w, tex_h );
}

void draw()
{
    EGL_DISPMANX_WINDOW_T nativewindow;

    init_egl(p_state);
    init_dispmanx(&nativewindow);

    egl_from_dispmanx(p_state, &nativewindow);
    //draw();
    pango();

    eglSwapBuffers(p_state->display, p_state->surface);

    sleep(30);
    eglTerminate(p_state->display);

    exit(0);
}

int main(int argc, char** argv) {

    draw();

    exit(0);
}
```

The program is run with pango.

This draws some text in red and black onto a gray rectangle against a white backdrop, as shown here:

FreeType

The previous programs basically did all the rendering calculations using the RPi's CPU and used the GPU only to render the resultant images. This is not using the GPU as you want: to manage as much as possible of the rendering calculations.

It is not possible to do everything on the GPU. It doesn't know about Linux file systems, for example. But more important, it doesn't know how or where information about text fonts and glyphs is stored. This information has to be built from an external library and massaged into a form suitable for the RPi's GPU.

The most common library in use on Linux systems now seems to be the FreeType library (www.freetype.org/). Here is the API reference: www.freetype.org/freetype2/docs/reference/ft2-basic_types.html#FT_Glyph_Format. There is also a tutorial at www.freetype.org/freetype2/docs/tutorial/index.html.

Font Faces

The highest-level object in FreeType is an FT_Library. There may be more than one of these, say, one for each thread in a multithreaded application. You will use only one. An FT_Face represents a single typeface in a particular style, such as Palatino Regular or Palatino Italic. An FT_Face is extracted from a font file that may contain many faces. Typically, face 0 is chosen. Once selected, a face may have characteristics such as the pixel size set.

The typical code to set up a face with a particular size is as follows:

```
void ft_init() {
    int err;

    err = FT_Init_FreeType(&ft_library);
    assert( !err);

    err = FT_New_Face(ft_library,
                      "/usr/share/ghostscript/9.05/Resource/CIDFSubst/DroidSansFallback.
ttf",
                     0,
                     &ft_face );
    assert( !err);

    int font_size = 128;
    err = FT_Set_Pixel_Sizes(ft_face, 0, font_size);
    assert( !err);
}
```

Where do you get the font files from? The file type required is usually TrueType with the extension .ttf, and many of these are found in the directory /usr/share/fonts/truetype/. These all contain the ASCII character set, and their appearance can be seen using gnome-font-viewer. For example, the font /usr/share/fonts/truetype/freefont/FreeMonoBold.ttf looks like this:

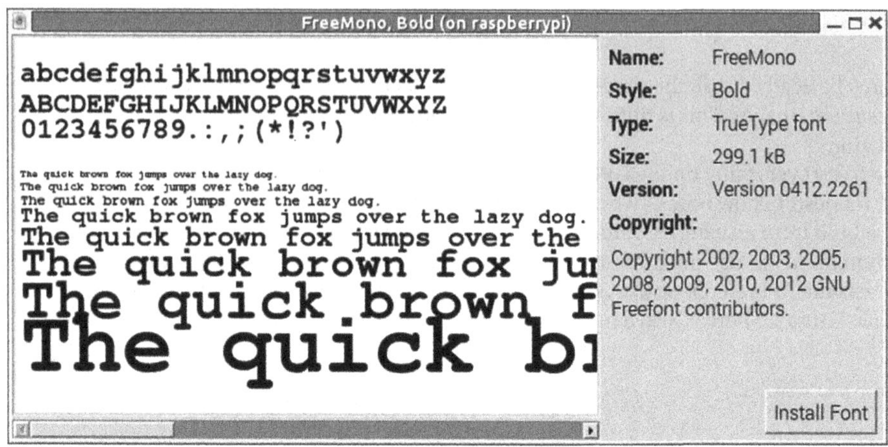

I want to be able to see Chinese characters as well as ASCII, and I know that with all the packages I have installed, Pango is able to find a font file containing them. Repolho shows at `https://repolinux.wordpress.com/2013/03/10/find-out-fallback-font-used-by-fontconfig-for-a-certain-character/` how to find a font file for any particular character. For example, for the Chinese character 好 (meaning "good"), the command is as follows:

```
FC_DEBUG=4 pango-view -t '好' 2>&1 |
 grep -o 'family: "[^"]\+' |
 cut -c 10- | tail -n 1
```

This will display the character in a small box, and when you quit (`q`) the box, it will print the name of a font file. On my system, it printed `Droid Sans Japanese`. This is the name of the font but not of the file. By using the following, the file `/usr/share/fonts/truetype/droid/DroidSansJapanese.ttf` was identified:

```
locate DroidSansJapanese
```

(I had installed the package `locate` and ran `updatedb` to build the database for the `locate` command.)

This need not be the only containing this character but is the one used by Pango. (That directory also contains `DroidSansFallbackFull.ttf`, which would probably be better if I wanted to draw more characters.) It may be that there is no such font file on your system. In that case, download the files `zysong.ttf` or `Cyberbit.ttf`.

Paths and Glyphs

Once a face has been set up, you can hand characters to it and get an index for each character in the face using `FT_Get_Char_Index(FT_Face face, FT_ULong charcode)`. Although the function name says `Char`, with Unicode characters these won't be 8-bit C `char`s except for the ASCII subset. You will need the full 16- or 32-bit Unicode character, which is an `FT_ULong` to the FreeType system. So, for example, the Chinese character 好 needs to be represented by its Unicode value of 0x597d. Unicode values can be found in many places, such as `scarfboy.com`.

Once an index is found, the character can be loaded into the face by `FT_Load_Glyph`. There are two things you can do with a loaded glyph: get it back as an *outline* or convert it to a *bitmap*. In each case there are different things that can done with OpenVG, which are discussed in the following sections.

Extracting the outline of a glyph from a face is easy, as shown here:

```
FT_Outline *outline = &ft_face->glyph->outline;
```

Getting a bitmap involves more steps: first you get the glyph as an `FT_Glyph` with `FT_Get_Glyph`. This is then converted to a bitmap with `FT_Glyph_To_Bitmap`. But a glyph has no field of a bitmap: it needs to be coerced into type `FT_BitmapGlyph`, and then the bitmap can be found.

```
FT_Glyph glyph;
FT_Get_Glyph(ft_face->glyph, &glyph);
FT_Glyph_To_Bitmap(&glyph, FT_RENDER_MODE_NORMAL, NULL, 1);
FT_BitmapGlyph bit_glyph = (FT_BitmapGlyph) glyph;
FT_Bitmap bitmap = bit_glyph->bitmap;
```

There are operations that can be performed on a glyph before extracting the bitmap. For example, using an `FT_Stroker`, the corners can be smoothed out. The code to do that is as follows:

```
FT_Glyph glyph;
FT_Get_Glyph(ft_face->glyph, &glyph);

// smooth the border
FT_Stroker ft_stroker;
FT_Stroker_New(ft_library, &ft_stroker);
FT_Stroker_Set(ft_stroker,
               200, // line_height_*border_thickness*64.0f,
               FT_STROKER_LINECAP_ROUND,
               FT_STROKER_LINEJOIN_ROUND,
               0);
FT_Glyph_StrokeBorder(&glyph, ft_stroker, 0, 1);

// now get the smoothed bitmap
FT_Glyph_To_Bitmap(&glyph, FT_RENDER_MODE_NORMAL, NULL, 1);
FT_BitmapGlyph bit_glyph = (FT_BitmapGlyph) glyph;
FT_Bitmap bitmap = bit_glyph->bitmap;
```

Drawing a FreeType Outline

The `FT_Outline` type from FreeType contains information about the outline of a glyph, but this is not the same form as used in an OpenVG path. It has to be converted from FreeType to OpenVG.

Functions to do this are in `/opt/vc/src/hello_pi/libs/vgfont/vgft.c`.

```
#define SEGMENTS_COUNT_MAX 256
#define COORDS_COUNT_MAX 1024

static VGuint segments_count;
static VGubyte segments[SEGMENTS_COUNT_MAX];
static VGuint coords_count;
static VGfloat coords[COORDS_COUNT_MAX];
```

```
static VGfloat float_from_26_6(FT_Pos x)
{
   return (VGfloat)x / 64.0f;
}

static void convert_contour(const FT_Vector *points,
                            const char *tags, short points_count)
{
   int first_coords = coords_count;

   int first = 1;
   char last_tag = 0;
   int c = 0;

   for (; points_count != 0; ++points, ++tags, --points_count) {
      ++c;

      char tag = *tags;
      if (first) {
         assert(tag & 0x1);
         assert(c==1); c=0;
         segments[segments_count++] = VG_MOVE_TO;
         first = 0;
      } else if (tag & 0x1) {
         /* on curve */

         if (last_tag & 0x1) {
            /* last point was also on -- line */
            assert(c==1); c=0;
            segments[segments_count++] = VG_LINE_TO;
         } else {
            /* last point was off -- quad or cubic */
            if (last_tag & 0x2) {
               /* cubic */
               assert(c==3); c=0;
               segments[segments_count++] = VG_CUBIC_TO;
            } else {
               /* quad */
               assert(c==2); c=0;
               segments[segments_count++] = VG_QUAD_TO;
            }
         }
      } else {
         /* off curve */

         if (tag & 0x2) {
            /* cubic */

            assert((last_tag & 0x1) || (last_tag & 0x2)); /* last either on or off and cubic */
         } else {
            /* quad */
```

```
            if (!(last_tag & 0x1)) {
               /* last was also off curve */

               assert(!(last_tag & 0x2)); /* must be quad */

               /* add on point half-way between */
               assert(c==2); c=1;
               segments[segments_count++] = VG_QUAD_TO;
               VGfloat x = (coords[coords_count - 2] + float_from_26_6(points->x)) * 0.5f;
               VGfloat y = (coords[coords_count - 1] + float_from_26_6(points->y)) * 0.5f;
               coords[coords_count++] = x;
               coords[coords_count++] = y;
            }
         }
      }
      last_tag = tag;

      coords[coords_count++] = float_from_26_6(points->x);
      coords[coords_count++] = float_from_26_6(points->y);
   }

   if (last_tag & 0x1) {
      /* last point was also on -- line (implicit with close path) */
      assert(c==0);
   } else {
      ++c;

      /* last point was off -- quad or cubic */
      if (last_tag & 0x2) {
         /* cubic */
         assert(c==3); c=0;
         segments[segments_count++] = VG_CUBIC_TO;
      } else {
         /* quad */
         assert(c==2); c=0;
         segments[segments_count++] = VG_QUAD_TO;
      }
      coords[coords_count++] = coords[first_coords + 0];
      coords[coords_count++] = coords[first_coords + 1];
   }

   segments[segments_count++] = VG_CLOSE_PATH;
}

static void convert_outline(const FT_Vector *points,
                            const char *tags, const short *contours,
                            short contours_count, short points_count)
{
   segments_count = 0;
   coords_count = 0;
```

```
   short last_contour = 0;
   for (; contours_count != 0; ++contours, --contours_count) {
      short contour = *contours + 1;
      convert_contour(points + last_contour, tags + last_contour, contour - last_contour);
      last_contour = contour;
   }
   assert(last_contour == points_count);

   assert(segments_count <= SEGMENTS_COUNT_MAX); /* oops... we overwrote some me
mory */
   assert(coords_count <= COORDS_COUNT_MAX);
}
```

These functions are used to set segments and coordinates suitable for adding to an OpenVG path created using vgCreatePath with vgAppendPath. The resultant path can then be drawn with vgDrawpath.

```
FT_Outline *outline = &ft_face->glyph->outline;
if (outline->n_contours != 0) {
   vg_path = vgCreatePath(VG_PATH_FORMAT_STANDARD,
                          VG_PATH_DATATYPE_F, 1.0f,
                          0.0f, 0, 0, VG_PATH_CAPABILITY_ALL);
   assert(vg_path != VG_INVALID_HANDLE);

   convert_outline(outline->points, outline->tags,
                   outline->contours, outline->n_contours,
                   outline->n_points);
   vgAppendPathData(vg_path, segments_count, segments, coords);
} else {
   vg_path = VG_INVALID_HANDLE;
}

VGfloat strokeColor[4] = {0.4, 0.1, 1.0, 1.0}; // purple
VGfloat fillColor[4] = {1.0, 1.0, 0.0, 1.0}; // yellow

setfill(fillColor);
setstroke(strokeColor, 5);

vgDrawPath(vg_path, VG_FILL_PATH | VG_STROKE_PATH);
```

It looks like this:

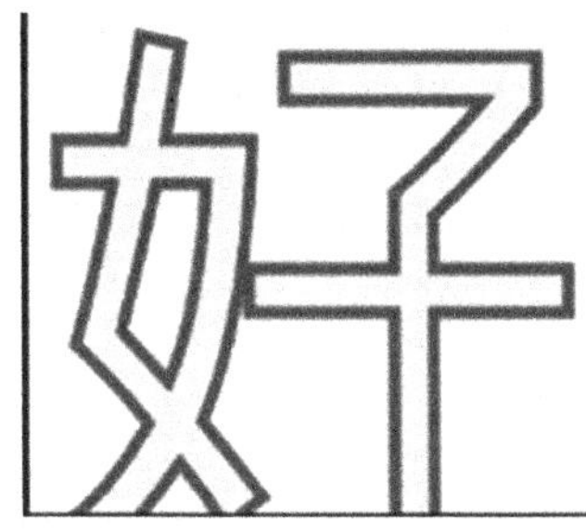

The complete program is text-outline.c.

Drawing a FreeType Bitmap

In an earlier section, you derived an FT_Bitmap for a glyph. The bitmap contains fields for the stride (which it calls *pitch*), width, and height, plus a buffer of data. The format is grayscale with 256 levels. What this becomes in OpenVG I'm not sure. VG_sL_8 seems to work OK for me, while omxplayer uses VG_A_8.

The only real wrinkle is that the bitmap image is upside down in OpenVG coordinates. So, you can't just use vgCreateImage because that draws the bitmap upside down. Instead, you have to add the data to a VGImage using vgImageSubData. The program omxplayer uses a neat trick to insert the data upside down. While vgImageSubData expects to load data from bottom to top, increasing the stride by one each time, you hand the data in from the top, decreasing the stride each time.

The image is drawn using vgSetPixels to set the location on the screen. The relevant code is as follows:

```
FT_Glyph glyph;
FT_Get_Glyph(ft_face->glyph, &glyph);

FT_Stroker ft_stroker;
FT_Stroker_New(ft_library, &ft_stroker);
FT_Stroker_Set(ft_stroker,
               200, // line_height_*border_thickness*64.0f,
               FT_STROKER_LINECAP_ROUND,
               FT_STROKER_LINEJOIN_ROUND,
               0);
FT_Glyph_StrokeBorder(&glyph, ft_stroker, 0, 1);

FT_Glyph_To_Bitmap(&glyph, FT_RENDER_MODE_NORMAL, NULL, 1);
FT_BitmapGlyph bit_glyph = (FT_BitmapGlyph) glyph;
FT_Bitmap bitmap = bit_glyph->bitmap;
printf("Bitmap mode is %d (2 is gray) with %d levels\n",
       bitmap.pixel_mode, bitmap.num_grays);

int tex_s = bitmap.pitch;
int tex_w = bitmap.width;
int tex_h = bitmap.rows;

VGImage image = vgCreateImage(VG_sL_8, tex_w, tex_h,
                              VG_IMAGE_QUALITY_NONANTIALIASED);

// invert bitmap image
vgImageSubData(image,
               bitmap.buffer + tex_s*(tex_h-1),
               -tex_s,
               VG_sL_8, //VG_A_8, VG_sL_8 seem to be okay too
               0, 0, tex_w, tex_h);

vgSetPixels(600, 600, image, 0, 0, tex_w, tex_h);
```

The complete program is `text-bitmap.c`, and it looks like this:

There is no border or color control.

VGFonts and Multiple Characters

If you want to draw multiple characters, you need to get multiple outline paths or bitmaps from FreeType and convert them. If characters are repeated or drawn multiple times, then this can lead to redundant processing. OpenVG has a type called `VGFont`, which can essentially act as a cache for the paths and bitmaps, so that you load each one into the font and then can draw them multiple times without having to repeat the loading process.

A font is created using `vgCreateFont`, which takes a parameter for the expected size or 0. Paths are added to a font with `vgSetGlyphToPath`, while bitmaps are added using `vgSetGlyphToImage`. These take as parameters the font, the index used for the glyph, the path or image, and two other parameters describing the geometry of the path/image. These are discussed next.

VGDrawGlyph

There is then a bonus function: instead of extracting the path or image from the font and then drawing it, the function `vgDrawGlyph` will perform the drawing for you, stroking or filling the path as required, as follows:

```
vgDrawGlyph(font, ch,
           VG_STROKE_PATH | VG_FILL_PATH,
           VG_FALSE);
```

Where it draws the path/image depends on the value of the parameter `VG_GLYPH_ORIGIN`. This is initially set to an (x, y) location, as follows:

```
VGfloat origin[2] = {300.0f, 300.0f};
vgSetfv(VG_GLYPH_ORIGIN, 2, origin);
```

Each time a glyph is drawn, this location is updated by the width and height of the glyph so that successive glyphs are drawn next to each other.

A pseudo-code algorithm for drawing a string is as follows:

```
for each ch in the string
    if ch is not in the font
    then
        add the glyph to the font, using ch as index
    draw the glyph using ch as index
```

There is only one glitch in this: there is no way of telling whether a glyph is in the font! If you are just adding a small set of glyphs such as the ASCII characters, this is not a problem: just loop through them all adding each one. But if it is a large set, such as the 120,000+ Unicode characters, then you won't want to do that. Instead, you would like to add them on an as-needed basis, say, as you loop through a piece of text.

To avoid repeated additions of a character, you need to keep a list *outside* of OpenVG. I used the hash table from GLib: add it to the hash table when first seen and add it the font, but do not add it if it's already there. The code to add an outline glyph is as follows:

```
void add_glyph(int *ch) {
    if (g_hash_table_lookup(glyphs_seen, GINT_TO_POINTER(ch)) != NULL) {
        printf("Glyph %d already seen\n", *ch);
        return;
    }
    g_hash_table_insert(glyphs_seen,
                        GINT_TO_POINTER(ch),
                        GINT_TO_POINTER(ch));
    printf("Inserting %d\n", *ch);

    int glyph_index = FT_Get_Char_Index(ft_face, *ch);
    if (glyph_index == 0) printf("No glyph found\n");
    FT_Load_Glyph(ft_face, glyph_index,
                  FT_LOAD_NO_HINTING| FT_LOAD_LINEAR_DESIGN);

    VGPath vg_path;

    FT_Outline *outline = &ft_face->glyph->outline;
    if (outline->n_contours != 0) {
        vg_path = vgCreatePath(VG_PATH_FORMAT_STANDARD,
                               VG_PATH_DATATYPE_F, 1.0f,
                               0.0f, 0, 0, VG_PATH_CAPABILITY_ALL);
        assert(vg_path != VG_INVALID_HANDLE);

        convert_outline(outline->points, outline->tags,
                        outline->contours, outline->n_contours,
                        outline->n_points);
        vgAppendPathData(vg_path, segments_count, segments, coords);
    } else {
        vg_path = VG_INVALID_HANDLE;
    }
```

```
    VGfloat glyphOrigin[2] = {0.0f, 0.0f};
    printf("Width %d, height %d advance %d\n",
           ft_face->glyph->metrics.width,
           ft_face->glyph->metrics.height,
           ft_face->glyph->linearHoriAdvance);
    VGfloat escapement[2] = {ft_face->glyph->linearHoriAdvance,
                             0.0f};

    vgSetGlyphToPath(font, *ch,
                     vg_path, 0,
                     glyphOrigin,
                     escapement);
}
```

You can then draw a set of glyphs as follows:

```
void draw() {
    EGL_DISPMANX_WINDOW_T nativewindow;
    glyphs_seen = g_hash_table_new(g_int_hash, g_int_equal);

    init_egl(p_state);
    init_dispmanx(&nativewindow);
 "/usr/share/fonts/truetype/droid/DroidSansFallbackFull.ttf",
    egl_from_dispmanx(p_state, &nativewindow);

    vgClear(0, 0, 1920, 1080);

    ft_init();

    font = vgCreateFont(0);
    //font_border = vgCreateFont(0);

    VGfloat strokeColor[4] = {0.4, 0.1, 1.0, 1.0}; // purple
    VGfloat fillColor[4] = {1.0, 1.0, 0.0, 1.0}; // yellowish

    setfill(fillColor);
    setstroke(strokeColor, 5);

    VGfloat origin[2] = {300.0f, 300.0f};
    vgSetfv(VG_GLYPH_ORIGIN, 2, origin);

    int ch;

    ch = 0x597d;
    add_glyph(&ch);
    vgDrawGlyph(font, ch,
                VG_STROKE_PATH,
                VG_FALSE);

    ch = 'A';
    add_glyph(&ch);
```

```
    vgDrawGlyph(font, ch,
                VG_STROKE_PATH,
                VG_FALSE);

    ch = 'b';
    add_glyph(&ch);
    vgDrawGlyph(font, ch,
                VG_STROKE_PATH | VG_FILL_PATH,
                VG_FALSE);
    ch = 'g';
    add_glyph(&ch);
    vgDrawGlyph(font, ch,
                VG_STROKE_PATH | VG_FILL_PATH,
                VG_FALSE);
    ch = 'b';
    add_glyph(&ch);
    vgDrawGlyph(font, ch,
                VG_STROKE_PATH,
                VG_FALSE);

    vgFlush();

    eglSwapBuffers(p_state->display, p_state->surface);
}
```

with appearance

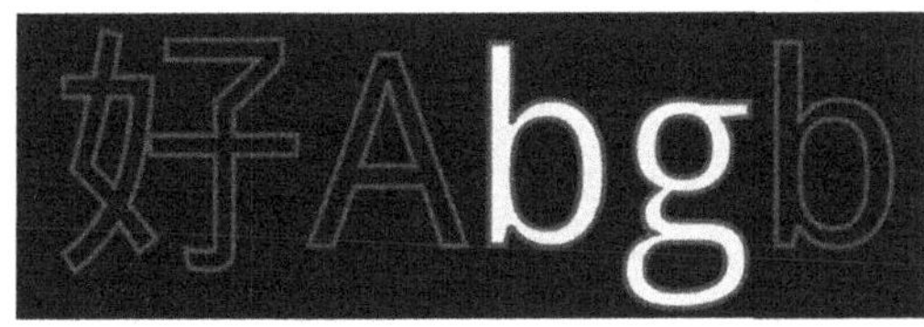

The complete program is `text-font.c`.

vgDrawGlyphs

In the previous example, you drew some glyphs filled and some in outline only. You could have made other changes too, such as different color fills. However, if *all* the characters are to be drawn using the same parameters, then the convenience function `vgDrawGlyphs` can be used.

Glyph Metrics

The geometry of a glyph is complex, and I have ignored it so far. For FreeType, the geometry is discussed in the tutorial "Managing Glyphs" at `www.freetype.org/freetype2/docs/tutorial/step2.html`. The metrics are accessible from the structure `face->glyph->metrics` and are as follows:

> `width`: This is the width of the glyph image's bounding box. It is independent of the layout direction.

> `height`: This is the height of the glyph image's bounding box. It is independent of the layout direction. Be careful not to confuse it with the `height` field in the `FT_Size_Metrics` structure.
>
> `horiBearingX`: For horizontal text layouts, this is the horizontal distance from the current cursor position to the leftmost border of the glyph image's bounding box.
>
> `horiBearingY`: For horizontal text layouts, this is the vertical distance from the current cursor position (on the baseline) to the topmost border of the glyph image's bounding box.
>
> `horiAdvance`: For horizontal text layouts, this is the horizontal distance to increment the pen position when the glyph is drawn as part of a string of text.
>
> `vertBearingX`: For vertical text layouts, this is the horizontal distance from the current cursor position to the leftmost border of the glyph image's bounding box.
>
> `vertBearingY`: For vertical text layouts, this is the vertical distance from the current cursor position (on the baseline) to the topmost border of the glyph image's bounding box.
>
> `vertAdvance`: For vertical text layouts, this is the vertical distance used to increment the pen position when the glyph is drawn as part of a string of text.

For horizontal layout,

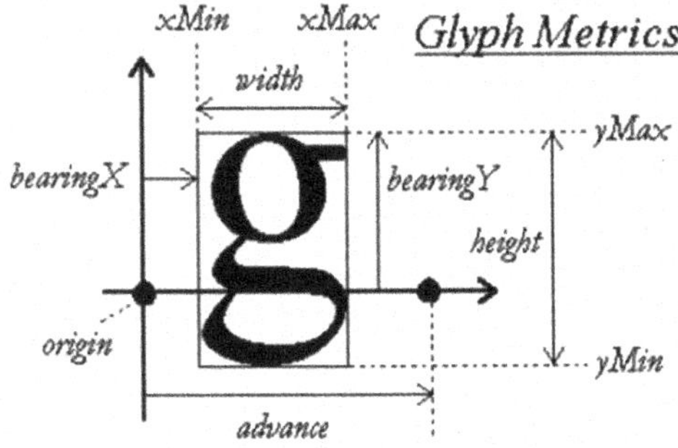

OpenVG does not seem to go into so much detail. The only diagram that is shown is as follows:

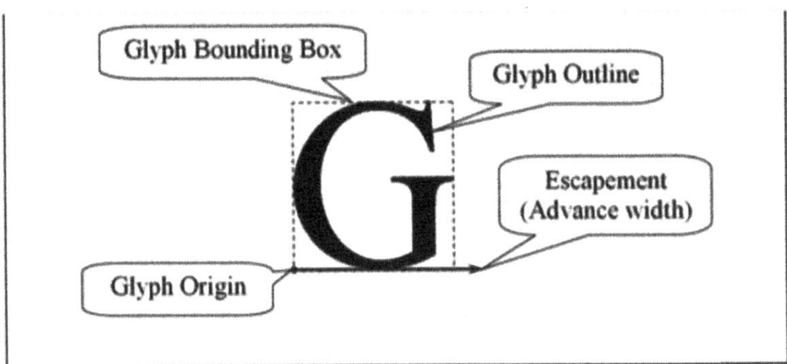

When adding a glyph to a font, only the glyph origin and the escapement are set, each as an (x, y) vector. The best values seem to be as follows:

```
VGfloat glyphOrigin[2] = {0.0f, 0.0f};
VGfloat escapement[2] = {ft_face->glyph->linearHoriAdvance,
                         0.0f};
```

Conclusion

This chapter has looked at a number of different ways of drawing text using OpenVG. There are trade-offs between ease of use and utilizing the GPU.

Resources

- *Text in Cairo*: `http://zetcode.com/gfx/cairo/cairotext/`
- *Pango Text Attribute Markup Language*: `https://developer.gnome.org/pango/stable/PangoMarkupFormat.html`

CHAPTER 19

Overlays on the Raspberry Pi

OpenMAX is concerned with displaying images and videos. Both OpenGL ES and OpenVG are concerned with drawing shapes in various ways. Subtitles are systems where dynamic text is drawn on top of images, static or dynamic. This chapter investigates issues in displaying overlays from one system on top of another on the Raspberry Pi.

Introduction

In Chapter 15, you considered drawing OpenMAX images/videos on top of an OpenGL ES surface. In this chapter, you'll consider the converse problem of using an OpenMAX surface and drawing on top of *that* surface.

Drawing OpenMAX onto an OpenGL ES surface required the use of a special function `eglCreateImageKHR` to link a Dispmanx OpenGL ES surface to an `EGLImage` that OpenMAX draws into. The converse is simpler.

1. Create and render OpenMAX images/videos.
2. Create and draw into OpenGL ES/OpenVG surfaces.

For this to work, the GPU must place the OpenMAX surface under the OpenGL ES/OpenVG surface so that drawing into OpenGL ES/OpenVG is drawing into an overlay.

While this allows basic overlays, drawing text adds a substantial layer of complexity. Text is *extremely* complex, even in English, and adding additional languages with their own wrinkles (such as Chinese hieroglyphics or Arabic right-to-left writing) makes this an area where you do *not* want to roll your own solutions. The principal library to manage this currently seems to be Pango built on top of Cairo, so this is investigated in this chapter.

Dispmanx Layers

To ensure that the OpenGL ES/OpenVG surface is drawn on top of the OpenMAX surface, they must be placed in different *layers*. The layer is controlled at the Dispmanx level in the function described in `/opt/vc/include/interface/vmcs_host/vc_dispmanx.h`.

```
vc_dispmanx_element_add (DISPMANX_UPDATE_HANDLE_T update,
                         DISPMANX_DISPLAY_HANDLE_T display,
                         int32_t layer,
                         const VC_RECT_T *dest_rect,
                         DISPMANX_RESOURCE_HANDLE_T src,
                         const VC_RECT_T *src_rect,
                         DISPMANX_PROTECTION_T protection,
```


J. Newmarch, *Raspberry Pi GPU Audio Video Programming*, DOI 10.1007/978-1-4842-2472-4_19

```
                        VC_DISPMANX_ALPHA_T *alpha,
                        DISPMANX_CLAMP_T *clamp,
                        DISPMANX_TRANSFORM_T transform );
```

The default for OpenMAX appears to be layer 0. For the overlays in this chapter, you will choose layer 1.

In OpenGL ES, you used the utilities from the *OpenGL ES 2.0 Programming Guide* by Aaftab Munshi, Dan Ginsburg, and Dave Shreiner. This creates the Dispmanx window at layer 0 in `esUtils.h`. This needs to be modified for the OpenGL ES examples discussed here, to set the Dispmanx window at layer 1.

```
dispman_element = vc_dispmanx_element_add ( dispman_update, dispman_display,
      1/*layer - changed*/, &dst_rect, 0/*src*/,
      &src_rect, DISPMANX_PROTECTION_NONE,
      0 /*alpha*/, 0/*clamp*/, 0/*transform*/);
```

Drawing OpenGL ESon Top of an OpenMAX Image

In Chapter 6, I discussed the program `Hello Triangle.c`, which draws a triangle into an OpenGL ES surface. In Chapter 12, I showed a program in the section "Rendering an Image Using Tunneling" to render an image using OpenMAX.

Drawing a triangle on top of the image is surprisingly easy: just combine the two sets of code, using the change to the layer of the OpenGL ES Dispmanx window above. With minor adjustments to the two sets of code, the program is `il_render_image_gles_overlay.c`.

```
#include <stdio.h>
#include <stdlib.h>
#include <sys/stat.h>

#include <assert.h>

#include <OMX_Core.h>
#include <OMX_Component.h>

#include <EGL/egl.h>
#include <EGL/eglext.h>
#include <GLES2/gl2.h>

#include <bcm_host.h>
#include <ilclient.h>
#include "esUtil.h"

typedef struct
{
    EGLDisplay display;
    EGLSurface surface;
    EGLContext context;
    EGLConfig config;
} EGL_STATE_T;
```

```
EGL_STATE_T state, *p_state = &state;

typedef struct
{
   // Handle to a program object
   GLuint programObject;

} UserData;

///
// Create a shader object, load the shader source, and
// compile the shader.
//
GLuint LoadShader ( GLenum type, const char *shaderSrc )
{
   GLuint shader;
   GLint compiled;

   // Create the shader object
   shader = glCreateShader ( type );

   if ( shader == 0 )
        return 0;

   // Load the shader source
   glShaderSource ( shader, 1, &shaderSrc, NULL );

   // Compile the shader
   glCompileShader ( shader );

   // Check the compile status
   glGetShaderiv ( shader, GL_COMPILE_STATUS, &compiled );

   if ( !compiled )
   {
      GLint infoLen = 0;

      glGetShaderiv ( shader, GL_INFO_LOG_LENGTH, &infoLen );

      if ( infoLen > 1 )
      {
         char* infoLog = malloc (sizeof(char) * infoLen );

         glGetShaderInfoLog ( shader, infoLen, NULL, infoLog );
         esLogMessage ( "Error compiling shader:\n%s\n", infoLog );

         free ( infoLog );
      }

      glDeleteShader ( shader );
      return 0;
   }
```

```
   return shader;

}

///
// Initialize the shader and program object
//
int Init ( ESContext *esContext )
{
   esContext->userData = malloc(sizeof(UserData));

   UserData *userData = esContext->userData;
   GLbyte vShaderStr[] =
      "attribute vec4 vPosition;    \n"
      "void main()                  \n"
      "{                            \n"
      "   gl_Position = vPosition;  \n"
      "}                            \n";

   GLbyte fShaderStr[] =
      "precision mediump float;\n"\
      "void main()                                  \n"
      "{                                            \n"
      "  gl_FragColor = vec4 ( 1.0, 0.0, 0.0, 0.0 );\n"
      "}                                            \n";

   GLuint vertexShader;
   GLuint fragmentShader;
   GLuint programObject;
   GLint linked;

   // Load the vertex/fragment shaders
   vertexShader = LoadShader ( GL_VERTEX_SHADER, vShaderStr );
   fragmentShader = LoadShader ( GL_FRAGMENT_SHADER, fShaderStr );

   // Create the program object
   programObject = glCreateProgram ( );

   if ( programObject == 0 )
      return 0;

   glAttachShader ( programObject, vertexShader );
   glAttachShader ( programObject, fragmentShader );

   // Bind vPosition to attribute 0
   glBindAttribLocation ( programObject, 0, "vPosition" );

   // Link the program
   glLinkProgram ( programObject );

   // Check the link status
   glGetProgramiv ( programObject, GL_LINK_STATUS, &linked );
```

```
   if ( !linked )
   {
      GLint infoLen = 0;

      glGetProgramiv ( programObject, GL_INFO_LOG_LENGTH, &infoLen );

      if ( infoLen > 1 )
      {
         char* infoLog = malloc (sizeof(char) * infoLen );

         glGetProgramInfoLog ( programObject, infoLen, NULL, infoLog );
         esLogMessage ( "Error linking program:\n%s\n", infoLog );

         free ( infoLog );
      }

      glDeleteProgram ( programObject );
      return GL_FALSE;
   }

   // Store the program object
   userData->programObject = programObject;

   glClearColor ( 0.0f, 0.0f, 0.0f, 0.0f );
   return GL_TRUE;
}

///
// Draw a triangle using the shader pair created in Init()
//
float incr = 0.0f;
void Draw ( ESContext *esContext )
{
   UserData *userData = esContext->userData;
   GLfloat vVertices[] = {  0.0f-incr,  0.5f-incr, 0.0f+incr,
                           -0.5f, -0.5f, 0.0f,
                            0.5f, -0.5f, 0.0f };

   incr += 0.01f;
   // Set the viewport
   glViewport ( 0, 0, esContext->width, esContext->height );

   // Clear the color buffer
   glClear ( GL_COLOR_BUFFER_BIT );

   // Use the program object
   glUseProgram ( userData->programObject );

   // Load the vertex data
   glVertexAttribPointer ( 0, 3, GL_FLOAT, GL_FALSE, 0, vVertices );
   glEnableVertexAttribArray ( 0 );
```

```
    glDrawArrays ( GL_TRIANGLES, 0, 3 );
}

#define IMG  "cimg0135.jpg"
//#define IMG "hype.jpg"

void printState(OMX_HANDLETYPE handle) {
    // elided
}

char *err2str(int err) {
    return "error elided";
}

void eos_callback(void *userdata, COMPONENT_T *comp, OMX_U32 data) {
    fprintf(stderr, "Got eos event\n");
}

void error_callback(void *userdata, COMPONENT_T *comp, OMX_U32 data) {
    fprintf(stderr, "OMX error %s\n", err2str(data));
}

int get_file_size(char *fname) {
    struct stat st;

    if (stat(fname, &st) == -1) {
            perror("Stat'ing img file");
            return -1;
        }
    return(st.st_size);
}

unsigned int uWidth;
unsigned int uHeight;

OMX_ERRORTYPE read_into_buffer_and_empty(FILE *fp,
                                         COMPONENT_T *component,
                                         OMX_BUFFERHEADERTYPE *buff_header,
                                         int *toread) {
    OMX_ERRORTYPE r;

    int buff_size = buff_header->nAllocLen;
    int nread = fread(buff_header->pBuffer, 1, buff_size, fp);

    printf("Read %d\n", nread);

    buff_header->nFilledLen = nread;
    *toread -= nread;
    if (*toread <= 0) {
        printf("Setting EOS on input\n");
        buff_header->nFlags |= OMX_BUFFERFLAG_EOS;
    }
```

```
    r = OMX_EmptyThisBuffer(ilclient_get_handle(component),
                            buff_header);
    if (r != OMX_ErrorNone) {
        fprintf(stderr, "Empty buffer error %s\n",
                err2str(r));
    }
    return r;
}

static void set_image_decoder_input_format(COMPONENT_T *component) {
   // set input image format
    printf("Setting image decoder format\n");
    OMX_IMAGE_PARAM_PORTFORMATTYPE imagePortFormat;
    //setHeader(&imagePortFormat,  sizeof(OMX_IMAGE_PARAM_PORTFORMATTYPE));
    memset(&imagePortFormat, 0, sizeof(OMX_IMAGE_PARAM_PORTFORMATTYPE));
    imagePortFormat.nSize = sizeof(OMX_IMAGE_PARAM_PORTFORMATTYPE);
    imagePortFormat.nVersion.nVersion = OMX_VERSION;

    imagePortFormat.nPortIndex = 320;
    imagePortFormat.eCompressionFormat = OMX_IMAGE_CodingJPEG;
    OMX_SetParameter(ilclient_get_handle(component),
                     OMX_IndexParamImagePortFormat, &imagePortFormat);

}

void setup_decodeComponent(ILCLIENT_T  *handle,
                           char *decodeComponentName,
                           COMPONENT_T **decodeComponent) {
    int err;

    err = ilclient_create_component(handle,
                                    decodeComponent,
                                    decodeComponentName,
                                    ILCLIENT_DISABLE_ALL_PORTS
                                        |
                                        ILCLIENT_ENABLE_INPUT_BUFFERS
                                        /* |
                                        ILCLIENT_ENABLE_OUTPUT_BUFFERS
                                        */
                                    );
    if (err == -1) {
        fprintf(stderr, "DecodeComponent create failed\n");
        exit(1);
    }
    printState(ilclient_get_handle(*decodeComponent));

    err = ilclient_change_component_state(*decodeComponent,
                                          OMX_StateIdle);
    if (err < 0) {
        fprintf(stderr, "Couldn't change state to Idle\n");
        exit(1);
    }
```

```
    printState(ilclient_get_handle(*decodeComponent));

    // must be before we enable buffers
    set_image_decoder_input_format(*decodeComponent);
}

void setup_renderComponent(ILCLIENT_T  *handle,
                              char *renderComponentName,
                              COMPONENT_T **renderComponent) {
    int err;

   err = ilclient_create_component(handle,
                                renderComponent,
                                renderComponentName,
                                ILCLIENT_DISABLE_ALL_PORTS
                                   /* |
                                    ILCLIENT_ENABLE_INPUT_BUFFERS
                                   */
                                );
    if (err == -1) {
        fprintf(stderr, "RenderComponent create failed\n");
        exit(1);
    }
    printState(ilclient_get_handle(*renderComponent));

    err = ilclient_change_component_state(*renderComponent,
                                          OMX_StateIdle);
    if (err < 0) {
        fprintf(stderr, "Couldn't change state to Idle\n");
        exit(1);
    }
    printState(ilclient_get_handle(*renderComponent));
}

void overlay_gles() {

    ESContext esContext;
    UserData  userData;

    //sleep(5);

    esInitContext ( &esContext );
    esContext.userData = &userData;

    esCreateWindow ( &esContext, "Hello Triangle", 1920, 1080, ES_WINDOW_RGB );

    if ( !Init ( &esContext ) )
        return 0;

    esRegisterDrawFunc ( &esContext, Draw );
```

```
    esMainLoop ( &esContext );
}

draw_openmax_image() {

    int i;
    char *decodeComponentName;
    char *renderComponentName;
    int err;
    ILCLIENT_T  *handle;
    COMPONENT_T *decodeComponent;
    COMPONENT_T *renderComponent;
    FILE *fp = fopen(IMG, "r");
    int toread = get_file_size(IMG);
    OMX_BUFFERHEADERTYPE *buff_header;

    decodeComponentName = "image_decode";
    renderComponentName = "video_render";

    //sleep(5);

    bcm_host_init();

    handle = ilclient_init();
    if (handle == NULL) {
        fprintf(stderr, "IL client init failed\n");
        exit(1);
    }

   if (OMX_Init() != OMX_ErrorNone) {
        ilclient_destroy(handle);
        fprintf(stderr, "OMX init failed\n");
        exit(1);
    }

   ilclient_set_error_callback(handle,
                               error_callback,
                               NULL);
   ilclient_set_eos_callback(handle,
                             eos_callback,
                             NULL);

   setup_decodeComponent(handle, decodeComponentName, &decodeComponent);
   setup_renderComponent(handle, renderComponentName, &renderComponent);
   // both components now in Idle state, no buffers, ports disabled

    // input port
    ilclient_enable_port_buffers(decodeComponent, 320,
                                 NULL, NULL, NULL);
    ilclient_enable_port(decodeComponent, 320);
```

```
err = ilclient_change_component_state(decodeComponent,
                                      OMX_StateExecuting);
if (err < 0) {
    fprintf(stderr, "Couldn't change state to Executing\n");
    exit(1);
}
printState(ilclient_get_handle(decodeComponent));

// Read the first block so that the decodeComponent can get
// the dimensions of the image and call port settings
// changed on the output port to configure it
buff_header =
    ilclient_get_input_buffer(decodeComponent,
                              320,
                              1 /* block */);
if (buff_header != NULL) {
    read_into_buffer_and_empty(fp,
                               decodeComponent,
                               buff_header,
                               &toread);

    // If all the file has been read in, then
    // we have to re-read this first block.
    // Broadcom bug?
    if (toread <= 0) {
        printf("Rewinding\n");
        // wind back to start and repeat
        fp = freopen(IMG, "r", fp);
        toread = get_file_size(IMG);
    }
}

// wait for first input block to set params for output port
err = ilclient_wait_for_event(decodeComponent,
                              OMX_EventPortSettingsChanged,
                              321, 0, 0, 1,
                              ILCLIENT_EVENT_ERROR | ILCLIENT_PARAMETER_CHANGED,
                              5);
if (err < 0) {
    fprintf(stderr, "No port settings changed\n");
}
printf("Port settings changed\n");

TUNNEL_T tunnel;
set_tunnel(&tunnel, decodeComponent, 321, renderComponent, 90);
if ((err = ilclient_setup_tunnel(&tunnel, 0, 0)) < 0) {
    fprintf(stderr, "Error setting up tunnel %X\n", err);
    exit(1);
} else {
    printf("Tunnel set up ok\n");
}
```

```
    // Okay to go back to processing data
    // enable the decode output ports

    OMX_SendCommand(ilclient_get_handle(decodeComponent),
                    OMX_CommandPortEnable, 321, NULL);

    ilclient_enable_port(decodeComponent, 321);

    // enable the render output ports
    /*
    OMX_SendCommand(ilclient_get_handle(renderComponent),
                    OMX_CommandPortEnable, 90, NULL);
    */
    ilclient_enable_port(renderComponent, 90);

    // set both components to executing state
     err = ilclient_change_component_state(decodeComponent,
                                           OMX_StateExecuting);
     if (err < 0) {
         fprintf(stderr, "Couldn't change state to Idle\n");
         exit(1);
     }
     err = ilclient_change_component_state(renderComponent,
                                           OMX_StateExecuting);
     if (err < 0) {
         fprintf(stderr, "Couldn't change state to Idle\n");
         exit(1);
     }

     // now work through the file
     while (toread > 0) {
         OMX_ERRORTYPE r;

         // do we have a decode input buffer we can fill and empty?
         buff_header =
             ilclient_get_input_buffer(decodeComponent,
                                       320,
                                       1 /* block */);
         if (buff_header != NULL) {
             read_into_buffer_and_empty(fp,
                                     decodeComponent,
                                     buff_header,
                                     &toread);
         }
     }

     ilclient_wait_for_event(renderComponent,
                             OMX_EventBufferFlag,
                             90, 0, OMX_BUFFERFLAG_EOS, 0,
                             ILCLIENT_BUFFER_FLAG_EOS, 10000);
     printf("EOS on render\n");
}
```

```
int main(int argc, char** argv) {

    draw_openmax_image();
    overlay_gles();

    exit(0);
}
```

One of the changes is to let the coordinates of one of the triangle vertices change on each draw. The triangle acts as an overlay, with the image behind preserved on each change. I don't see how to make the overlay OpenGL ES surface transparent at all, though.

It looks like this:

The commands I used to build this (with considerable redundancy!) are as follows:

```
cc -g -Wall -pthread -I/usr/include/gtk-3.0 -I/usr/include/atk-1.0 -I/usr/include/at-spi2-
atk/2.0 -I/usr/include/pango-1.0 -I/usr/include/gio-unix-2.0/ -I/usr/include/cairo -I/usr/
include/gdk-pixbuf-2.0 -I/usr/include/glib-2.0 -I/usr/lib/arm-linux-gnueabihf/glib-2.0/
include -I/usr/include/harfbuzz -I/usr/include/freetype2 -I/usr/include/pixman-1 -I/usr/
include/libpng12   -I../../openGLES/Common/ -DSTANDALONE -D__STDC_CONSTANT_MACROS -D__
STDC_LIMIT_MACROS -DTARGET_POSIX -D_LINUX -fPIC -DPIC -D_REENTRANT -D_LARGEFILE64_SOURCE
-D_FILE_OFFSET_BITS=64 -U_FORTIFY_SOURCE -Wall -g -DHAVE_LIBOPENMAX=2 -DOMX -DOMX_SKIP64BIT
-ftree-vectorize -pipe -DUSE_EXTERNAL_OMX -DHAVE_LIBBCM_HOST -DUSE_EXTERNAL_LIBBCM_HOST
-DUSE_VCHIQ_ARM -Wno-psabi -I/opt/vc/include/ -I /opt/vc/include/IL -I/opt/vc/include/
interface/vcos/pthreads  -I/opt/vc/include/interface/vcos/ -I/opt/vc/include/interface/
vmcs_host/linux -I./ -I/opt/vc/src/hello_pi/libs/ilclient -I/opt/vc/src/hello_pi/libs/vgfont
-I/opt/vc/src/hello_pi/libs/ilclient -DRASPBERRY_PI   -c -o il_render_image_gles_overlay.o
il_render_image_gles_overlay.c
```

```
cc il_render_image_gles_overlay.o  ../openGLES/Common/esUtil.o ../openGLES/Common/
esTransform.o ../openGLES/Common/esShader.o ../openGLES/Common/esShapes.o  -Wl,--whole-
archive -lilclient -L/opt/vc/lib/ -lopenmaxil -lbcm_host -lvcos -lvchiq_arm -lpthread -lrt
-L/opt/vc/src/hello_pi/libs/ilclient -L/opt/vc/src/hello_pi/libs/vgfont -Wl,--no-whole-
archive -rdynamic  -lavutil -lavcodec -lavformat -lavresample  -lEGL -lGLESv2  -lm -lgtk-3
-lgdk-3 -latk-1.0 -lgio-2.0 -lpangocairo-1.0 -lgdk_pixbuf-2.0 -lcairo-gobject -lpango-1.0
-lcairo -lgobject-2.0 -lglib-2.0    -o il_render_image_gles_overlay
```

Drawing OpenGL ES on Top of an OpenMAX Video

The same basic techniques are applied to overlaying on top of a video, with the relatively minor complexity that both the video and the OpenGL ES rendering take place at the same time, requiring the use of threads. You can take `il_render_video.c` from Chapter 13 and combine this with the `Hello Triangle.c` program. You can place the `Hello Triangle.c` code in its own `pthread`, to run asynchronously with OpenMAX running in the main thread. It could be the other way around, of course.

The resulting program is `il_render_video_gles_overlay.c`.

```
#include <stdio.h>
#include <stdlib.h>
#include <sys/stat.h>

#include <OMX_Core.h>
#include <OMX_Component.h>

#include <bcm_host.h>
#include <ilclient.h>

#include <EGL/egl.h>
#include <EGL/eglext.h>
#include <GLES2/gl2.h>
#include "esUtil.h"

#define IMG "/opt/vc/src/hello_pi/hello_video/test.h264"

pthread_t tid;

typedef struct
{
    EGLDisplay display;
    EGLSurface surface;
    EGLContext context;
    EGLConfig config;
} EGL_STATE_T;

EGL_STATE_T state, *p_state = &state;

typedef struct
{
```

```
   // Handle to a program object
   GLuint programObject;

} UserData;

///
// Create a shader object, load the shader source, and
// compile the shader.
//
GLuint LoadShader ( GLenum type, const char *shaderSrc )
{
   GLuint shader;
   GLint compiled;

   // Create the shader object
   shader = glCreateShader ( type );

   if ( shader == 0 )
        return 0;

   // Load the shader source
   glShaderSource ( shader, 1, &shaderSrc, NULL );

   // Compile the shader
   glCompileShader ( shader );

   // Check the compile status
   glGetShaderiv ( shader, GL_COMPILE_STATUS, &compiled );

   if ( !compiled )
   {
      GLint infoLen = 0;

      glGetShaderiv ( shader, GL_INFO_LOG_LENGTH, &infoLen );

      if ( infoLen > 1 )
      {
         char* infoLog = malloc (sizeof(char) * infoLen );

         glGetShaderInfoLog ( shader, infoLen, NULL, infoLog );
         esLogMessage ( "Error compiling shader:\n%s\n", infoLog );

         free ( infoLog );
      }

      glDeleteShader ( shader );
      return 0;
   }

   return shader;

}
```

```
///
// Initialize the shader and program object
//
int Init ( ESContext *esContext )
{
   esContext->userData = malloc(sizeof(UserData));

   UserData *userData = esContext->userData;
   GLbyte vShaderStr[] =
      "attribute vec4 vPosition;    \n"
      "void main()                  \n"
      "{                            \n"
      "   gl_Position = vPosition;  \n"
      "}                            \n";

   GLbyte fShaderStr[] =
      "precision mediump float;\n"\
      "void main()                                  \n"
      "{                                            \n"
      "  gl_FragColor = vec4 ( 1.0, 0.0, 0.0, 0.0 );\n"
      "}                                            \n";

   GLuint vertexShader;
   GLuint fragmentShader;
   GLuint programObject;
   GLint linked;

   // Load the vertex/fragment shaders
   vertexShader = LoadShader ( GL_VERTEX_SHADER, vShaderStr );
   fragmentShader = LoadShader ( GL_FRAGMENT_SHADER, fShaderStr );

   // Create the program object
   programObject = glCreateProgram ( );

   if ( programObject == 0 )
      return 0;

   glAttachShader ( programObject, vertexShader );
   glAttachShader ( programObject, fragmentShader );

   // Bind vPosition to attribute 0
   glBindAttribLocation ( programObject, 0, "vPosition" );

   // Link the program
   glLinkProgram ( programObject );

   // Check the link status
   glGetProgramiv ( programObject, GL_LINK_STATUS, &linked );

   if ( !linked )
   {
      GLint infoLen = 0;
```

```
      glGetProgramiv ( programObject, GL_INFO_LOG_LENGTH, &infoLen );

      if ( infoLen > 1 )
      {
         char* infoLog = malloc (sizeof(char) * infoLen );

         glGetProgramInfoLog ( programObject, infoLen, NULL, infoLog );
         esLogMessage ( "Error linking program:\n%s\n", infoLog );

         free ( infoLog );
      }

      glDeleteProgram ( programObject ) ;
      return GL_FALSE;
   }

   // Store the program object
   userData->programObject = programObject;

   glClearColor ( 0.0f, 0.0f, 0.0f, 0.0f );
   return GL_TRUE;
}

///
// Draw a triangle using the shader pair created in Init()
//
float incr = 0.0f;
void Draw ( ESContext *esContext )
{
   UserData *userData = esContext->userData;
   GLfloat vVertices[] = {  0.0f-incr,  0.5f-incr, 0.0f+incr,
                           -0.5f, -0.5f, 0.0f,
                            0.5f, -0.5f, 0.0f };

   incr += 0.01f;
   // Set the viewport
   glViewport ( 0, 0, esContext->width, esContext->height );

   // Clear the color buffer
   glClear ( GL_COLOR_BUFFER_BIT );

   // Use the program object
   glUseProgram ( userData->programObject );

   // Load the vertex data
   glVertexAttribPointer ( 0, 3, GL_FLOAT, GL_FALSE, 0, vVertices );
   glEnableVertexAttribArray ( 0 );

   glDrawArrays ( GL_TRIANGLES, 0, 3 );
}
```

```
void printState(OMX_HANDLETYPE handle) {
    // elided
}

char *err2str(int err) {
    return "error elided";
}

void eos_callback(void *userdata, COMPONENT_T *comp, OMX_U32 data) {
    fprintf(stderr, "Got eos event\n");
}

void error_callback(void *userdata, COMPONENT_T *comp, OMX_U32 data) {
    fprintf(stderr, "OMX error %s\n", err2str(data)) ;
}

int get_file_size(char *fname) {
    struct stat st;

    if (stat(fname, &st) == -1) {
        perror("Stat'ing img file");
        return -1;
    }
    return(st.st_size);
}

unsigned int uWidth;
unsigned int uHeight;

OMX_ERRORTYPE read_into_buffer_and_empty(FILE *fp,
                                          COMPONENT_T *component,
                                          OMX_BUFFERHEADERTYPE *buff_header,
                                          int *toread) {
    OMX_ERRORTYPE r;

    int buff_size = buff_header->nAllocLen;
    int nread = fread(buff_header->pBuffer, 1, buff_size, fp);

    buff_header->nFilledLen = nread;
    *toread -= nread;
    printf("Read %d, %d still left\n", nread, *toread);

    if (*toread <= 0) {
        printf("Setting EOS on input\n");
        buff_header->nFlags |= OMX_BUFFERFLAG_EOS;
    }
    r = OMX_EmptyThisBuffer(ilclient_get_handle(component),
                            buff_header);
    if (r != OMX_ErrorNone) {
        fprintf(stderr, "Empty buffer error %s\n",
                err2str(r));
```

```
    }
    return r;
}

static void set_video_decoder_input_format(COMPONENT_T *component) {
    int err;

    // set input video format
    printf("Setting video decoder format\n");
    OMX_VIDEO_PARAM_PORTFORMATTYPE videoPortFormat;

    memset(&videoPortFormat, 0, sizeof(OMX_VIDEO_PARAM_PORTFORMATTYPE));
    videoPortFormat.nSize = sizeof(OMX_VIDEO_PARAM_PORTFORMATTYPE);
    videoPortFormat.nVersion.nVersion = OMX_VERSION;

    videoPortFormat.nPortIndex = 130;
    videoPortFormat.eCompressionFormat = OMX_VIDEO_CodingAVC;

    err = OMX_SetParameter(ilclient_get_handle(component),
                           OMX_IndexParamVideoPortFormat, &videoPortFormat);
    if (err != OMX_ErrorNone) {
        fprintf(stderr, "Error setting video decoder format %s\n", err2str(err));
        return err;
    } else {
        printf("Video decoder format set up ok\n") ;
    }

}

void setup_decodeComponent(ILCLIENT_T  *handle,
                           char *decodeComponentName,
                           COMPONENT_T **decodeComponent) {
    int err;

    err = ilclient_create_component(handle,
                                    decodeComponent,
                                    decodeComponentName,
                                    ILCLIENT_DISABLE_ALL_PORTS
                                    |
                                    ILCLIENT_ENABLE_INPUT_BUFFERS
                                    |
                                    ILCLIENT_ENABLE_OUTPUT_BUFFERS
                                    );
    if (err == -1) {
        fprintf(stderr, "DecodeComponent create failed\n");
        exit(1);
    }
    printState(ilclient_get_handle(*decodeComponent));

    err = ilclient_change_component_state(*decodeComponent,
                                          OMX_StateIdle);
```

```
    if (err < 0) {
        fprintf(stderr, "Couldn't change state to Idle\n");
        exit(1);
    }
    printState(ilclient_get_handle(*decodeComponent));

    // must be before we enable buffers
    set_video_decoder_input_format(*decodeComponent);
}

void setup_renderComponent(ILCLIENT_T  *handle,
                           char *renderComponentName,
                           COMPONENT_T **renderComponent) {
    int err;

    err = ilclient_create_component(handle,
                                    renderComponent,
                                    renderComponentName,
                                    ILCLIENT_DISABLE_ALL_PORTS
                                    |
                                    ILCLIENT_ENABLE_INPUT_BUFFERS
                                    );
    if (err == -1) {
        fprintf(stderr, "RenderComponent create failed\n");
        exit(1);
    }
    printState(ilclient_get_handle(*renderComponent)) ;

    err = ilclient_change_component_state(*renderComponent,
                                          OMX_StateIdle);
    if (err < 0) {
        fprintf(stderr, "Couldn't change state to Idle\n");
        exit(1);
    }
    printState(ilclient_get_handle(*renderComponent));
}

void *overlay_gles(void *args) {

    ESContext esContext;
    UserData  userData;

    //sleep(10);

    esInitContext ( &esContext );
    esContext.userData = &userData;

    esCreateWindow ( &esContext, "Hello Triangle", 1920, 1080, ES_WINDOW_RGB );

    if ( !Init ( &esContext ) )
        return 0;
```

```
    esRegisterDrawFunc ( &esContext, Draw );

    esMainLoop ( &esContext );
}

void draw_openmax_video(int argc, char** argv) {
    int i;
    char *decodeComponentName;
    char *renderComponentName;
    int err;
    ILCLIENT_T  *handle;
    COMPONENT_T *decodeComponent;
    COMPONENT_T *renderComponent;
    FILE *fp; // = fopen(IMG, "r");
    int toread = get_file_size(IMG);
    OMX_BUFFERHEADERTYPE *buff_header;

    decodeComponentName = "video_decode";
    renderComponentName = "video_render";

    bcm_host_init();

    if (argc > 1) {
        fp = fopen(argv[1], "r");
    } else {
        fp =fopen(IMG, "r");
    }
    if (fp == NULL) {
        fprintf(stderr, "Can't open file\n") ;
        exit(1);
    }

    handle = ilclient_init();
    if (handle == NULL) {
        fprintf(stderr, "IL client init failed\n");
        exit(1);
    }

    if (OMX_Init() != OMX_ErrorNone) {
        ilclient_destroy(handle);
        fprintf(stderr, "OMX init failed\n");
        exit(1);
    }

    ilclient_set_error_callback(handle,
                                error_callback,
                                NULL);
    ilclient_set_eos_callback(handle,
                              eos_callback,
                              NULL);
```

```
setup_decodeComponent(handle, decodeComponentName, &decodeComponent);
setup_renderComponent(handle, renderComponentName, &renderComponent);
// both components now in Idle state, no buffers, ports disabled

// input port
ilclient_enable_port_buffers(decodeComponent, 130,
                             NULL, NULL, NULL);
ilclient_enable_port(decodeComponent, 130);

err = ilclient_change_component_state(decodeComponent,
                                      OMX_StateExecuting);
if (err < 0) {
    fprintf(stderr, "Couldn't change state to Executing\n");
    exit(1);
}
printState(ilclient_get_handle(decodeComponent));

// Read the first block so that the decodeComponent can get
// the dimensions of the video and call port settings
// changed on the output port to configure it
while (toread > 0) {
    buff_header =
        ilclient_get_input_buffer(decodeComponent,
                                  130,
                                  1 /* block */);
    if (buff_header != NULL) {
        read_into_buffer_and_empty(fp,
                                   decodeComponent,
                                   buff_header,
                                   &toread);

        // If all the file has been read in, then
        // we have to re-read this first block.
        // Broadcom bug?
        if (toread <= 0) {
            printf("Rewinding\n");
            // wind back to start and repeat
            fp = freopen(IMG, "r", fp);
            toread = get_file_size(IMG);
        }
    }

    if (toread > 0 && ilclient_remove_event(decodeComponent,
                                            OMX_EventPortSettingsChanged,
                                            131, 0, 0, 1) == 0) {
        printf("Removed port settings event\n");
        break;
    } else {
        printf("No portr settting seen yet\n");
    }
```

```
        // wait for first input block to set params for output port
        if (toread == 0) {
            // wait for first input block to set params for output port
            err = ilclient_wait_for_event(decodeComponent,
                                          OMX_EventPortSettingsChanged,
                                          131, 0, 0, 1,
                                          ILCLIENT_EVENT_ERROR | ILCLIENT_PARAMETER_CHANGED,
                                          2000);
            if (err < 0) {
                fprintf(stderr, "No port settings change\n");
                //exit(1);
            } else {
                printf("Port settings changed\n");
                break;
            }
        }
    }

    // set up the tunnel between decode and render ports
    TUNNEL_T tunnel;
    set_tunnel(&tunnel, decodeComponent, 131, renderComponent, 90);
    if ((err = ilclient_setup_tunnel(&tunnel, 0, 0)) < 0) {
        fprintf(stderr, "Error setting up tunnel %X\n", err);
        exit(1);
    } else {
        printf("Tunnel set up ok\n");
    }

    // Okay to go back to processing data
    // enable the decode output ports

    OMX_SendCommand(ilclient_get_handle(decodeComponent),
                    OMX_CommandPortEnable, 131, NULL);

    ilclient_enable_port(decodeComponent, 131);

    // enable the render output ports

    OMX_SendCommand(ilclient_get_handle(renderComponent),
                    OMX_CommandPortEnable, 90, NULL);

    ilclient_enable_port(renderComponent, 90);

    // set both components to executing state
    err = ilclient_change_component_state(decodeComponent,
                                          OMX_StateExecuting);
```

```
    if (err < 0) {
        fprintf(stderr, "Couldn't change state to Idle\n");
        exit(1);
    }
    err = ilclient_change_component_state(renderComponent,
                                          OMX_StateExecuting);
    if (err < 0) {
        fprintf(stderr, "Couldn't change state to Idle\n");
        exit(1);
    }

    // now work through the file
    while (toread > 0) {
        OMX_ERRORTYPE r;

        // do we have a decode input buffer we can fill and empty?
        buff_header =
            ilclient_get_input_buffer(decodeComponent,
                                      130,
                                      1 /* block */);
        if (buff_header != NULL) {
            read_into_buffer_and_empty(fp,
                                       decodeComponent,
                                       buff_header,
                                       &toread);
        }
    }

    ilclient_wait_for_event(renderComponent,
                            OMX_EventBufferFlag,
                            90, 0, OMX_BUFFERFLAG_EOS, 0,
                            ILCLIENT_BUFFER_FLAG_EOS, 10000);
    printf("EOS on render\n");
}

int main(int argc, char** argv) {

    pthread_create(&tid, NULL, overlay_gles, NULL);
    draw_openmax_video(argc, argv);

    exit(0);
}
```

A snapshot (taken using raspi2png) is shown here:

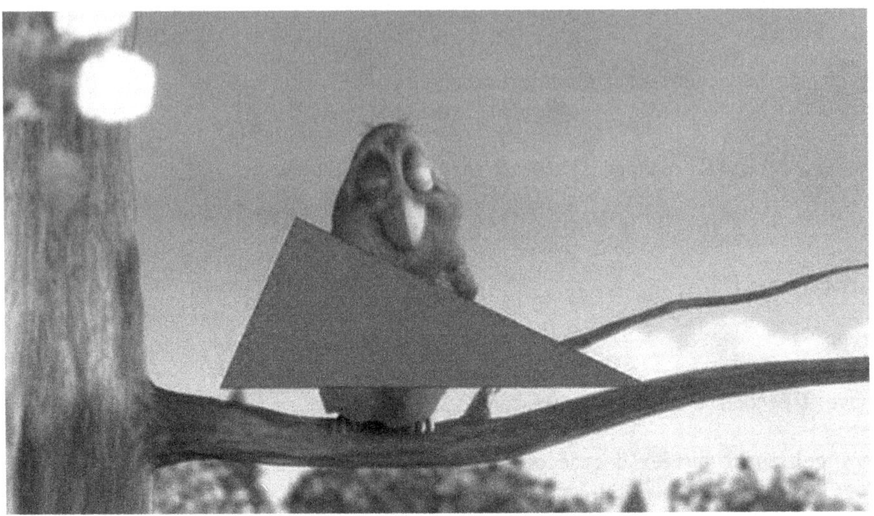

Drawing OpenVG on Top of an OpenMAX Image

The same principle holds for drawing OpenVG rather than OpenGL ES overlaid on top of an OpenMAX image. You create a Dispmanx layer above the OpenMAX layer and draw into each. A program to draw an ellipse on top of an image is il_render_image_vg_overlay.c, as shown here:

```
#include <stdio.h>
#include <stdlib.h>
#include <sys/stat.h>

#include <assert.h>

#include <OMX_Core.h>
#include <OMX_Component.h>

#include <EGL/egl.h>
#include <EGL/eglext.h>
#include <GLES2/gl2.h>
#include <VG/openvg.h>
#include <VG/vgu.h>

#include <bcm_host.h>
#include <ilclient.h>

typedef struct
{
    EGLDisplay display;
    EGLSurface surface;
    EGLContext context;
```

```
    EGLConfig config;
} EGL_STATE_T;

EGL_STATE_T state, *p_state = &state;

void init_egl(EGL_STATE_T *state)
{
    EGLint num_configs;
    EGLBoolean result;

    //bcm_host_init();

    static const EGLint attribute_list[] =
        {
            EGL_RED_SIZE, 8,
            EGL_GREEN_SIZE, 8,
            EGL_BLUE_SIZE, 8,
            EGL_ALPHA_SIZE, 8,
            EGL_SURFACE_TYPE, EGL_WINDOW_BIT,
            EGL_RENDERABLE_TYPE, EGL_OPENVG_BIT,
            EGL_NONE
        };

    static const EGLint context_attributes[] =
        {
            EGL_CONTEXT_CLIENT_VERSION, 2,
            EGL_NONE
        };

    // get an EGL display connection
    state->display = eglGetDisplay(EGL_DEFAULT_DISPLAY);

    // initialize the EGL display connection
    result = eglInitialize(state->display, NULL, NULL);

    /* if we want all configs:
    EGLConfig *configs;
    eglGetConfigs(state->display, NULL, 0, &num_configs);
    printf("EGL has %d configs\n", num_configs);

    configs = calloc(num_configs, sizeof *configs);
    eglGetConfigs(state->display, configs, num_configs, &num_configs);
    */

    // get an appropriate EGL frame buffer configuration
    result = eglChooseConfig(state->display, attribute_list, &state->config, 1, &num_
    configs);
    assert(EGL_FALSE != result);

    // Choose the OpenGL ES API
    //result = eglBindAPI(EGL_OPENGL_ES_API);
```

```
    result = eglBindAPI(EGL_OPENVG_API);
    assert(EGL_FALSE != result);

    // create an EGL rendering context
    state->context = eglCreateContext(state->display,
                                      state->config, EGL_NO_CONTEXT,
                                      NULL);
    //context_attributes);
                                      // breaks if we use this: context_attributes);
    assert(state->context!=EGL_NO_CONTEXT);
}

void init_dispmanx(EGL_DISPMANX_WINDOW_T *nativewindow) {
    int32_t success = 0;
    uint32_t screen_width;
    uint32_t screen_height;

    DISPMANX_ELEMENT_HANDLE_T dispman_element;
    DISPMANX_DISPLAY_HANDLE_T dispman_display;
    DISPMANX_UPDATE_HANDLE_T dispman_update;
    VC_RECT_T dst_rect;
    VC_RECT_T src_rect;

    bcm_host_init();

    // create an EGL window surface
    success = graphics_get_display_size(0 /* LCD */,
                                        &screen_width,
                                        &screen_height);
    assert( success >= 0 );

    dst_rect.x = 0;
    dst_rect.y = 0;
    dst_rect.width = screen_width;
    dst_rect.height = screen_height;

    src_rect.x = 0;
    src_rect.y = 0;
    src_rect.width = screen_width << 16;
    src_rect.height = screen_height << 16;

    dispman_display = vc_dispmanx_display_open( 0 /* LCD */);
    dispman_update = vc_dispmanx_update_start( 0 );

    dispman_element =
        vc_dispmanx_element_add(dispman_update, dispman_display,
                                1/*layer*/, &dst_rect, 0/*src*/,
                                &src_rect, DISPMANX_PROTECTION_NONE,
                                0 /*alpha*/, 0/*clamp*/, 0/*transform*/);

    // Build an EGL_DISPMANX_WINDOW_T from the Dispmanx window
    nativewindow->element = dispman_element;
```

```
    nativewindow->width = screen_width;
    nativewindow->height = screen_height;
    vc_dispmanx_update_submit_sync(dispman_update);

    printf("Got a Dispmanx window\n");
}

void egl_from_dispmanx(EGL_STATE_T *state,
                       EGL_DISPMANX_WINDOW_T *nativewindow) {
    EGLBoolean result;

    state->surface = eglCreateWindowSurface(state->display,
                                            state->config,
                                            nativewindow, NULL );
    assert(state->surface != EGL_NO_SURFACE);

    // connect the context to the surface
    result = eglMakeCurrent(state->display, state->surface, state->surface, state->context);
    assert(EGL_FALSE != result);
}

// setfill sets the fill color
void setfill(float color[4]) {
    VGPaint fillPaint = vgCreatePaint();
    vgSetParameteri(fillPaint, VG_PAINT_TYPE, VG_PAINT_TYPE_COLOR);
    vgSetParameterfv(fillPaint, VG_PAINT_COLOR, 4, color);
    vgSetPaint(fillPaint, VG_FILL_PATH);
    vgDestroyPaint(fillPaint);
}

// setstroke sets the stroke color and width
void setstroke(float color[4], float width) {
    VGPaint strokePaint = vgCreatePaint();
    vgSetParameteri(strokePaint, VG_PAINT_TYPE, VG_PAINT_TYPE_COLOR);
    vgSetParameterfv(strokePaint, VG_PAINT_COLOR, 4, color);
    vgSetPaint(strokePaint, VG_STROKE_PATH);
    vgSetf(VG_STROKE_LINE_WIDTH, width);
    vgSeti(VG_STROKE_CAP_STYLE, VG_CAP_BUTT);
    vgSeti(VG_STROKE_JOIN_STYLE, VG_JOIN_MITER);
    vgDestroyPaint(strokePaint);
}

// Ellipse makes an ellipse at the specified location and dimensions, applying style
void Ellipse(float x, float y, float w, float h, float sw, float fill[4], float stroke[4]) {
    VGPath path = vgCreatePath(VG_PATH_FORMAT_STANDARD, VG_PATH_DATATYPE_F, 1.0f, 0.0f, 0,
0, VG_PATH_CAPABILITY_ALL);
    vguEllipse(path, x, y, w, h);
    setfill(fill);
    setstroke(stroke, sw);
    vgDrawPath(path, VG_FILL_PATH | VG_STROKE_PATH);
    vgDestroyPath(path);
}
```

```
void draw(){
    VGfloat color[4] = {0.4, 0.1, 1.0, 1.0};

    vgClear(0, 0, 1920, 1080);
    Ellipse(1920/2, 1080/2, 600, 400, 0, color, color);
    vgFlush();
}

void overlay_vg()
{
    EGL_DISPMANX_WINDOW_T nativewindow;

    init_egl(p_state);
    init_dispmanx(&nativewindow);

    egl_from_dispmanx(p_state, &nativewindow);
    draw();
    eglSwapBuffers(p_state->display, p_state->surface);

    sleep(10);
    eglTerminate(p_state->display);

    exit(0);
}

#define IMG  "cimg0135.jpg"

void printState(OMX_HANDLETYPE handle) {
    // elided
}

char *err2str(int err) {
    return "error elided";
}

void eos_callback(void *userdata, COMPONENT_T *comp, OMX_U32 data) {
    fprintf(stderr, "Got eos event\n");
}

void error_callback(void *userdata, COMPONENT_T *comp, OMX_U32 data) {
    fprintf(stderr, "OMX error %s\n", err2str(data));
}

int get_file_size(char *fname) {
    struct stat st;

    if (stat(fname, &st) == -1) {
            perror("Stat'ing img file");
            return -1;
        }
    return(st.st_size);
}
```

```
unsigned int uWidth;
unsigned int uHeight;

OMX_ERRORTYPE read_into_buffer_and_empty(FILE *fp,
                                         COMPONENT_T *component,
                                         OMX_BUFFERHEADERTYPE *buff_header,
                                         int *toread) {
    OMX_ERRORTYPE r;

    int buff_size = buff_header->nAllocLen;
    int nread = fread(buff_header->pBuffer, 1, buff_size, fp);

    printf("Read %d\n", nread);

    buff_header->nFilledLen = nread;
    *toread -= nread;
    if (*toread <= 0) {
        printf("Setting EOS on input\n");
        buff_header->nFlags |= OMX_BUFFERFLAG_EOS;
    }
    r = OMX_EmptyThisBuffer(ilclient_get_handle(component),
                            buff_header);
    if (r != OMX_ErrorNone) {
        fprintf(stderr, "Empty buffer error %s\n",
                err2str(r));
    }
    return r;
}

static void set_image_decoder_input_format(COMPONENT_T *component) {
   // set input image format
    printf("Setting image decoder format\n");
    OMX_IMAGE_PARAM_PORTFORMATTYPE imagePortFormat;
    //setHeader(&imagePortFormat,  sizeof(OMX_IMAGE_PARAM_PORTFORMATTYPE));
    memset(&imagePortFormat, 0, sizeof(OMX_IMAGE_PARAM_PORTFORMATTYPE));
    imagePortFormat.nSize = sizeof(OMX_IMAGE_PARAM_PORTFORMATTYPE);
    imagePortFormat.nVersion.nVersion = OMX_VERSION;

    imagePortFormat.nPortIndex = 320;
    imagePortFormat.eCompressionFormat = OMX_IMAGE_CodingJPEG;
    OMX_SetParameter(ilclient_get_handle(component),
                     OMX_IndexParamImagePortFormat, &imagePortFormat);

}

void setup_decodeComponent(ILCLIENT_T  *handle,
                           char *decodeComponentName,
                           COMPONENT_T **decodeComponent) {
    int err;
```

```
    err = ilclient_create_component(handle,
                                    decodeComponent,
                                    decodeComponentName,
                                    ILCLIENT_DISABLE_ALL_PORTS
                                        |
                                        ILCLIENT_ENABLE_INPUT_BUFFERS
                                        /* |
                                        ILCLIENT_ENABLE_OUTPUT_BUFFERS
                                        */
                                    );
    if (err == -1) {
        fprintf(stderr, "DecodeComponent create failed\n");
        exit(1);
    }
    printState(ilclient_get_handle(*decodeComponent));

    err = ilclient_change_component_state(*decodeComponent,
                                          OMX_StateIdle);
    if (err < 0) {
        fprintf(stderr, "Couldn't change state to Idle\n");
        exit(1);
    }
    printState(ilclient_get_handle(*decodeComponent));

    // must be before we enable buffers
    set_image_decoder_input_format(*decodeComponent);
}

void setup_renderComponent(ILCLIENT_T  *handle,
                              char *renderComponentName,
                              COMPONENT_T **renderComponent) {
    int err;

   err = ilclient_create_component(handle,
                                    renderComponent,
                                    renderComponentName,
                                    ILCLIENT_DISABLE_ALL_PORTS
                                       /* |
                                        ILCLIENT_ENABLE_INPUT_BUFFERS
                                       */
                                    );
    if (err == -1) {
        fprintf(stderr, "RenderComponent create failed\n");
        exit(1);
    }
    printState(ilclient_get_handle(*renderComponent));

    err = ilclient_change_component_state(*renderComponent,
                                          OMX_StateIdle);
    if (err < 0) {
        fprintf(stderr, "Couldn't change state to Idle\n");
        exit(1);
```

```
    }
    printState(ilclient_get_handle(*renderComponent));
}

void draw_openmax_image() {
    int i;
    char *decodeComponentName;
    char *renderComponentName;
    int err;
    ILCLIENT_T  *handle;
    COMPONENT_T *decodeComponent;
    COMPONENT_T *renderComponent;
    FILE *fp = fopen(IMG, "r");
    int toread = get_file_size(IMG);
    OMX_BUFFERHEADERTYPE *buff_header;

    decodeComponentName = "image_decode";
    renderComponentName = "video_render";

    bcm_host_init();

    handle = ilclient_init();
    if (handle == NULL) {
        fprintf(stderr, "IL client init failed\n");
        exit(1);
    }

   if (OMX_Init() != OMX_ErrorNone) {
        ilclient_destroy(handle);
        fprintf(stderr, "OMX init failed\n");
        exit(1);
    }

   ilclient_set_error_callback(handle,
                               error_callback,
                               NULL);
   ilclient_set_eos_callback(handle,
                             eos_callback,
                             NULL);

   setup_decodeComponent(handle, decodeComponentName, &decodeComponent);
   setup_renderComponent(handle, renderComponentName, &renderComponent);
   // both components now in Idle state, no buffers, ports disabled

    // input port
    ilclient_enable_port_buffers(decodeComponent, 320,
                                 NULL, NULL, NULL);
    ilclient_enable_port(decodeComponent, 320);

    err = ilclient_change_component_state(decodeComponent,
                                          OMX_StateExecuting);
```

```
    if (err < 0) {
        fprintf(stderr, "Couldn't change state to Executing\n");
        exit(1);
    }
    printState(ilclient_get_handle(decodeComponent));

    // Read the first block so that the decodeComponent can get
    // the dimensions of the image and call port settings
    // changed on the output port to configure it
    buff_header =
        ilclient_get_input_buffer(decodeComponent,
                                  320,
                                  1 /* block */);
    if (buff_header != NULL) {
        read_into_buffer_and_empty(fp,
                                   decodeComponent,
                                   buff_header,
                                   &toread);

        // If all the file has been read in, then
        // we have to re-read this first block.
        // Broadcom bug?
        if (toread <= 0) {
            printf("Rewinding\n");
            // wind back to start and repeat
            fp = freopen(IMG, "r", fp);
            toread = get_file_size(IMG);
        }
    }

    // wait for first input block to set params for output port
    err = ilclient_wait_for_event(decodeComponent,
                                  OMX_EventPortSettingsChanged,
                                  321, 0, 0, 1,
                                  ILCLIENT_EVENT_ERROR | ILCLIENT_PARAMETER_CHANGED,
                                  5);
    if (err < 0) {
        fprintf(stderr, "No port settings changed\n");
    }
    printf("Port settings changed\n");

    TUNNEL_T tunnel;
    set_tunnel(&tunnel, decodeComponent, 321, renderComponent, 90);
    if ((err = ilclient_setup_tunnel(&tunnel, 0, 0)) < 0) {
        fprintf(stderr, "Error setting up tunnel %X\n", err);
        exit(1);
    } else {
        printf("Tunnel set up ok\n");
    }

   // Okay to go back to processing data
   // enable the decode output ports
```

```
    OMX_SendCommand(ilclient_get_handle(decodeComponent),
                    OMX_CommandPortEnable, 321, NULL);

    ilclient_enable_port(decodeComponent, 321);

    // enable the render output ports
    /*
    OMX_SendCommand(ilclient_get_handle(renderComponent),
                    OMX_CommandPortEnable, 90, NULL);
    */
    ilclient_enable_port(renderComponent, 90);

    // set both components to executing state
     err = ilclient_change_component_state(decodeComponent,
                                           OMX_StateExecuting);
     if (err < 0) {
         fprintf(stderr, "Couldn't change state to Idle\n");
         exit(1);
     }
     err = ilclient_change_component_state(renderComponent,
                                           OMX_StateExecuting);
     if (err < 0) {
         fprintf(stderr, "Couldn't change state to Idle\n");
         exit(1);
     }

     // now work through the file
     while (toread > 0) {
         OMX_ERRORTYPE r;

         // do we have a decode input buffer we can fill and empty?
         buff_header =
             ilclient_get_input_buffer(decodeComponent,
                                       320,
                                       1 /* block */);
         if (buff_header != NULL) {
             read_into_buffer_and_empty(fp,
                                        decodeComponent,
                                        buff_header,
                                        &toread);
         }
     }

    ilclient_wait_for_event(renderComponent,
                            OMX_EventBufferFlag,
                            90, 0, OMX_BUFFERFLAG_EOS, 0,
                            ILCLIENT_BUFFER_FLAG_EOS, 10000);
    printf("EOS on render\n");
}

int main(int argc, char** argv) {
```

```
    draw_openmax_image();
    overlay_vg();
    sleep(100);

    exit(0);
}
```

It looks like this:

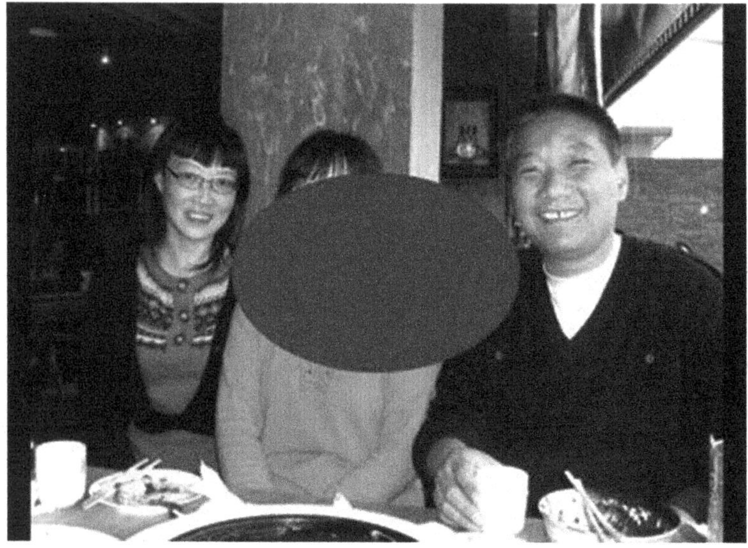

Drawing OpenVG on Top of an OpenMAX Video

The same technique works again. You draw the video in one thread on one surface and the OpenVG in another thread on another surface.

This program draws a shrinking ellipse on top of an OpenMAX video. It is `il_render_video_vg_overlay.c`.

```
#include <stdio.h>
#include <stdlib.h>
#include <sys/stat.h>

#include <OMX_Core.h>
#include <OMX_Component.h>

#include <bcm_host.h>
#include <ilclient.h>

#include <assert.h>
#include <EGL/egl.h>
#include <EGL/eglext.h>
```

```
#include <GLES2/gl2.h>
#include <VG/openvg.h>
#include <VG/vgu.h>
//#include "esUtil.h"

//#define IMG  "cimg0135.jpg"
#define IMG "/opt/vc/src/hello_pi/hello_video/test.h264"
//#define IMG "/home/pi/timidity/short.mpg"
//#define IMG "small.ogv"

pthread_t tid;

typedef struct
{
    EGLDisplay display;
    EGLSurface surface;
    EGLContext context;
    EGLConfig config;
} EGL_STATE_T;

EGL_STATE_T state, *p_state = &state;

typedef struct
{
   // Handle to a program object
   GLuint programObject;

} UserData;

void printState(OMX_HANDLETYPE handle) {
    // elided
}

char *err2str(int err) {
    return "error elided";
}

void eos_callback(void *userdata, COMPONENT_T *comp, OMX_U32 data) {
    fprintf(stderr, "Got eos event\n");
}

void error_callback(void *userdata, COMPONENT_T *comp, OMX_U32 data) {
    fprintf(stderr, "OMX error %s\n", err2str(data));
}

int get_file_size(char *fname) {
    struct stat st;

    if (stat(fname, &st) == -1) {
        perror("Stat'ing img file");
        return -1;
    }
```

```
    return(st.st_size);
}

unsigned int uWidth;
unsigned int uHeight;

OMX_ERRORTYPE read_into_buffer_and_empty(FILE *fp,
                                         COMPONENT_T *component,
                                         OMX_BUFFERHEADERTYPE *buff_header,
                                         int *toread) {
    OMX_ERRORTYPE r;

    int buff_size = buff_header->nAllocLen;
    int nread = fread(buff_header->pBuffer, 1, buff_size, fp);

    buff_header->nFilledLen = nread;
    *toread -= nread;
    printf("Read %d, %d still left\n", nread, *toread);

    if (*toread <= 0) {
        printf("Setting EOS on input\n");
        buff_header->nFlags |= OMX_BUFFERFLAG_EOS;
    }
    r = OMX_EmptyThisBuffer(ilclient_get_handle(component),
                            buff_header);
    if (r != OMX_ErrorNone) {
        fprintf(stderr, "Empty buffer error %s\n",
                err2str(r));
    }
    return r;
}

static void set_video_decoder_input_format(COMPONENT_T *component) {
    int err;

    // set input video format
    printf("Setting video decoder format\n");
    OMX_VIDEO_PARAM_PORTFORMATTYPE videoPortFormat;

    memset(&videoPortFormat, 0, sizeof(OMX_VIDEO_PARAM_PORTFORMATTYPE));
    videoPortFormat.nSize = sizeof(OMX_VIDEO_PARAM_PORTFORMATTYPE);
    videoPortFormat.nVersion.nVersion = OMX_VERSION;

    videoPortFormat.nPortIndex = 130;
    videoPortFormat.eCompressionFormat = OMX_VIDEO_CodingAVC;

    err = OMX_SetParameter(ilclient_get_handle(component),
                           OMX_IndexParamVideoPortFormat, &videoPortFormat);
    if (err != OMX_ErrorNone) {
        fprintf(stderr, "Error setting video decoder format %s\n", err2str(err));
        return err;
```

```
    } else {
        printf("Video decoder format set up ok\n");
    }

}

void setup_decodeComponent(ILCLIENT_T  *handle,
                           char *decodeComponentName,
                           COMPONENT_T **decodeComponent) {
    int err;

    err = ilclient_create_component(handle,
                                    decodeComponent,
                                    decodeComponentName,
                                    ILCLIENT_DISABLE_ALL_PORTS
                                    |
                                    ILCLIENT_ENABLE_INPUT_BUFFERS
                                    |
                                    ILCLIENT_ENABLE_OUTPUT_BUFFERS
                                    );
    if (err == -1) {
        fprintf(stderr, "DecodeComponent create failed\n");
        exit(1);
    }
    printState(ilclient_get_handle(*decodeComponent));

    err = ilclient_change_component_state(*decodeComponent,
                                          OMX_StateIdle);
    if (err < 0) {
        fprintf(stderr, "Couldn't change state to Idle\n");
        exit(1);
    }
    printState(ilclient_get_handle(*decodeComponent));

    // must be before we enable buffers
    set_video_decoder_input_format(*decodeComponent);
}

void setup_renderComponent(ILCLIENT_T  *handle,
                           char *renderComponentName,
                           COMPONENT_T **renderComponent) {
    int err;

    err = ilclient_create_component(handle,
                                    renderComponent,
                                    renderComponentName,
                                    ILCLIENT_DISABLE_ALL_PORTS
                                    |
                                    ILCLIENT_ENABLE_INPUT_BUFFERS
                                    );
    if (err == -1) {
```

```
        fprintf(stderr, "RenderComponent create failed\n");
        exit(1);
    }
    printState(ilclient_get_handle(*renderComponent));

    err = ilclient_change_component_state(*renderComponent,
                                          OMX_StateIdle);
    if (err < 0) {
        fprintf(stderr, "Couldn't change state to Idle\n");
        exit(1);
    }
    printState(ilclient_get_handle(*renderComponent));
}

void init_egl(EGL_STATE_T *state)
{
    EGLint num_configs;
    EGLBoolean result;

    //bcm_host_init();

    static const EGLint attribute_list[] =
        {
            EGL_RED_SIZE, 8,
            EGL_GREEN_SIZE, 8,
            EGL_BLUE_SIZE, 8,
            EGL_ALPHA_SIZE, 8,
            EGL_SURFACE_TYPE, EGL_WINDOW_BIT,
            EGL_SAMPLES, 1,
            EGL_NONE
        };

    static const EGLint context_attributes[] =
        {
            EGL_CONTEXT_CLIENT_VERSION, 2,
            EGL_NONE
        };

    // get an EGL display connection
    state->display = eglGetDisplay(EGL_DEFAULT_DISPLAY);

    // initialize the EGL display connection
    result = eglInitialize(state->display, NULL, NULL);

    // get an appropriate EGL frame buffer configuration
    result = eglChooseConfig(state->display, attribute_list, &state->config, 1, &num_
    configs);
    assert(EGL_FALSE != result);

    // Choose the OpenVG API
    result = eglBindAPI(EGL_OPENVG_API);
    assert(EGL_FALSE != result);
```

```
    // create an EGL rendering context
    state->context = eglCreateContext(state->display,
                                      state->config,
                                      NULL, // EGL_NO_CONTEXT,
                                      NULL);
                                      // breaks if we use this: context_attributes);
    assert(state->context!=EGL_NO_CONTEXT);
}

void init_dispmanx(EGL_DISPMANX_WINDOW_T *nativewindow) {
    int32_t success = 0;
    uint32_t screen_width;
    uint32_t screen_height;

    DISPMANX_ELEMENT_HANDLE_T dispman_element;
    DISPMANX_DISPLAY_HANDLE_T dispman_display;
    DISPMANX_UPDATE_HANDLE_T dispman_update;
    VC_RECT_T dst_rect;
    VC_RECT_T src_rect;

    bcm_host_init();

    // create an EGL window surface
    success = graphics_get_display_size(0 /* LCD */,
                                        &screen_width,
                                        &screen_height);
    assert( success >= 0 );

    dst_rect.x = 0;
    dst_rect.y = 0;
    dst_rect.width = screen_width;
    dst_rect.height = screen_height;

    src_rect.x = 0;
    src_rect.y = 0;
    src_rect.width = screen_width << 16;
    src_rect.height = screen_height << 16;

    dispman_display = vc_dispmanx_display_open( 0 /* LCD */);
    dispman_update = vc_dispmanx_update_start( 0 );

    dispman_element =
        vc_dispmanx_element_add(dispman_update, dispman_display,
                                1 /*layer*/, &dst_rect, 0 /*src*/,
                                &src_rect, DISPMANX_PROTECTION_NONE,
                                0 /*alpha*/, 0 /*clamp*/, 0 /*transform*/);

    // Build an EGL_DISPMANX_WINDOW_T from the Dispmanx window
    nativewindow->element = dispman_element;
    nativewindow->width = screen_width;
    nativewindow->height = screen_height;
    vc_dispmanx_update_submit_sync(dispman_update);
```

```
    printf("Got a Dispmanx window\n");
}

void egl_from_dispmanx(EGL_STATE_T *state,
                       EGL_DISPMANX_WINDOW_T *nativewindow) {
    EGLBoolean result;
    static const EGLint attribute_list[] =
        {
            EGL_RENDER_BUFFER, EGL_SINGLE_BUFFER,
            EGL_NONE
        };

    state->surface = eglCreateWindowSurface(state->display,
                                            state->config,
                                            nativewindow,
                                            NULL );
                                            //attribute_list);
    assert(state->surface != EGL_NO_SURFACE);

    // connect the context to the surface
    result = eglMakeCurrent(state->display, state->surface, state->surface, state->context);
    assert(EGL_FALSE != result);
}

// setfill sets the fill color
void setfill(float color[4]) {
    VGPaint fillPaint = vgCreatePaint();
    vgSetParameteri(fillPaint, VG_PAINT_TYPE, VG_PAINT_TYPE_COLOR);
    vgSetParameterfv(fillPaint, VG_PAINT_COLOR, 4, color);
    vgSetPaint(fillPaint, VG_FILL_PATH);
    vgDestroyPaint(fillPaint);
}

// setstroke sets the stroke color and width
void setstroke(float color[4], float width) {
    VGPaint strokePaint = vgCreatePaint();
    vgSetParameteri(strokePaint, VG_PAINT_TYPE, VG_PAINT_TYPE_COLOR);
    vgSetParameterfv(strokePaint, VG_PAINT_COLOR, 4, color);
    vgSetPaint(strokePaint, VG_STROKE_PATH);
    vgSetf(VG_STROKE_LINE_WIDTH, width);
    vgSeti(VG_STROKE_CAP_STYLE, VG_CAP_BUTT);
    vgSeti(VG_STROKE_JOIN_STYLE, VG_JOIN_MITER);
    vgDestroyPaint(strokePaint);
}

// Ellipse makes an ellipse at the specified location and dimensions, applying style
void Ellipse(float x, float y, float w, float h, float sw, float fill[4], float stroke[4]) {
    VGPath path = vgCreatePath(VG_PATH_FORMAT_STANDARD, VG_PATH_DATATYPE_F, 1.0f, 0.0f, 0,
0, VG_PATH_CAPABILITY_ALL);
    vguEllipse(path, x, y, w, h);
    setfill(fill);
    setstroke(stroke, sw);
```

```
    vgDrawPath(path, VG_FILL_PATH | VG_STROKE_PATH);
    vgDestroyPath(path);
}

void draw() {
    EGL_DISPMANX_WINDOW_T nativewindow;

    init_egl(p_state);
    init_dispmanx(&nativewindow);

    egl_from_dispmanx(p_state, &nativewindow);

    vgClear(0, 0, 1920, 1080);

    VGfloat color[4] = {0.4, 0.1, 1.0, 1.0}; // purple
    float c = 1.0;
    float clearColor[4] = {c, c, c, 0.0}; // white transparent
    vgSetfv(VG_CLEAR_COLOR, 4, clearColor);
    vgClear(0, 0, 1920, 1080);

    vgSeti(VG_BLEND_MODE, VG_BLEND_SRC_OVER);
    static float ht = 400;
    while (1) {
        vgClear(0, 0, 1920, 1080);
        Ellipse(1920/2, 1080/2, 600, ht--, 0, color, color);
        if (ht <= 0) {
            ht = 200;
        }
        eglSwapBuffers(p_state->display, p_state->surface);
    }
    assert(vgGetError() == VG_NO_ERROR);

    vgFlush();

    eglSwapBuffers(p_state->display, p_state->surface);
}

void *overlay_vg(void *args) {
    draw();
}

void draw_openmax_video(int argc, char** argv) {
    int i;
    char *decodeComponentName;
    char *renderComponentName;
    int err;
    ILCLIENT_T  *handle;
    COMPONENT_T *decodeComponent;
    COMPONENT_T *renderComponent;
    FILE *fp; // = fopen(IMG, "r");
    int toread = get_file_size(IMG);
    OMX_BUFFERHEADERTYPE *buff_header;
```

```
    decodeComponentName = "video_decode";
    renderComponentName = "video_render";

    bcm_host_init();

    if (argc > 1) {
        fp = fopen(argv[1], "r");
    } else {
        fp =fopen(IMG, "r");
    }
    if (fp == NULL) {
        fprintf(stderr, "Can't open file\n");
        exit(1);
    }

    handle = ilclient_init();
    if (handle == NULL) {
        fprintf(stderr, "IL client init failed\n");
        exit(1);
    }

    if (OMX_Init() != OMX_ErrorNone) {
        ilclient_destroy(handle);
        fprintf(stderr, "OMX init failed\n");
        exit(1);
    }

    ilclient_set_error_callback(handle,
                                error_callback,
                                NULL);
    ilclient_set_eos_callback(handle,
                              eos_callback,
                              NULL);

    setup_decodeComponent(handle, decodeComponentName, &decodeComponent);
    setup_renderComponent(handle, renderComponentName, &renderComponent);
    // both components now in Idle state, no buffers, ports disabled

    // input port
    ilclient_enable_port_buffers(decodeComponent, 130,
                                 NULL, NULL, NULL);
    ilclient_enable_port(decodeComponent, 130);

    err = ilclient_change_component_state(decodeComponent,
                                          OMX_StateExecuting);
    if (err < 0) {
        fprintf(stderr, "Couldn't change state to Executing\n");
        exit(1);
    }
    printState(ilclient_get_handle(decodeComponent));
```

```
// Read the first block so that the decodeComponent can get
// the dimensions of the video and call port settings
// changed on the output port to configure it
while (toread > 0) {
    buff_header =
        ilclient_get_input_buffer(decodeComponent,
                                   130,
                                   1 /* block */);
    if (buff_header != NULL) {
        read_into_buffer_and_empty(fp,
                                    decodeComponent,
                                    buff_header,
                                    &toread);

        // If all the file has been read in, then
        // we have to re-read this first block.
        // Broadcom bug?
        if (toread <= 0) {
            printf("Rewinding\n");
            // wind back to start and repeat
            fp = freopen(IMG, "r", fp);
            toread = get_file_size(IMG);
        }
    }

    if (toread > 0 && ilclient_remove_event(decodeComponent,
                                             OMX_EventPortSettingsChanged,
                                             131, 0, 0, 1) == 0) {
        printf("Removed port settings event\n");
        break;
    } else {
        printf("No portr settting seen yet\n");
    }
    // wait for first input block to set params for output port
    if (toread == 0) {
        // wait for first input block to set params for output port
        err = ilclient_wait_for_event(decodeComponent,
                                       OMX_EventPortSettingsChanged,
                                       131, 0, 0, 1,
                                       ILCLIENT_EVENT_ERROR | ILCLIENT_PARAMETER_CHANGED,
                                       2000);
        if (err < 0) {
            fprintf(stderr, "No port settings change\n");
            //exit(1);
        } else {
            printf("Port settings changed\n");
            break;
        }
    }
}
```

```
    // set up the tunnel between decode and render ports
    TUNNEL_T tunnel;
    set_tunnel(&tunnel, decodeComponent, 131, renderComponent, 90);
    if ((err = ilclient_setup_tunnel(&tunnel, 0, 0)) < 0) {
        fprintf(stderr, "Error setting up tunnel %X\n", err);
        exit(1);
    } else {
        printf("Tunnel set up ok\n");
    }

    // Okay to go back to processing data
    // enable the decode output ports

    OMX_SendCommand(ilclient_get_handle(decodeComponent),
                    OMX_CommandPortEnable, 131, NULL);

    ilclient_enable_port(decodeComponent, 131);

    // enable the render output ports

    OMX_SendCommand(ilclient_get_handle(renderComponent),
                    OMX_CommandPortEnable, 90, NULL);

    ilclient_enable_port(renderComponent, 90);

    // set both components to executing state
    err = ilclient_change_component_state(decodeComponent,
                                          OMX_StateExecuting);
    if (err < 0) {
        fprintf(stderr, "Couldn't change state to Idle\n");
        exit(1);
    }
    err = ilclient_change_component_state(renderComponent,
                                          OMX_StateExecuting);
    if (err < 0) {
        fprintf(stderr, "Couldn't change state to Idle\n");
        exit(1);
    }

    // now work through the file
    while (toread > 0) {
        OMX_ERRORTYPE r;

        // do we have a decode input buffer we can fill and empty?
        buff_header =
            ilclient_get_input_buffer(decodeComponent,
                                      130,
                                      1 /* block */);
        if (buff_header != NULL) {
            read_into_buffer_and_empty(fp,
                                       decodeComponent,
```

```
                                                buff_header,
                                                &toread);
        }
    }

    ilclient_wait_for_event(renderComponent,
                            OMX_EventBufferFlag,
                            90, 0, OMX_BUFFERFLAG_EOS, 0,
                            ILCLIENT_BUFFER_FLAG_EOS, 10000);
    printf("EOS on render\n");
}

int main(int argc, char** argv) {

    pthread_create(&tid, NULL, overlay_vg, NULL);
    draw_openmax_video(argc, argv);

    exit(0);
}
```

A snapshot is as follows:

Drawing Text on Top of an OpenMAX Video Using Pango

Drawing text can be made with any of the native OpenVG, Cairo, or Pango APIs. The easiest is probably Pango as it looks after many layout issues for you. In Chapter 18, you used Pango to lay out text on an OpenVG surface. That copies across to here, drawing Pango text onto an OpenVG surface, which is overlaid onto an OpenMAX surface playing a video.

The program il_render_video_pango_overlay.c draws a series of colored text lines on a video.

```
#include <stdio.h>
#include <stdlib.h>
#include <sys/stat.h>

#include <OMX_Core.h>
#include <OMX_Component.h>

#include <bcm_host.h>
#include <ilclient.h>

#include <assert.h>
#include <EGL/egl.h>
#include <EGL/eglext.h>
#include <GLES2/gl2.h>
#include <VG/openvg.h>
#include <VG/vgu.h>

#include <gtk/gtk.h>

#define IMG "/opt/vc/src/hello_pi/hello_video/test.h264"

pthread_t tid;

typedef struct
{
    EGLDisplay display;
    EGLSurface surface;
    EGLContext context;
    EGLConfig config;
} EGL_STATE_T;

EGL_STATE_T state, *p_state = &state;

typedef struct
{
   // Handle to a program object
   GLuint programObject;

} UserData;

void printState(OMX_HANDLETYPE handle) {
    // elided
}

char *err2str(int err) {
    return "error elided";
}
```

```
void eos_callback(void *userdata, COMPONENT_T *comp, OMX_U32 data) {
    fprintf(stderr, "Got eos event\n");
}

void error_callback(void *userdata, COMPONENT_T *comp, OMX_U32 data) {
    fprintf(stderr, "OMX error %s\n", err2str(data));
}

int get_file_size(char *fname) {
    struct stat st;

    if (stat(fname, &st) == -1) {
        perror("Stat'ing img file");
        return -1;
    }
    return(st.st_size);
}

unsigned int uWidth;
unsigned int uHeight;

OMX_ERRORTYPE read_into_buffer_and_empty(FILE *fp,
                                       COMPONENT_T *component,
                                       OMX_BUFFERHEADERTYPE *buff_header,
                                       int *toread) {
    OMX_ERRORTYPE r;

    int buff_size = buff_header->nAllocLen;
    int nread = fread(buff_header->pBuffer, 1, buff_size, fp);

    buff_header->nFilledLen = nread;
    *toread -= nread;
    printf("Read %d, %d still left\n", nread, *toread);

    if (*toread <= 0) {
        printf("Setting EOS on input\n");
        buff_header->nFlags |= OMX_BUFFERFLAG_EOS;
    }
    r = OMX_EmptyThisBuffer(ilclient_get_handle(component),
                            buff_header);
    if (r != OMX_ErrorNone) {
        fprintf(stderr, "Empty buffer error %s\n",
                err2str(r));
    }
    return r;
}

static void set_video_decoder_input_format(COMPONENT_T *component) {
    int err;

    // set input video format
    printf("Setting video decoder format\n");
```

```
    OMX_VIDEO_PARAM_PORTFORMATTYPE videoPortFormat;

    memset(&videoPortFormat, 0, sizeof(OMX_VIDEO_PARAM_PORTFORMATTYPE));
    videoPortFormat.nSize = sizeof(OMX_VIDEO_PARAM_PORTFORMATTYPE);
    videoPortFormat.nVersion.nVersion = OMX_VERSION;

    videoPortFormat.nPortIndex = 130;
    videoPortFormat.eCompressionFormat = OMX_VIDEO_CodingAVC;

    err = OMX_SetParameter(ilclient_get_handle(component),
                           OMX_IndexParamVideoPortFormat, &videoPortFormat);
    if (err != OMX_ErrorNone) {
        fprintf(stderr, "Error setting video decoder format %s\n", err2str(err));
        return err;
    } else {
        printf("Video decoder format set up ok\n");
    }

}

void setup_decodeComponent(ILCLIENT_T  *handle,
                           char *decodeComponentName,
                           COMPONENT_T **decodeComponent) {
    int err;

    err = ilclient_create_component(handle,
                                    decodeComponent,
                                    decodeComponentName,
                                    ILCLIENT_DISABLE_ALL_PORTS
                                    |
                                    ILCLIENT_ENABLE_INPUT_BUFFERS
                                    |
                                    ILCLIENT_ENABLE_OUTPUT_BUFFERS
                                    );
    if (err == -1) {
        fprintf(stderr, "DecodeComponent create failed\n");
        exit(1);
    }
    printState(ilclient_get_handle(*decodeComponent));

    err = ilclient_change_component_state(*decodeComponent,
                                          OMX_StateIdle);
    if (err < 0) {
        fprintf(stderr, "Couldn't change state to Idle\n");
        exit(1);
    }
    printState(ilclient_get_handle(*decodeComponent));

    // must be before we enable buffers
    set_video_decoder_input_format(*decodeComponent);
}
```

```
void setup_renderComponent(ILCLIENT_T  *handle,
                           char *renderComponentName,
                           COMPONENT_T **renderComponent) {
    int err;

    err = ilclient_create_component(handle,
                                    renderComponent,
                                    renderComponentName,
                                    ILCLIENT_DISABLE_ALL_PORTS
                                    |
                                    ILCLIENT_ENABLE_INPUT_BUFFERS
                                    );
    if (err == -1) {
        fprintf(stderr, "RenderComponent create failed\n");
        exit(1);
    }
    printState(ilclient_get_handle(*renderComponent));

    err = ilclient_change_component_state(*renderComponent,
                                          OMX_StateIdle);
    if (err < 0) {
        fprintf(stderr, "Couldn't change state to Idle\n");
        exit(1);
    }
    printState(ilclient_get_handle(*renderComponent));
}

void init_egl(EGL_STATE_T *state)
{
    EGLint num_configs;
    EGLBoolean result;

    //bcm_host_init();

    static const EGLint attribute_list[] =
        {
            EGL_RED_SIZE, 8,
            EGL_GREEN_SIZE, 8,
            EGL_BLUE_SIZE, 8,
            EGL_ALPHA_SIZE, 8,
            EGL_SURFACE_TYPE, EGL_WINDOW_BIT,
            EGL_SAMPLES, 1,
            EGL_NONE
        };

    static const EGLint context_attributes[] =
        {
            EGL_CONTEXT_CLIENT_VERSION, 2,
            EGL_NONE
        };
```

```
    // get an EGL display connection
    state->display = eglGetDisplay(EGL_DEFAULT_DISPLAY);

    // initialize the EGL display connection
    result = eglInitialize(state->display, NULL, NULL);

    // get an appropriate EGL frame buffer configuration
    result = eglChooseConfig(state->display, attribute_list, &state->config, 1, &num_
    configs);
    assert(EGL_FALSE != result);

    // Choose the OpenVG API
    result = eglBindAPI(EGL_OPENVG_API);
    assert(EGL_FALSE != result);

    // create an EGL rendering context
    state->context = eglCreateContext(state->display,
                                      state->config,
                                      NULL, // EGL_NO_CONTEXT,
                                      NULL);
                                      // breaks if we use this: context_attributes);
    assert(state->context!=EGL_NO_CONTEXT);
}

void init_dispmanx(EGL_DISPMANX_WINDOW_T *nativewindow) {
    int32_t success = 0;
    uint32_t screen_width;
    uint32_t screen_height;

    DISPMANX_ELEMENT_HANDLE_T dispman_element;
    DISPMANX_DISPLAY_HANDLE_T dispman_display;
    DISPMANX_UPDATE_HANDLE_T dispman_update;
    VC_RECT_T dst_rect;
    VC_RECT_T src_rect;

    bcm_host_init();

    // create an EGL window surface
    success = graphics_get_display_size(0 /* LCD */,
                                        &screen_width,
                                        &screen_height);
    assert( success >= 0 );

    dst_rect.x = 0;
    dst_rect.y = 0;
    dst_rect.width = screen_width;
    dst_rect.height = screen_height;

    src_rect.x = 0;
    src_rect.y = 0;
    src_rect.width = screen_width << 16;
    src_rect.height = screen_height << 16;
```

```
    dispman_display = vc_dispmanx_display_open( 0 /* LCD */);
    dispman_update = vc_dispmanx_update_start( 0 );

    dispman_element =
        vc_dispmanx_element_add(dispman_update, dispman_display,
                                1 /*layer*/, &dst_rect, 0 /*src*/,
                                &src_rect, DISPMANX_PROTECTION_NONE,
                                0 /*alpha*/, 0 /*clamp*/, 0 /*transform*/);

    // Build an EGL_DISPMANX_WINDOW_T from the Dispmanx window
    nativewindow->element = dispman_element;
    nativewindow->width = screen_width;
    nativewindow->height = screen_height;
    vc_dispmanx_update_submit_sync(dispman_update);

    printf("Got a Dispmanx window\n");
}

void egl_from_dispmanx(EGL_STATE_T *state,
                       EGL_DISPMANX_WINDOW_T *nativewindow) {
    EGLBoolean result;
    static const EGLint attribute_list[] =
        {
            EGL_RENDER_BUFFER, EGL_SINGLE_BUFFER,
            EGL_NONE
        };

    state->surface = eglCreateWindowSurface(state->display,
                                            state->config,
                                            nativewindow,
                                            NULL );
                                            //attribute_list);
    assert(state->surface != EGL_NO_SURFACE);

    // connect the context to the surface
    result = eglMakeCurrent(state->display, state->surface, state->surface, state->context);
    assert(EGL_FALSE != result);
}

// setfill sets the fill color
void setfill(float color[4]) {
    VGPaint fillPaint = vgCreatePaint();
    vgSetParameteri(fillPaint, VG_PAINT_TYPE, VG_PAINT_TYPE_COLOR);
    vgSetParameterfv(fillPaint, VG_PAINT_COLOR, 4, color);
    vgSetPaint(fillPaint, VG_FILL_PATH);
    vgDestroyPaint(fillPaint);
}


// setstroke sets the stroke color and width
void setstroke(float color[4], float width) {
    VGPaint strokePaint = vgCreatePaint();
```

```
    vgSetParameteri(strokePaint, VG_PAINT_TYPE, VG_PAINT_TYPE_COLOR);
    vgSetParameterfv(strokePaint, VG_PAINT_COLOR, 4, color);
    vgSetPaint(strokePaint, VG_STROKE_PATH);
    vgSetf(VG_STROKE_LINE_WIDTH, width);
    vgSeti(VG_STROKE_CAP_STYLE, VG_CAP_BUTT);
    vgSeti(VG_STROKE_JOIN_STYLE, VG_JOIN_MITER);
    vgDestroyPaint(strokePaint);
}

// Ellipse makes an ellipse at the specified location and dimensions, applying style
void Ellipse(float x, float y, float w, float h, float sw, float fill[4], float stroke[4]) {
    VGPath path = vgCreatePath(VG_PATH_FORMAT_STANDARD, VG_PATH_DATATYPE_F, 1.0f, 0.0f, 0,
0, VG_PATH_CAPABILITY_ALL);
    vguEllipse(path, x, y, w, h);
    setfill(fill);
    setstroke(stroke, sw);
    vgDrawPath(path, VG_FILL_PATH | VG_STROKE_PATH);
    vgDestroyPath(path);
}

char *get_markup_text() {
    static int n = 0;
    int num_texts = 6;
    gchar *texts[] = {
 "<span foreground=\"red\" font='96'></span>\
<span foreground=\"black\" font='96'>hello</span>",
         "<span foreground=\"red\" font='96'>h</span>\
<span foreground=\"black\" font='96'>ello</span>",
         "<span foreground=\"red\" font='96'>he</span>\
<span foreground=\"black\" font='96'>llo</span>",
         "<span foreground=\"red\" font='96'>hel</span>\
<span foreground=\"black\" font='96'>lo</span>",
         "<span foreground=\"red\" font='96'>hell</span>\
<span foreground=\"black\" font='96'>o</span>",
         "<span foreground=\"red\" font='96'>hello</span>\
<span foreground=\"black\" font='96'></span>"
    };

    return texts[n++ % num_texts];
}

void pango() {
    int width = 500;
    int height = 200;

    cairo_surface_t *surface = cairo_image_surface_create (CAIRO_FORMAT_ARGB32,
                                                           width, height);
    cairo_t *cr = cairo_create(surface);

    // Pango marked up text, half red, half black
    gchar *markup_text = get_markup_text();
    PangoAttrList *attrs;
```

```
    gchar *text;

    pango_parse_markup (markup_text, -1, 0, &attrs, &text, NULL, NULL);

    // draw Pango text
    PangoLayout *layout;
    PangoFontDescription *desc;

    cairo_move_to(cr, 30.0, 150.0);
    cairo_scale(cr, 1.0f, -1.0f);
    layout = pango_cairo_create_layout (cr);
    pango_layout_set_text (layout, text, -1);
    pango_layout_set_attributes(layout, attrs);
    pango_cairo_update_layout (cr, layout);
    pango_cairo_show_layout (cr, layout);

    int tex_w = cairo_image_surface_get_width(surface);
    int tex_h = cairo_image_surface_get_height(surface);
    int tex_s = cairo_image_surface_get_stride(surface);
    cairo_surface_flush(surface);
    cairo_format_t type =
        cairo_image_surface_get_format (surface);
    printf("Format is (0 is ARGB32) %d\n", type);
    unsigned char* data = cairo_image_surface_get_data(surface);

    PangoLayoutLine *layout_line =
        pango_layout_get_line(layout, 0);
    PangoRectangle ink_rect;
    PangoRectangle logical_rect;
    pango_layout_line_get_pixel_extents (layout_line,
                                         &ink_rect,
                                         &logical_rect);
    printf("Layout line rectangle is %d, %d, %d, %d\n",
           ink_rect.x, ink_rect.y, ink_rect.width, ink_rect.height);

    int left = 1920/2 - ink_rect.width/2;
    vgWritePixels(data, tex_s,
                  //VG_lBGRA_8888,
                  VG_sARGB_8888 ,
                  left, 300, tex_w, tex_h );

}

void draw()
{
    EGL_DISPMANX_WINDOW_T nativewindow;

    init_egl(p_state);
    init_dispmanx(&nativewindow);

    egl_from_dispmanx(p_state, &nativewindow);
```

```
    VGfloat color[4] = {0.4, 0.1, 1.0, 1.0}; // purple
    float c = 1.0;
    float clearColor[4] = {c, c, c, 0.0}; // white transparent
    vgSetfv(VG_CLEAR_COLOR, 4, clearColor);

    while (1) {
        vgClear(0, 0, 1920, 1080);
        pango();
        eglSwapBuffers(p_state->display, p_state->surface);
        sleep(1);
    }

    sleep(30);
    eglTerminate(p_state->display);

    exit(0);
}

#if 0
void draw() {
    EGL_DISPMANX_WINDOW_T nativewindow;

    init_egl(p_state);
    init_dispmanx(&nativewindow);

    egl_from_dispmanx(p_state, &nativewindow);

    vgClear(0, 0, 1920, 1080);

    VGfloat color[4] = {0.4, 0.1, 1.0, 1.0}; // purple
    float c = 1.0;
    float clearColor[4] = {c, c, c, 0.0}; // white transparent
    vgSetfv(VG_CLEAR_COLOR, 4, clearColor);
    vgClear(0, 0, 1920, 1080);

    vgSeti(VG_BLEND_MODE, VG_BLEND_SRC_OVER);
    static float ht = 400;
    while (1) {
        vgClear(0, 0, 1920, 1080);
        Ellipse(1920/2, 1080/2, 600, ht--, 0, color, color);
        if (ht <= 0) {
            ht = 200;
        }
        eglSwapBuffers(p_state->display, p_state->surface);
    }
    assert(vgGetError() == VG_NO_ERROR);

    vgFlush();

    eglSwapBuffers(p_state->display, p_state->surface);
}
```

```
#endif

void *overlay_vg(void *args) {
    draw();
}

void draw_openmax_video(int argc, char** argv) {
    int i;
    char *decodeComponentName;
    char *renderComponentName;
    int err;
    ILCLIENT_T  *handle;
    COMPONENT_T *decodeComponent;
    COMPONENT_T *renderComponent;
    FILE *fp; // = fopen(IMG, "r");
    int toread = get_file_size(IMG);
    OMX_BUFFERHEADERTYPE *buff_header;

    decodeComponentName = "video_decode";
    renderComponentName = "video_render";

    bcm_host_init();

    if (argc > 1) {
        fp = fopen(argv[1], "r");
    } else {
        fp =fopen(IMG, "r");
    }
    if (fp == NULL) {
        fprintf(stderr, "Can't open file\n");
        exit(1);
    }

    handle = ilclient_init();
    if (handle == NULL) {
        fprintf(stderr, "IL client init failed\n");
        exit(1);
    }

    if (OMX_Init() != OMX_ErrorNone) {
        ilclient_destroy(handle);
        fprintf(stderr, "OMX init failed\n");
        exit(1);
    }

    ilclient_set_error_callback(handle,
                                error_callback,
                                NULL);
    ilclient_set_eos_callback(handle,
                              eos_callback,
                              NULL);
```

```
    setup_decodeComponent(handle, decodeComponentName, &decodeComponent);
    setup_renderComponent(handle, renderComponentName, &renderComponent);
    // both components now in Idle state, no buffers, ports disabled

    // input port
    ilclient_enable_port_buffers(decodeComponent, 130,
                                 NULL, NULL, NULL);
    ilclient_enable_port(decodeComponent, 130);

    err = ilclient_change_component_state(decodeComponent,
                                          OMX_StateExecuting);
    if (err < 0) {
        fprintf(stderr, "Couldn't change state to Executing\n");
        exit(1);
    }
    printState(ilclient_get_handle(decodeComponent));

    // Read the first block so that the decodeComponent can get
    // the dimensions of the video and call port settings
    // changed on the output port to configure it
    while (toread > 0) {
        buff_header =
            ilclient_get_input_buffer(decodeComponent,
                                      130,
                                      1 /* block */);
        if (buff_header != NULL) {
            read_into_buffer_and_empty(fp,
                                       decodeComponent,
                                       buff_header,
                                       &toread);

            // If all the file has been read in, then
            // we have to re-read this first block.
            // Broadcom bug?
            if (toread <= 0) {
                printf("Rewinding\n");
                // wind back to start and repeat
                fp = freopen(IMG, "r", fp);
                toread = get_file_size(IMG);
            }
        }

        if (toread > 0 && ilclient_remove_event(decodeComponent,
                                                OMX_EventPortSettingsChanged,
                                                131, 0, 0, 1) == 0) {
            printf("Removed port settings event\n");
            break;
        } else {
            printf("No portr settting seen yet\n");
        }
        // wait for first input block to set params for output port
```

```
    if (toread == 0) {
        // wait for first input block to set params for output port
        err = ilclient_wait_for_event(decodeComponent,
                                      OMX_EventPortSettingsChanged,
                                      131, 0, 0, 1,
                                      ILCLIENT_EVENT_ERROR | ILCLIENT_PARAMETER_CHANGED,
                                      2000);
        if (err < 0) {
            fprintf(stderr, "No port settings change\n");
            //exit(1);
        } else {
            printf("Port settings changed\n");
            break;
        }
    }
}

// set up the tunnel between decode and render ports
TUNNEL_T tunnel;
set_tunnel(&tunnel, decodeComponent, 131, renderComponent, 90);
if ((err = ilclient_setup_tunnel(&tunnel, 0, 0)) < 0) {
    fprintf(stderr, "Error setting up tunnel %X\n", err);
    exit(1);
} else {
    printf("Tunnel set up ok\n");
}

// Okay to go back to processing data
// enable the decode output ports

OMX_SendCommand(ilclient_get_handle(decodeComponent),
                OMX_CommandPortEnable, 131, NULL);

ilclient_enable_port(decodeComponent, 131);

// enable the render output ports

OMX_SendCommand(ilclient_get_handle(renderComponent),
                OMX_CommandPortEnable, 90, NULL);

ilclient_enable_port(renderComponent, 90);

// set both components to executing state
err = ilclient_change_component_state(decodeComponent,
                                      OMX_StateExecuting);
if (err < 0) {
    fprintf(stderr, "Couldn't change state to Idle\n");
    exit(1);
}
err = ilclient_change_component_state(renderComponent,
                                      OMX_StateExecuting);
```

```
    if (err < 0) {
        fprintf(stderr, "Couldn't change state to Idle\n");
        exit(1);
    }

    // now work through the file
    while (toread > 0) {
        OMX_ERRORTYPE r;

        // do we have a decode input buffer we can fill and empty?
        buff_header =
            ilclient_get_input_buffer(decodeComponent,
                                      130,
                                      1 /* block */);
        if (buff_header != NULL) {
            read_into_buffer_and_empty(fp,
                                       decodeComponent,
                                       buff_header,
                                       &toread);
        }
    }

    ilclient_wait_for_event(renderComponent,
                            OMX_EventBufferFlag,
                            90, 0, OMX_BUFFERFLAG_EOS, 0,
                            ILCLIENT_BUFFER_FLAG_EOS, 10000);
    printf("EOS on render\n");
}

int main(int argc, char** argv) {

    pthread_create(&tid, NULL, overlay_vg, NULL);
    draw_openmax_video(argc, argv);

    exit(0);
}
```

A snapshot looks like this:

Conclusion

This chapter looked at overlaying drawings and text on top of images and videos. A number of scenarios were dealt with, all using the concept of different layers at the Dispmanx level to implement them.

Index

J. Newmarch, *Raspberry Pi GPU Audio Video Programming*, DOI 10.1007/978-1-4842-2472-4

F

G

H

I

J

K

L

M, N

O

W, X, Y, Z

Printed in the United States
By Bookmasters